MODERN ASPECTS OF ELECTROCHEMISTRY

No. 13

LIST OF CONTRIBUTORS

J. AUGUSTYNSKI
Chemistry Department
University of Geneva
Geneva, Switzerland

LUCETTE BALSENC
Chemistry Department
University of Geneva
Geneva, Switzerland

J. BARTHEL
Department of Chemistry
University of Regensburg
Regensburg, West Germany

J.-P. FARGES
School of Chemistry
Macquarie University
Nth. Ryde, NSW 2113, Australia

E. D. GODDARD
Union Carbide Corporation
Tarrytown Technical Center
Tarrytown, New York

H.-J. GORES
Department of Chemistry
University of Regensburg
Regensburg, West Germany

F. GUTMANN
School of Chemistry
Macquarie University
Nth. Ryde, NSW 2113, Australia

P. SOMASUNDARAN
Henry Krumb School of Mines
Columbia University
New York, New York

SERGIO TRASATTI
Laboratory of Electrochemistry
The University
Milan, Italy

R. WACHTER
Department of Chemistry
University of Regensburg
Regensburg, West Germany

A Continuation Order Plan is available for this series. A continuation order will bring delivery of each new volume immediately upon publication. Volumes are billed only upon actual shipment. For further information please contact the publisher.

MODERN ASPECTS OF ELECTROCHEMISTRY

No. 13

Edited by

B. E. CONWAY
Department of Chemistry
University of Ottawa
Ottawa, Canada

and

J. O'M. BOCKRIS
Department of Chemistry
Texas A & M University
College Station, Texas

PLENUM PRESS • NEW YORK AND LONDON

Library of Congress cataloged the first volume of this title as follows:

Modern aspects of electrochemistry. no. [1]
Washington, Butterworths, 1954-
   v. illus. 23 cm.
   No. 1-2 issued as Modern aspects series of chemistry.
   Editors: no. 1- J. Bockris (with B. E. Conway, no. 3-
Imprint varies: no. 1, New York, Academic Press.—No. 2,
London, Butterworths.

1. Electrochemistry—Collected works.	I. Bockris, John O'M.,
ed. II. Conway, B. E. ed.	(Series: Modern as-
pects series of chemistry)	
QD552.M6	54-12732 rev

Library of Congress Catalog Card Number 54-12732
ISBN 978-1-4615-7457-6   ISBN 978-1-4615-7455-2 (eBook)
DOI 10.1007/978-1-4615-7455-2

© 1979 Plenum Press, New York
Softcover reprint of the hardcover 1st edition 1979
A Division of Plenum Publishing Corporation
227 West 17th Street, New York, N.Y. 10011

All rights reserved

No part of this book may be reproduced, stored in a retrieval
system, or transmitted, in any form or by any means, electronic,
mechanical, photocopying, microfilming, recording, or otherwise,
without written permission from the Publisher

Preface

The present volume contains five chapters covering areas of contemporary interest in the fields of electrolyte solutions, the state of solvent molecules at electrode surfaces, charged colloid interfaces, surface chemistry of oxide electrodes and electrochemistry, and bioelectrochemistry of charge transfer complexes.

The first chapter, by Barthel, Wachter, and Gores, covers the topic of conductance of nonaqueous protic and aprotic electrolyte solutions. This field is not only of intrinsic interest in itself, illustrating the important departures of ion-transport behavior in organic solvents from that, more well known, in water, but the information and extensive new data presented in this chapter will be of interest to those working with nonaqueous alkali-metal batteries where the conductivity and ion-association behavior of electrolytes in various solvents other than water is of great importance.

The second chapter is devoted to a very fundamental and ubiquitous aspect of electrochemistry of electrodes: the state of solvent molecules, adsorbed and oriented, at their surfaces. The role of solvent adsorption and orientation in double-layer properties, it will be recalled, remained poorly understood until the early 1960s. This chapter, by Trasatti, gives a thorough account of the present state of knowledge of solvent orientation at electrode interfaces and of the unsuspected (until recent years) role it plays in properties of the double layer and in determining the potential profile at charged metal surfaces in solution.

In Chapter 3, Somasundaran and Goddard give a useful account of electrochemical aspects of adsorption at mineral solids. This field pertains to the important technology of mineral flotation for concentration of ores, many of which (e.g., the sulfides) are semimetallic substances. The adsorption behavior of various surfactants at such materials is therefore closely connected with problems of adsorption at electrodes and with the stability of colloids.

The use of modern techniques of surface science for the study of electrodes has been a rapidly developing area of electrochemistry in recent years. Augustynski and Balsenc give, in Chapter 4, a self-contained account of the principles of Auger and X-ray photoelectron spectroscopy, with applications to the study of oxide films on noble-metal surfaces. Further knowledge of the chemical valence states of elements in such films is desirable for improved understanding of electrocatalysis of reactions such as electrolytic chlorine and oxygen evolution at these metals.

The final chapter, by Gutmann and Farges, is complementary to a previous contribution by these authors to the *Modern Aspects* series (No. 12) and describes the electrochemistry of charge transfer complexes. These are of interest in bioelectrochemistry and more recently in photoelectrochemistry as substances that can be involved in solar energy transfer and utilization.

Ottawa, Canada
August, 1979

B. E. Conway
J. O'M. Bockris

Contents

Chapter 1

TEMPERATURE DEPENDENCE OF CONDUCTANCE OF ELECTROLYTES IN NONAQUEOUS SOLUTIONS

J. Barthel, R. Wachter, and H.-J. Gores

I. Introduction	1
II. Experimental Aspects	4
1. Temperature Control	4
2. Conductance Measurements	6
3. Organic Solvents	10
III. Dilute Electrolyte Solutions	14
1. Conductance Equations	14
2. Thermodynamics of Ion-Pair Formation	18
3. Ionic Conductance	25
4. Analysis of Experimental Data	27
5. Results and Discussion	37
IV. Concentrated Electrolyte Solutions	52
1. Survey	52
2. Analysis of Experimental Data	55
3. Electrolytes in Pure Organic Solvents	55
4. Organic Solvent Mixtures	66
References	71

Chapter 2

SOLVENT ADSORPTION AND DOUBLE-LAYER POTENTIAL DROP AT ELECTRODES

Sergio Trasatti

I. Introduction 81
   1. Scope of the Chapter 81
   2. Historical Survey 84
II. Water Dipole Contribution to the Potential of
    Zero Charge 90
   1. Basic Double-Layer Model 90
   2. Relative Values of Surface Potential of Water
     at Various Metals 95
   3. Surface Potential of Water at Mercury 100
   4. Surface Potential of Water at the Free Surface ... 105
   5. Surface Potential of Water at Other Metals 107
III. Relation of Surface Potential to Strength of Water
    Adsorption 108
   1. Hydrophilicity of Metals 109
   2. The Interaction Parameter of Metal Surfaces 115
   3. Additional Evidence for the Hydrophilicity
     Scale 117
   4. Dielectric Aspects 122
IV. Description of the Structure of Interfacial Water 126
   1. Theoretical Structural Models 127
   2. Comparison with Experiments 133
   3. The Idea of Water Clusters 142
V. Electric Field Effects 145
   1. Charge Dependence of Water Dipole
     Orientation 145
   2. The Capacity Hump 149
   3. The Charge for Zero Net Dipole Orientation 151
VI. Temperature Effects 155
   1. Temperature Dependence of Water Dipole
     Orientation 155
   2. Surface Excess Entropy in the Inner Layer 164
VII. Suggestions for a Possible Alternative Model 168

VIII. Nonaqueous Solvents	174
1. Qualitative Aspects	174
2. Quantitative Aspects	184
References	193

Chapter 3

ELECTROCHEMICAL ASPECTS OF ADSORPTION ON MINERAL SOLIDS
P. Somasundaran and E. D. Goddard

I. Introduction	207
II. Basic Principles	208
III. Charge Generation	209
IV. Electrostatic Adsorption	219
V. Lateral Interaction between Adsorbates	225
VI. Chemical Forces	233
VII. Chemical State of the Adsorbate	233
VIII. Miscellaneous	236
IX. Application of Boundary Tension and Contact Angle Measurements	239
X. Adsorption Kinetics	240
XI. Current Research Trends	242
References	248

Chapter 4

APPLICATION OF AUGER AND PHOTOELECTRON SPECTROSCOPY TO ELECTROCHEMICAL PROBLEMS
J. Augustynski and Lucette Balsenc

I. Introduction	251
1. Methods	251
2. Surface Analysis	252

II. Auger Electron Spectroscopy 253
   1. Physical Process 253
   2. The Auger Spectrum 254
   3. Instrumentation 255
   4. Qualitative Analysis 257
   5. Quantitative Analysis 258
   6. Chemical Effects 263
   7. Background Subtraction and Deconvolution 269
   8. Electron Beam Effects 274
III. X-Ray-Excited Photoelectron Spectroscopy 277
   1. Physical Process 277
   2. The Photoelectron Spectrum 277
   3. Instrumentation 283
   4. Elemental Analysis 284
   5. Quantitative Analysis 284
   6. Chemical Effects 287
   7. Determination of Valence Band Density of States 294
   8. Deconvolution Methods 296
   9. X-Ray-Excited AES 296
   10. X-Ray-Induced Damage 298
IV. AES and XPS for Surface Analysis 299
   1. Escape Depth 300
   2. Distinction between Surface and Bulk Contributions 302
   3. Profile 303
   4. Surface Roughness 309
   5. Structure 310
V. Surface Analysis Applied to Electrocatalysts 312
   1. Oxide Layers on Noble Metals 313
   2. RuO_2-Based Film Electrodes 325
   3. Characterization of Electrode Surfaces in Connection with Adsorption Studies 330
VI. Application of AES and XPS to Passivity and Corrosion Studies 331
   1. AES Studies 332
   2. XPS Studies 334
VII. Concluding Remarks 342
References 347

Chapter 5

AN INTRODUCTION TO THE ELECTROCHEMISTRY OF CHARGE TRANSFER COMPLEXES II
F. Gutmann and J.-P. Farges

I.	Introduction	361
II.	Ionics	362
	1. Donicity	362
	2. Conductivities	364
	3. Charge Carriers.	369
	4. Solvent Interactions	376
	5. Colloid and Surface Complexations	379
	6. Stochastic Processes	381
III.	Electrodics	387
	1. Complex Formation as an Electrode Reaction	387
	2. Electrocatalysis and Heterogeneous Catalysis	388
	3. Electrochemical Methods	401
IV.	Ternary and Proton Transfer Complexes.	407
	1. Ternary Complexes	407
	2. Proton Complexes	410
V.	Charge Transfer Complexes as Electrochemical Energy Storage Devices	412
References.		413

Index 423

1

Temperature Dependence of Conductance of Electrolytes in Nonaqueous Solutions

J. Barthel, R. Wachter, and H.-J. Gores

Department of Chemistry, University of Regensburg, West Germany

I. INTRODUCTION

NOTE. All equations in this chapter can be used either with SI or cgs units.[1] For this purpose, appropriate physical constants and conversion factors are summarized in Table 1. Concentration c is used throughout at mol liter^{-1}. This requires a conversion factor 10^n with $n(\text{SI}) = +3$; $n(\text{cgs}) = -3$.

The importance of nonaqueous electrolyte solutions, both in fundamental research and in technology, is manifested by the steady growth of publications in this field in the last decade. Monographs and handbooks[2-11] especially devoted to nonaqueous electrolyte solutions and review articles[12-26] provide a comprehensive survey of investigations and results. A perusal of published data on the temperature dependence of nonaqueous electrolyte solution properties, however, yields few items of information; this kind of research especially has been scarcely applied to transport phenomena in spite of the associated technological interest. One of the technical aspects of these investigations in concentrated electrolyte solutions is the optimization of supporting electrolytes of low-temperature nonaqueous cells and high-energy batteries.[26-30,256,270]

Table 1
Physical Constants and Conversion Factors

Quantity and symbol	Values with estimated uncertainty[a]	
	SI	cgs
Avogadro constant, N_A	6.02252×10^{23} mol^{-1} $\pm 0.00028 \times 10^{23}$ mol^{-1}	
Planck constant, h	6.6256×10^{-34} J s $\pm 0.0005 \times 10^{-34}$ J s	6.6256×10^{-27} erg s $\pm 0.0005 \times 10^{-27}$ erg s
Boltzmann constant, k	1.38054×10^{-23} J K^{-1} $\pm 0.00009 \times 10^{-23}$ J K^{-1}	1.38054×10^{-16} erg K^{-1} $\pm 0.00009 \times 10^{-16}$ erg K^{-1}
Charge of proton, e	1.60210×10^{-19} C $\pm 0.00007 \times 10^{-19}$ C	4.8029×10^{-10} esu $\pm 0.0002 \times 10^{-10}$ esu
Permittivity of vacuum, ε_0	8.854185×10^{-12} J^{-1} C^2 m^{-1} $\pm 0.000018 \times 10^{-12}$ J^{-1} C^2 m^{-1}	$(4\pi)^{-1}$ (exactly)
Conversion factor,[b] Ξ	1 (exactly)	9.98755×10^{11}

[a] From Ref. 1.
[b] From Ref. 8.

The reason for the lack of reliable conductance data, especially in dilute solutions, is obvious. Evaluation of measurements within the framework of present electrolyte theory requires an accuracy in the measurement of data which is difficult to obtain over a sufficiently large temperature range. In order to obtain conductance data which can be evaluated by a complete conductance equation, e.g., Eqs. (1), it is necessary to set each temperature of the temperature program quickly and reproducibly within 10^{-3} K, taking into account a temperature coefficient of conductance up to 0.1 K^{-1} (cf. Section III). Moreover, short as well as long time deviations beyond 10^{-3} K cannot be permitted. Figure 1 shows a thermostat assembly which fulfils these requirements[31,32] within a sufficiently large temperature range (cf. Section II).

According to modern electrolyte theory, conductance data obtained for dilute solutions can successfully be analyzed with the help of the conductance equation in the form of the following set of equations (cf. Section III):

$$\Lambda = \alpha[\Lambda^\infty - S(\alpha c)^{1/2} + E\alpha c \log \alpha c + J_1 \alpha c + J_2(\alpha c)^{3/2}] \quad (1a)$$

$$K_A = \frac{1-\alpha}{\alpha^2 c} \frac{y_A}{y'^2_\pm} \quad (1b)$$

$$y'_\pm = \exp\left(-\frac{\kappa q}{1+\kappa R_y}\right) \tag{1c}$$

$$y_A = 1 \tag{1d}$$

As usual, Λ is the experimentally determined molar conductance of the solution at an electrolyte concentration c and Λ^∞ the molar conductance at infinite dilution ($c \to 0$). The coefficients S, E, J_1, and J_2 depend on the pair-distribution functions of the ions and the boundary conditions on which a special conductance theory is based (cf. Ref. 8). K_A is the equilibrium constant for the formation of ion pairs from the free ions, α is the degree of dissociation, $y'_\pm$ the mean activity coefficient on the molar scale of the dissociated part of the electrolyte, and y_A that of the associated part. The Debye parameter κ is given by Eq. (2); the Bjerrum distance q by Eq. (3). R_y is the distance parameter for the activity coefficient (cf. Section III):

$$\kappa^2 = 16\pi q N_A(\alpha c) \times 10^n \tag{2}$$

$$q = \frac{z^2 e^2}{8\pi\varepsilon_0\varepsilon_r kT} \tag{3}$$

Equally well-founded equations are not so far known for conductance data for highly concentrated electrolyte solutions. Bruno and Della Monica[33-35] have extended Angell's model of mass transport in aqueous solutions,[36] which is based on the fused-salt theory,[37] to salts dissolved in media other than water. Extensions of Eqs. (1) with the help of empirical parameters,[38-41] e.g., the viscosity of the solution, become doubtful at high concentrations because of the approximations underlying Eq. (1a) as established for low concentrations and the validity of the ion-pair concept itself at high concentrations.

A noteworthy exact extension of Eq. (1a) for nonassociating electrolytes ($\alpha = 1$) into the region of moderate concentrations ($c \approx 0.25$ mol liter^{-1}) is due to Ebeling, Geisler, Kraeft, and Sändig.[42] They based their conductance equation on a statistical-mechanical treatment[42-45] which takes into account interaction potentials other than those for hard spheres. A recent further improvement of this model yields a conductance equation[46] valid up to still higher concentrations.[46,47] Data of nonaqueous solutions, however, have not been treated up to now.

Compilation of conductance data for highly concentrated solutions can best be achieved in an unambiguous manner by means of empirical functions which do not have a theoretical basis. A comprehensive study in this field then allows rules to be expressed which give a first insight into the effects of the conductance-determining factors and are useful for technological applications.

Our experience with different empirical functions has led to the application of functions which relate the specific conductance* κ of the solution to the maximum κ_{max} of κ and the appropriate concentration μ in the molal scale as adaptation parameters.[48] An equation published by Casteel and Amis,[49]

$$\frac{\kappa}{\kappa_{max}} = \left(\frac{m}{\mu}\right)^a \exp\left[b(m-\mu)^2 - \frac{a}{\mu}(m-\mu)\right] \qquad (4)$$

has been shown to provide a good fit of data for many solutions over a wide concentration range around μ and at various temperatures (cf. Section IV). The quantities κ_{max}, μ, and the constants a and b of Eq. (4) are adjusted by a least-squares method. Equation (4) fulfills the conditions $\kappa = \kappa_{max}$ if $m = \mu$, and $\kappa = 0$ if $m = 0$ and $a > 0$.

II. EXPERIMENTAL ASPECTS

1. Temperature Control

Figure 1 shows the thermostat assembly with a three-electrode conductance cell immersed in the controlled temperature bath as used in the author's laboratory. The range of operation of a thermostat is limited at high temperatures by the vapor pressure and at low temperatures by the viscosity of the bath liquid. With silicone oil, Baysilon M5® (Bayer), as the bath fluid a temperature range of −60 to +50°C is covered which allows the temperature dependence of the properties of numerous nonaqueous electrolyte solutions and their solvents to be investigated.

---

* In accordance with the recommendations of the IUPAC Commission on Symbols, Technology, and Units[1] the symbol κ is used for specific conductance. Confusion with κ, Eq. (2), is easy to avoid. These two quantities are never used in the same context.

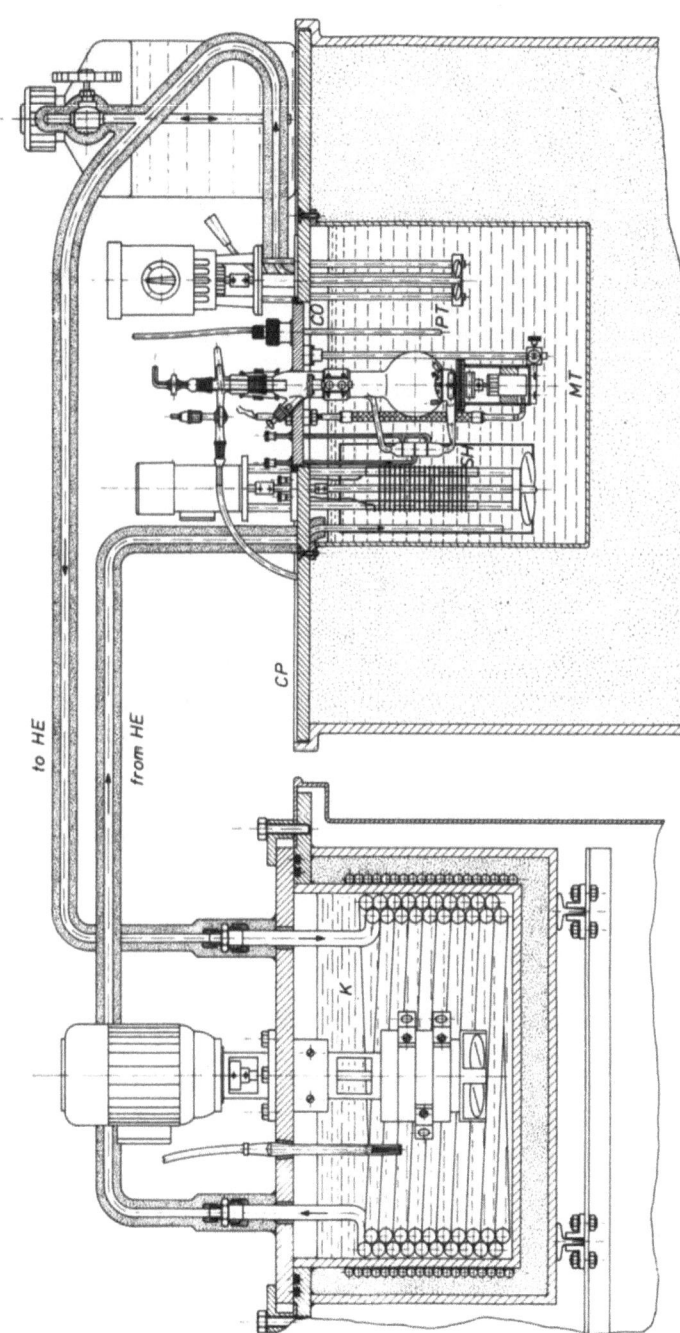

Figure 1. Thermostat assembly with cold bath (K), measurement thermostat (MT), and immersed conductivity cell. For explanation see text.

The measurement thermostat (MT) with a bath of about 60 liters is coupled to a cold bath (K) by means of a heat exchanger (HE). The temperature of the measurement thermostat is controlled by a PID controller joined to an a.c. bridge which contains a platinum resistance thermometer (PT). The error voltage of the bridge is used both for temperature measurement and, via the PID controller, for controlling the heating power of the source of heat (SH). The thermostat is hermetically sealed to prevent atmospheric moisture entering the bath. A circular opening (CO, ϕ 18.5 cm) in the cover plate (CP) permits the immersion or connection of measuring cells (conductance, permittivity, viscosity, density, solubility, etc.) which for their part are supplied with assembly plates guaranteeing hermetical sealing of the opening (CO). For a detailed description, see Refs. 31 and 32.

2. Conductance Measurements

The experimental methods for determining the concentration dependence of conductance are as follows:

(i) preparation of each solution by mixing the solvent and electrolyte compounds by weight (usual method);
(ii) stepwise dilution of a concentrated solution in the measuring cell by adding the pure solvent (cf. Refs. 12 and 50);
(iii) stepwise concentration by successive additions of weighed samples of the electrolyte compound, starting from the pure solvent (cf. Refs. 32, 51, and 52).

Each method requires appropriate measuring cells built according to the requirements of the electrolyte solution and the desired concentration range to be studied.

For the investigation of temperature dependence of conductance and for dilute solutions, method (iii) proves to be the most promising solution of the problem when a method of isologous sections is to be used to establish $\Lambda-c-T$ diagrams.[8,32,53] Figure 2 shows an appropriate type of conductance cell (cf. also Fig. 1).

For stepwise increase of concentration and isologous sections, the highly purified solvent is introduced into the mixing chamber (M) of the conductance cell (Fig. 2a) through the inlet (I) under protective gas [see Fig. 2b: cap of the conductance cell with gas inlet (G)]. The conductance of the solvent is measured at the various different temperatures of the program. Only then is the first electrolyte concentration prepared by

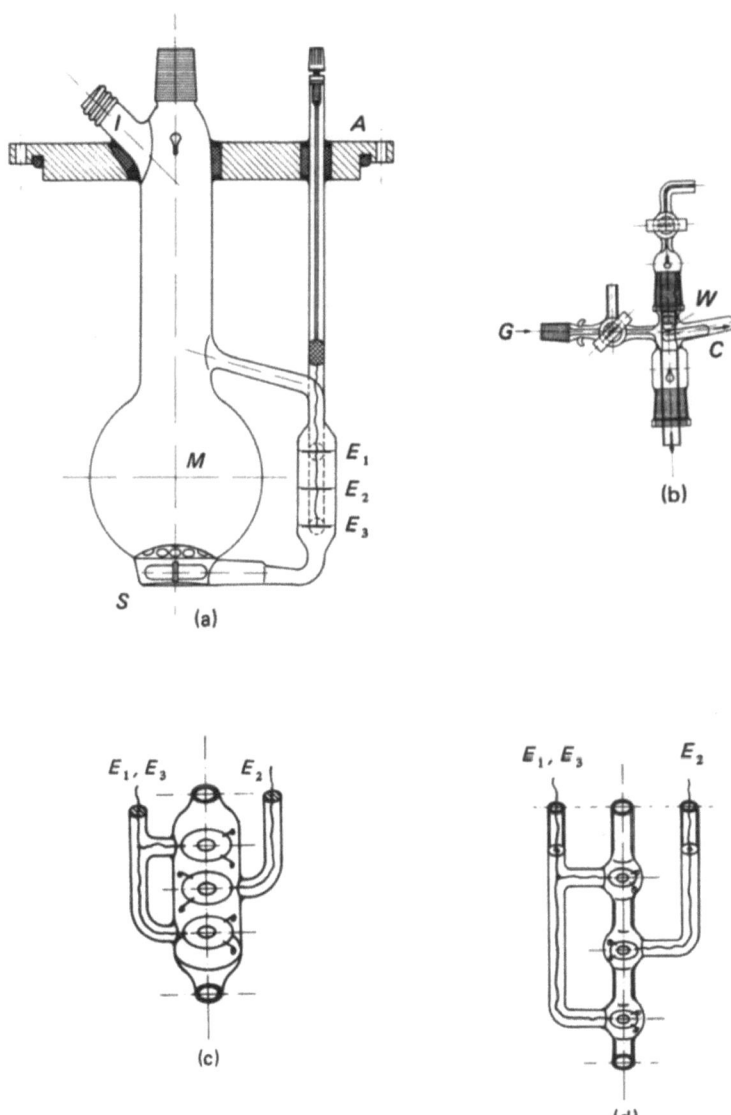

Figure 2. Three-electrode measuring cell and mixing chamber with assembly lid (A, Fig. a) for immersion in the temperature bath (cf. Fig. 1). Cell assembly (a); inlet cap (b); electrode assembly of low (c) and moderate (d) cell constant. For further explanation see text.

introducing the electrolyte in a weighing bottle (W) through the cap of the cell (Fig. 2b) with the help of a magnetic switch (C). After mixing thoroughly [stirrer (S)], the temperature program is repeated with this first solution. The highest electrolyte concentration envisaged is reached after eight or ten concentration steps. Addition of the electrolyte can be achieved alternatively by introducing a concentrated solution through (I) from a weighing burette or by a chemical reaction which produces a definite amount of the electrolyte compound in the measuring cell (cf. Refs. 53 and 54).

Conductance measurement in the measuring cell, Fig. 2, is accomplished with a three-electrode assembly in an arm of an a.c. bridge built on the classical lines of Jones[55] and Shedlovsky[56,57] according to present standards of technology. Current improvements are based on the application of highly stable resistors of low time constant ($<10^{-8}$ sec) and the use of lock-in amplifiers as null detectors providing a resolution up to 10^{-6} at a bridge supply voltage of 100 mV.[31,32]

The outer electrodes E_1 and E_3 of the measuring cell, Fig. 2, are connected and thus restrict the electric field to the interior of the measuring system. Figure 2c shows an electrode arrangement yielding low cell constants A ($0.05 < A/\text{cm}^{-1} < 2$); the arrangement in Fig. 2d permits higher values ($2 < A/\text{cm}^{-1} < 10$). The protected electrode assembly of Hawes and Kay[51] is an equally good system for guarding electrodes but is limited to low cell constants. In the range of high electrolyte concentrations, conductance cells with high cell constants are needed ($10 < A/\text{cm}^{-1} < 300$). Capillary cells (Fig. 3) can be advantageously used in connection with an appropriate filling device for isologous sections and by using method (i).[48] For a type of stirred capillary cell which possesses high cell constants, see Ref. 58. Surveys of further equipment and experimental techniques have been given by Hills[15] (1970), Braunstein and Robbins[59] (1971), and Evans and Matesich[60] (1973). For recent developments, see Refs. 61–64, 170.

One of the chief difficulties in a highly precise determination of the temperature dependence of electrolyte conductance is the temperature dependence of the cell constant A. At 25°C the cell constant A_{298} can be determined either by using standard solutions[65,66] or by means of conductance equations for dilute[67,68] or moderately concentrated[69,70] aqueous KCl solutions. A recent

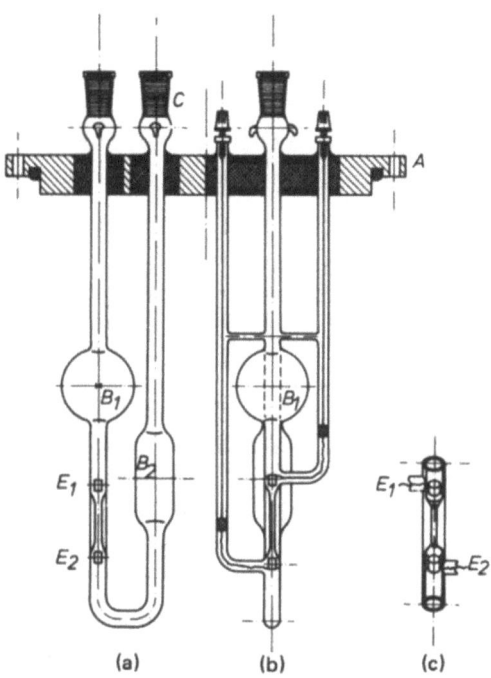

Figure 3. Capillary cell (a, b) with assembly lid (A) for immersion in the temperature bath (cf. Fig. 1); (c) electrode assembly of the cell. The arrangement of the bulbs B1 and B2 effects a replacement of the solution between the electrodes E1 and E2 at every temperature step with the help of pressure change within the cell. Inlet C permits a bubble-free filling under protective gas from an appropriate filling device.

survey compares the precision of these different methods.[71] In a determination of the temperature coefficient $\beta = A_{298}^{-1} (dA/dT)$, based on standard values at 25, 18, and 0°C,[65,66] only a narrow temperature range is available yielding β values of low accuracy. On the other hand, β is defined for well-designed capillary cells only by the linear expansivity of the cell material $[\alpha_{\text{Pyrex}} = 3.5 \times 10^{-6} \text{ K}^{-1}$, $\alpha_{\text{quartz}} = 0.5 \times 10^{-6} \text{ K}^{-1}$ (Ref. 72)].[71,73] A thorough experimental investigation yields β values from both methods within $\Delta\beta = 3 \times 10^{-6} \text{ K}^{-1}$.[32] The temperature-dependent cell constant is given by

$$A_T = A_{298}[1 + \beta(T - 298.15)] \qquad (5)$$

3. Organic Solvents

The degree of purity needed for a solvent is determined by the lowest electrolyte concentration of the planned series of measurements. Two classes of impurities must be distinguished:

(i) inert impurities which, by their presence, change only the physical properties of the solvent;
(ii) interfering impurities which react with the added electrolyte compound, and/or are involved in selective solvation effects.

A typical example of the first class of impurity is water in methanol–NaCl or methanol–NaOCH$_3$ solutions[74,75]; in the second class, CO_2 in methanol–KOCH$_3$ solutions (reacting to form potassium methyl carbonate)[12] is an example. For further examples, see Refs. 76–79. Investigations on conductance in binary solvent mixtures, especially in those containing small amounts of water, give further insight into these problems (e.g., see Ref. 80). Impurities of the first class are largely rendered ineffective by a correction to the specific conductance:

$$\kappa = \kappa_{solution} - \kappa_{solvent} \qquad (6)$$

Impurities of the second class must be reduced to a level at which they do not influence the measurements. Usually, the solvent should remain under an inert protective gas from the time of its chemical purification until its introduction into the measuring cell. Purification apparatus for this procedure has been described.[12,283] Figure 4 shows an apparatus for the complete purification of the solvent under an inert gas.

The analysis of data from temperature-dependent conductivity measurements by means of Eq. (1) requires also the knowledge of temperature-dependent solvent data (density, permittivity, viscosity). An inspection of the most comprehensive publication in this field[81] shows that data at low temperatures are lacking. A systematic study in the authors' laboratory helps to make accessible precise solvent data for temperatures in the range $-50 < \theta < +40°C$ which will complement the known temperature range.[84] Tables 2, 3, and 4 summarize the results which are used in the later text. Figure 5 shows the capacitor for

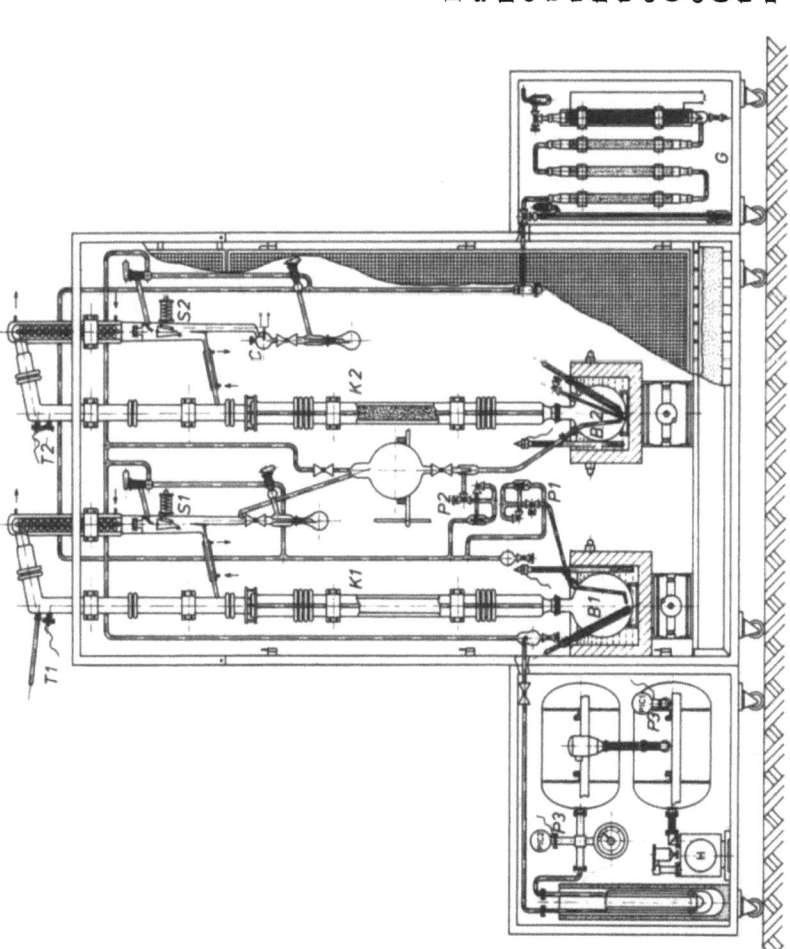

Figure 4. Apparatus for the purification of solvents in an inert gas atmosphere of arbitrary pressure ($2 < p$ (Torr) < 1000). Chemical purification of the solvent with appropriate reactants in B1 (approximately 6l); distillation through column K1 into B2; final distillation from B2 (packed column K2, about 40 theoretical plates) into fraction collector C controlled by a conductance measuring cell ($A \sim 0.01\,\text{cm}^{-1}$), S1, S2: electromagnetically controlled takeoff. P1, P2, P3: pressure control ($\Delta p \leq 0.2$ Torr). G: inlet for the purified protective gas. T1, T2: temperature control or registration. Grease-free throughflow valves are used in the distillation line.

Table 2
Density $\rho/\text{g ml}^{-1}$ of Nonaqueous Solvents as a Function of Temperature[a]

Solvent	$\dfrac{\text{g ml}^{-1}}{\rho} = a + b(T-273.15) + c(T-298.15)^2$					Experimental
	a	$b \times 10^3$	$c \times 10^6$	$10^5 \sigma(1/\rho)$	$\rho_{298.15}$	$\rho_{298.15}$
Methanol	1.23452	1.4283	1.52	6	0.78667	0.78664
Ethanol	1.24027	1.3031	1.39	7	0.78511	0.78507
n-Propanol	1.22050	1.173	1.28	40	0.79961	0.79954
Acetonitrile	1.24446	1.6458	2.92		0.77674	0.77676
Propylene carbonate	0.81527	0.7149	0.459		1.19986	1.19983

[a] Reference standard: $\rho_{298}(H_2O) = (0.9970751 \pm 7 \times 10^{-7})$ g ml^{-1} (Ref. 85).

Table 3
Relative Permittivity ε_r of Nonaqueous Solvents as a Function of Temperature[a]

Solvent	$\varepsilon_r = a + bT^{-1} + cT^{-2}$					Experimental
	a	$b \times 10^{-3}$	$c \times 10^{-6}$	$10^3 \sigma(\varepsilon)$	$\varepsilon_{298.15}$	$\varepsilon_{298.15}$
Methanol	−23.18	16.02	0.182	7	32.60	32.597
Ethanol	−24.33	15.48	−0.288	10	24.35	24.343
n-Proponal	−27.24	16.33	−0.629	14	20.45	20.421
Acetonitrile	−21.06	20.23	−0.963	3	35.96	35.958
Propylene carbonate	−33.09	36.65	−2.210	10	64.97	64.980

[a] Reference standard: $\varepsilon_{r,295}(Ar) = 1.0005173 \pm 4 \times 10^{-7}$ (Ref. 86).

Table 4
Viscosity η/cP of Nonaqueous Solvents as a Function of Temperature[a]

Solvent	$\ln \dfrac{\eta}{cP} = a + bT^{-1} + cT^{-2}$					Experimental
	a	$b \times 10^{-3}$	$c \times 10^{-6}$	$10^3 \sigma(\eta)$	$\eta_{298.15}$	$\eta_{298.15}$
Methanol	−5.0858	1.421	−0.0262	1.7	0.5417	0.5409
Ethanol	−6.5642	2.316	−0.099	2.0	1.089	1.087
n-Propanol	−6.2460	2.061	+0.0009	0.1	1.967	1.967
Acetonitrile	−3.5164	0.620	+0.0325	1.1	0.342	0.342
Propylene carbonate[b]	+5.3565	−4.234	+0.870	14	2.55	2.53

[a] Reference standard. kinematic viscosity $\nu_{295}(H_2O) = 1.0038$ cST (Ref. 87).
[b] Data of propylene carbonate are better represented by series expansion up to T^{-4} (Ref. 84).

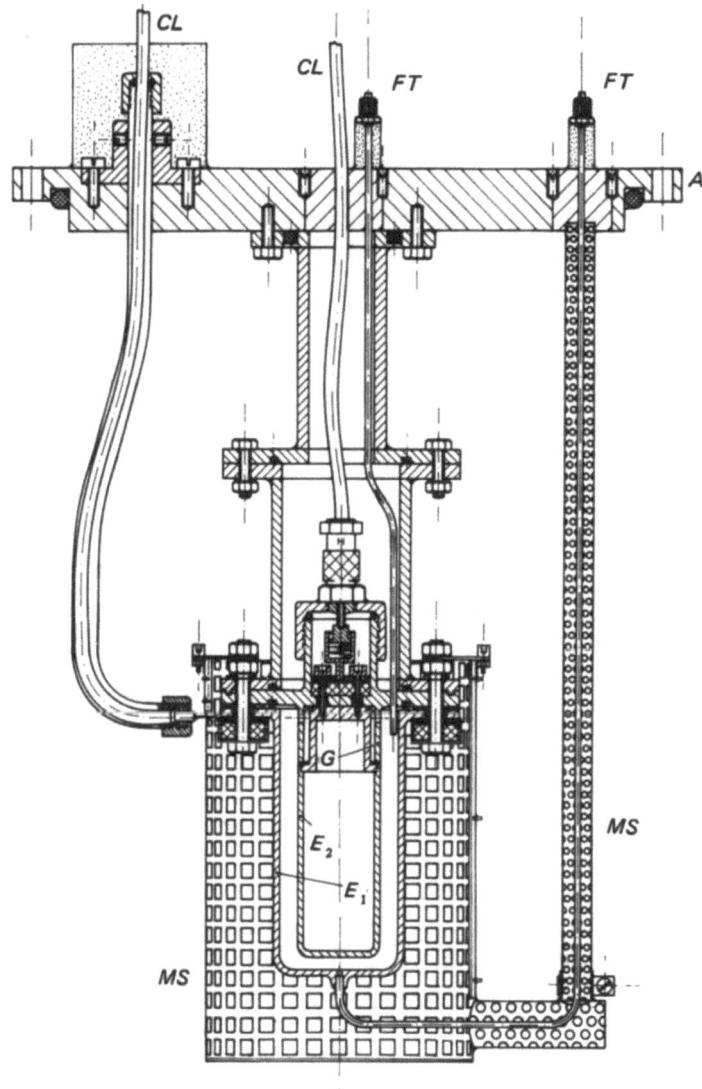

Figure 5. Capacitor for measuring permittivity of solvent. A: Assembly lid for immersion in the temperature bath, Fig. 1; CL: coaxial lines; E1: outer electrode E2: inner electrode; G: guardring, MS: metal shield; FT: filling tubes (stainless steel capillaries).

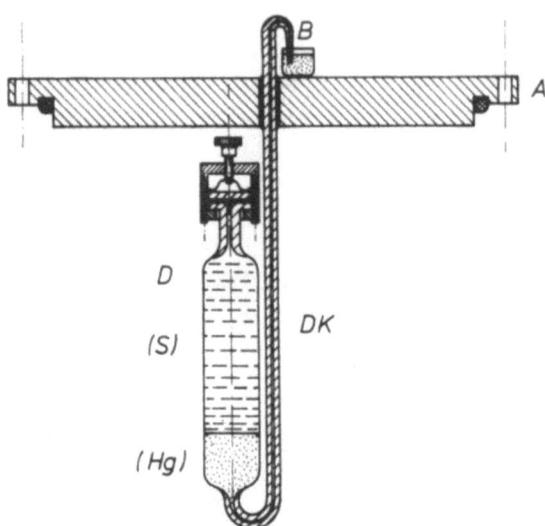

Figure 6. Dilatometer for determining temperature dependence of solvent density based on the lines of Gibson and Loeffler[82] and Burlow.[83] A: Assembly lid for immersion in the temperature bath, Fig. 1; D: dilatometer filled with solvent (S) and mercury (Hg); DK: dilatometer capillary; B: weighing bottle for the expelled mercury.

permittivity measurements on liquids, and Fig. 6 the dilatometer for density determination; viscosity is measured by means of the usual type of Ubbelohde viscosimeter.[84] These measuring devices are used in connection with the precise temperature bath shown in Fig. 1.

Further solvent data can be found in recent publications. A critical and comprehensive survey of organic solvent and mixed solvent data would go beyond the scope of this article.

III. DILUTE ELECTOLYTE SOLUTIONS

1. Conductance Equations

The theory of electrolytes is constructed around the pair-distribution functions $f_{ij}(\mathbf{r}_1, \mathbf{r}_2)$, Eq. (7), of the ions i, j in the solution.[43,88-93] A binary symmetrical electrolyte $Y = C^{z_+}A^{z_-}$, as considered in Section I, gives rise to ions i, $j = C^{z_+}$, A^{z_-} with densities ρ_i and ρ_j (ions/volume). The distribution of these ions depends on the forces acting between all the particles, viz., ions and solvent molecules, i.e., polar, induction, dispersion, and repulsive forces (cf. Refs. 94–97). Chemical effects must also be taken into account. External forces are involved in the description of transport properties:

$$f_{ij}(\mathbf{r}_1, \mathbf{r}_{21}) = \rho_i(\mathbf{r}_1)\rho_{ij}(\mathbf{r}_1, \mathbf{r}_{21})$$
$$= \rho_j(\mathbf{r}_2)\rho_{ji}(\mathbf{r}_2, \mathbf{r}_{12}) = f_{ji}(\mathbf{r}_2, \mathbf{r}_{12}) \qquad (7)$$

In this equation, $\rho_{ij}(\mathbf{r}_1, \mathbf{r}_{21})$ represents the density of j ions at a distance $\mathbf{r}_{21}$ from an i ion situated at a point $P_1(\mathbf{r}_1)$. Orientation of a j ion with respect to the i ion which would require the introduction of Eulerian angles as further coordinates is not taken into account by Eq. (7).

In the discussion of conductance, we begin with a completely dissociated electrolyte Y. Only the theoretical features are considered in this context. Conductance theory is based on pair-distribution functions, Eq. (7), and Onsager's continuity equation[89]

$$-\frac{\partial f_{ij}}{\partial t} = \text{div}(f_{ij}\mathbf{v}_{ij}) + \text{div}(f_{ji}\mathbf{v}_{ji}) = -\frac{\partial f_{ji}}{\partial t} \qquad (8)$$

which can also be derived from the exact Liouville equation.[44] $\mathbf{v}_{ij} = \mathbf{v}_{ij}(\mathbf{r}_1, \mathbf{r}_{21})$ is the velocity of a j ion out of the vicinity of the i ion at $\mathbf{r}_1$. A local hydrodynamic flow in the solvent as a consequence of the forces acting on the ions yields the electrophoretic velocity.[44,92,98] The solvent is treated as a homogeneous medium of permittivity ε and viscosity η.

Conductance equations are obtained in the form

$$\Lambda = \Lambda^\infty - \Lambda^{\text{rel}} - \Lambda^{\text{el}} \qquad (9)$$

with a relaxation term Λ^{rel} and an electrophoretic term Λ^{el} relative to the underlying conductance-inhibiting effects, or as series expansions

$$\Lambda = \Lambda^\infty - Sc^{1/2} + Ec \log c + J_1 c + J_2 c^{3/2} \qquad (10)$$

in which the coefficients S, E, J_1, and J_2 contain contributions due both to the relaxation and electrophoretic effects.*

Restriction to the long-range coulombic forces in the pair-distribution functions and corresponding potentials yields the limiting law,[89,99]

$$\Lambda = \Lambda^\infty - Sc^{1/2} \qquad (11)$$

which has been the object of various semiempirical extensions (cf. Refs. 12 and 73) since Onsager derived it as the first exact conductance equation. Its validity has been repeatedly demonstrated subsequently, within the framework of statistical methods (cf. Ref. 100).

Short-range forces are taken into account by different models. In the theories of Fuoss and Onsager,[101–103] Pitts,[104] Falkenhagen et al.,[43,93,105,106] and Fuoss and Hsia,[69] and in related contributions,[39,92,107–110] rigid charged spheres are introduced through the potential of the Debye–Hückel theory and the appropriate boundary conditions. In contrast to this approach, a new statistical treatment[42–46,111–113] initially allows for the introduction of any interaction potential by the formulation of basic equations for it.[46] Hard-sphere models and additional short-range forces have been used up to now. A contribution of short-range forces to the relaxation force as an essentially new feature was introduced by Falkenhagen, Ebeling, and Kraeft.[44]

For comparison of equations, and for critical or detailed treatments, see Refs. 8, 12, 44, 93, 113–130.

In the following, Eq. (10) is used as the basic conductance equation. This type of equation, established by Fuoss and Onsager, was also adopted by many of the later authors. Some equations, originally established in another form, have been transformed to Eq. (10), e.g., the equation of Pitts by Fernández-Prini and Prue[131] and that of Fuoss and Hsia by Fernández-Prini.[119] A tabular survey of the coefficients appropriate to various equations has been cataloged.[8,12]

---

* For theoretical reasons, a further term of the order $c^{3/2} \log c$ would be expected[147] which always appears in the calculation of thermodynamic functions.[43]

The above brief survey is intended as information about the present state of theories leading to Eqs. (10) or (1a) when $\alpha = 1$. Conductance of nonaqueous solutions, however, implies significant ion-pair formation, i.e., $\alpha \neq 1$, as a further feature and requires the complete set of Eqs. (1) for analysis of the data. The main problem is to introduce an ion-pair concept which is not in contradiction to conductance models.

Without prejudice to the theoretical significance of attempts to explain the association constant in terms of the pair-distribution function and average potential,[42,45,88,132–136,161] this quantity may be still considered in the framework of Eq. (1) as the consequence of a chemical equilibrium between free ions and ion pairs according to the equation

$$C^{z_+} + A^{z_-} \rightleftarrows C^{z_+}A^{z_-} \qquad (12)$$

Then ion pairs must be considered as chemical entities of defined energy content. Their concentration $(1-\alpha)c$ and that of free ions, αc, yield the equilibrium constant, Eq. (1b), which introduces the ion-pair concept into the conductance equations (9), (10), or (11). This procedure enables conductance and association phenomena to be separated to an extent which permits an application of conductance measurements over a wide range, i.e., to determine ion-pair formation as a consequence of purely Coulombic ion–ion interaction as well as to evaluate stability constants of complexes or dissociation constants of molecules (cf. Ref. 137).

Ion pairs, as defined in a wide sense in terms of conductivity, are nonconducting paired states of oppositely charged ions. Ion pairs of equally charged ions can be neglected in dilute solutions. This concept is already the basis of applications of the limiting law, Eq. (11), to incompletely dissociated or associating electrolytes by the procedures of Fuoss and Kraus[138] or Shedlovsky.[139] Later on it was connected with Eq. (10) by Fuoss,[92,109,140] who used K_A^F (cf. Section III.2) as the association constant. Justice repeatedly stressed that only association constants on the basis of Bjerrum's treatment, K_A^B, are consistent with the set of Eqs. (1).[68,108,125,141] For critical remarks, see Refs. 128 and 129.

A treatment based on Bjerrum's concept yields the set of Eqs. (1) with the following specifications (cf. Refs. 88, 157, 161):

$$K_A = 4\pi N_A (10^n) \int_a^R r^2 \exp\left(\frac{2q}{r} - \frac{U_{+-}}{kT}\right) dr \quad (13a)$$

with

$$R = q \quad \text{for} \quad a < q; \quad R = a \quad \text{for} \quad a \geq q \quad (13b)$$

where a is the distance of closest approach of two ions and q is given by Eq. (3). U_{+-} is a constant, short-range interaction potential of the oppositely charged ions which satisfies the condition $U_{+-} = \infty$ if $r < a$. Experimental evidence has led to use of the set of Eqs. (1) in almost the same manner.[142–144,283] The condition, Eq. (13b), can be changed according to the requirements of the real solutions investigated (cf. Sections II.2–II.4). A recent revision of conductance theory by Fuoss,[109,145,146] which is based on a comparable concept, leads to a representation of the conductance equation in the form $\Lambda = \Lambda(\Lambda^\infty, R, K_A)$.

The thermodynamic basis of the association concept underlying Eq. (1b) is treated in the following section in a general way.

2. Thermodynamics of Ion-Pair Formation

Thermodynamic properties of a completely dissociated electrolyte, $Y = C^{z_+}A^{z_-}$, in solution are available (cf. Refs. 8, 91, 148) from knowledge of the chemical potential $\mu_Y = \mu_+ + \mu_-$ or

$$\mu_Y = \mu_Y^\infty + 2RT \ln c_\pm y_\pm \quad (14)$$

In Eq. (14), μ_Y is referred to the chemical potential μ_Y^∞ at infinite dilution, $c_\pm$ is the mean concentration of the electrolyte compound on the molar scale, and $y_\pm$ the corresponding mean activity coefficient.

If, however, an equilibrium of the type represented by Eq. (12) is assumed,

$$\mu_Y = \alpha(\mu'_+ + \mu'_-) + (1-\alpha)\mu'_A \quad (15)$$

μ'_+, μ'_-, and μ'_A are the chemical potentials of free cations, free anions, and the species $A = [C^{z_+}A^{z_-}]$ in solution, respectively, and α is the degree of dissociation. Equation (15) is subject to the equilibrium condition

$$\mu'_A - \mu'_+ - \mu'_- = 0 \qquad (16)$$

and yields, with $c'_\pm$ and $y'_\pm$ referred to the free ions in solution,

$$\mu_Y = \mu'^\infty_Y + 2RT \ln c'_\pm y'_\pm \qquad (17)$$

Taking into account the relationships $\mu'^\infty_Y = \mu^\infty_Y$ and $c'_\pm = \alpha c_\pm$, comparison of Eqs. (14) and (17) allows us to write

$$y_\pm = \alpha y'_\pm \qquad (18)$$

From the equilibrium condition, Eq. (16), we deduce

$$\mu^\infty_A - \mu^\infty_+ - \mu^\infty_- = -RT \ln \frac{c'_A}{c'^2_\pm} \frac{y_A}{y'^2_\pm} \qquad (19)$$

As the left side of Eq. (19) is the molar Gibbs' energy, ΔG°_A, required to form an ion pair in the infinitely dilute solution from the initially infinitely separated ions, and

$$K_A = \frac{c'_A}{c'^2_\pm} \frac{y_A}{y'^2_\pm} = \frac{1-\alpha}{\alpha^2 c} \frac{y_A}{y'^2_\pm} \qquad (20)$$

is the equilibrium constant of the equilibrium in Eq. (12), Eq. (19) can be written in the form

$$\Delta G^\circ_A = -RT \ln K_A \qquad (21)$$

The thermodynamic equilibrium constant K_A is independent of concentration and hence can be expressed by

$$K_A = \lim_{c \to 0} \frac{1-\alpha}{\alpha^2 c} \frac{y_A}{y'^2_\pm} = \lim_{c \to 0} \frac{1-\alpha}{c} \qquad (22)$$

Equation (22) is the link between thermodynamic and a statistical–mechanical treatment of the association problem.[144] An important fact can be illustrated if Eq. (18) is combined with Eq. (22) to give, for highly dilute solutions,

$$\ln y_\pm = -K_A c + \ln y'_\pm \qquad (23)$$

As the activity coefficient $y_\pm$ of the electrolyte compound Y is a well-defined quantity at each concentration, Eq. (23) indicates that the distinction between paired states of "free" ions and paired states of ions which are considered to be ion pairs is somewhat arbitrary.

In the preceding discussion no restriction is necessary regarding the nature of species $A = [C^{z_+}A^{z_-}]$. Species A can be an ion pair as defined in Section III.1, a complex compound or the undissociated molecule Y in solution. Accordingly, K_A is used as the association constant, stability constant, or reciprocal dissociation constant.

A further property of K_A is obvious. Like any thermodynamic potential, the quantity ΔG_A° is an additive function and thus can be divided, if ion pair formation is considered, into Coulombic ion–ion (ΔG_A^C) and residual (ΔG_A^*) energy terms,[8,31,144] leading to a formal separation of these contributions in the association constant of a single-stage equilibrium, i.e.,

$$K_A = K_A^C K_A^* = K_A^C \exp(-\Delta G_A^*/RT) \qquad (24)$$

This property has been repeatedly used for semiempirical extensions of K_A^C constants.[31,149–151] It is also implied in modern statistical treatments of the association problem. Association constants of the type K_A^C are obtained in the classical semiphenomenological treatment of the association problem from the expression

$$\frac{1-\alpha}{c} = \frac{4\pi}{c} \int_a^R r^2 \rho_{ij}^0(r)\, dr \qquad (25)$$

in which $\rho_{ij}^0(r)$ is a density function for spherical symmetry of ion–ion interaction potentials, taking into account only the long-range Coulombic forces within the region $a \leq r \leq R$, i.e.,

$$\rho_{ij}^0(r) = \rho_j^0 \exp\frac{2q}{r} \qquad (26)$$

a and R in Eq. (25) indicate the limits around the i ion, within which a paired state of oppositely charged ions i and j yields an ion pair; for the other symbols, see Eq. (3). As $\rho_j^0 = N_A c(10^n)$, for a symmetrical electrolyte, Y, the final equation

$$K_A^C = 4\pi N_A (10^n) \int_a^R r^2 \exp\frac{2q}{r}\, dr \qquad (27)$$

is obtained from Eqs. (22), (25), and (26). K_A^C, as given by Eq.

(27), is Prue's association constant, $K_A^P,^{76,152,153}$ from which the Bjerrum constant, $K_A^B,^{154}$ is obtained with an upper limit $R = q$ and from which the Fuoss constant, $K_A^F,^{150}$ can be formally calculated by setting $R = \frac{4}{3}a.^{155}$ A further constant, K_A^{FE}, derived by Falkenhagen and Ebeling,[156] is of the same type.

Short-range forces can be taken into account by an additional potential $U^*(r)$ which, as a first approximation, is considered constant[142,144,157] within $a \le r \le R$:

$$\rho_{ij}^0(r) = \rho_j^0 \exp\left(\frac{2q}{r} - \frac{U^*}{kT}\right) \qquad (28)$$

This assumption leads, with $N_A U^* = \Delta G_A^*$, to relations of the type of Eq. (24).[144] A further extension of Eq. (27)[144] is based on an interaction potential given by Kelbg,[158] yielding

$$\rho_{ij}^0(r) = \rho_j^0 \exp\left[\frac{2q}{r}\phi(r) - \frac{U^*(r)}{kT}\right] \qquad (29)$$

In Eq. (29), Coulombic forces are assumed to be modified by a function $\phi(r)$, e.g., by local permittivity $\varepsilon(r)$ or by a structural screening factor. Fernández-Prini and Prue[153] have used this concept setting $U^* = 0$ in connection with Booth's theory[159] of local permittivity.

A further separation of the distance- and angular-dependent functions determining the pair-distribution function is, at the present time, not possible. For detailed and critical discussions of association constants, see Refs. 8, 12, 88, 109, 137, 142–146, 155, 156, 160–162, 283, 285.

Treatment of equilibria in statistical thermodynamics is based on the reduced partition functions Q_i of reacting particles:

$$Q_i = \frac{1}{N_A} \frac{(2\pi m_i kT)^{3/2}}{h^3} Z_i \qquad (30)$$

Equation (30) shows the separation of the contributions due to translational $[(2\pi m_i kT)^{3/2}/h^3]$ and internal $[Z_i]$ degrees of freedom. The equilibrium constant of Eq. (12) is given by

$$K_A = \frac{Q_A}{Q_{C^+}^z Q_{A^-}^z} \exp\left(-\frac{\Delta E_0}{RT}\right) \qquad (31)$$

with

$$\Delta E_0 = E_0(C^{z_+}A^{z_-}) - E_0(C^{z_+}) - E_0(A^{z_-}) \quad (32)$$

as the difference in energy between the reacting species in their lowest energy levels. If A is a complex compound or the undissociated molecule Y itself, Eq. (31) applies in an unambiguous manner. An extension to ion pairs, as considered in the above semiphenomenological treatment, gives insight into the basic concept of association theory. If ions are supposed to yield no contributions from internal degrees of freedom $[Z_i(C^{z_+}) = Z_i(A^{z_-}) = 1]$ and the contribution to Z_i of the ion pair is that of an internal translation only,[155] derived from the motion of the ions within the sphere $a \le r \le R$, an ion pair built by the ions C^{z_+} and A^{z_-} in a distance between r and $r + \Delta r$ yields

$$Z_i(r, \Delta r) = \frac{(2\pi\mu kT)^{3/2}}{h^3} 4\pi r^2 \Delta r, \qquad \frac{1}{\mu} = \frac{1}{m_{C^{z_+}}} + \frac{1}{m_{A^{z_-}}} \quad (33)$$

Finally, ΔE_0 is given with the approximation of the most comprehensive of the semiphenomenological treatments, Eq. (29), as

$$\frac{\Delta E_0}{RT} = \frac{2q}{r}\phi(r) - \frac{U^*(r)}{kT} \quad (34)$$

From Eqs. (31) to (34) and taking into account the conversion to K_A as mol^{-1} liter, we obtain

$$K_A = 4\pi N_A(10^n) \int_a^R r^2 \exp\left[\frac{2q}{r}\phi(r) - \frac{U^*(r)}{kT}\right] dr \quad (35)$$

The way to derive Eq. (35) makes evident what the association theory is able to explain at present [cf. Eqs. (13a) and (13b)].

On the other hand, Eq. (31) has been applied to ion pairs by assuming rotational and vibrational degrees of freedom instead of internal translation contributing to $Z_i(C^{z_+}A^{z_-})$.[163–165] The limits of these treatments are obvious, too. It is essential to specify the model on which the discussion is based.

Conductance measurements, when used as a probe for incomplete dissociation or association of an electrolyte compound, only provide information on the total number of nonconducting species in the solution. From a thermodynamic point of view, the identification of $(1-\alpha)/c$ and the equilibrium constant K_A [cf. Eqs. (1) and (25)] must take into account the fact that generally more than one nonconducting species may exist in the solution. For example, for a multistep equilibrium (cf. Refs. 88, 161)

$$C^{z_+} + A^{z_-} \underset{K_I}{\rightleftarrows} [C^{z_+}A^{z_-}]_I \underset{K_{II}}{\rightleftarrows} [C^{z_+}A^{z_-}]_{II} \quad (36)$$

between free ions and ion pairs, complex compounds, or molecules $[C^{z_+}A^{z_-}]_I$ and $[C^{z_+}A^{z_-}]_{II}$ having different energy contents (cf. Fig. 7), it is necessary to write

$$K_I = \frac{c_I}{c_\pm'^2} \frac{y_I}{y_\pm'^2} \quad (37a)$$

$$K_{II} = \frac{c_{II}}{c_I} \frac{y_{II}}{y_I} \quad (37b)$$

where $c_\pm'$ is the mean concentration of the free ions, c_I the concentration of $[C^{z_+}A^{z_-}]_I$, and c_{II} that of $[C^{z_+}A^{z_-}]_{II}$; $y_\pm'$, y_I, and y_{II} are the appropriate activity coefficients on the molar scale. As usual, $y_I = y_{II} = 1$ in dilute solutions, and hence

$$K_I K_{II} + K_I = \frac{c_I + c_{II}}{c_\pm'^2 y_\pm'^2} \quad (38)$$

where $c_I + c_{II} = c_A'$ is the total concentration of nonconducting

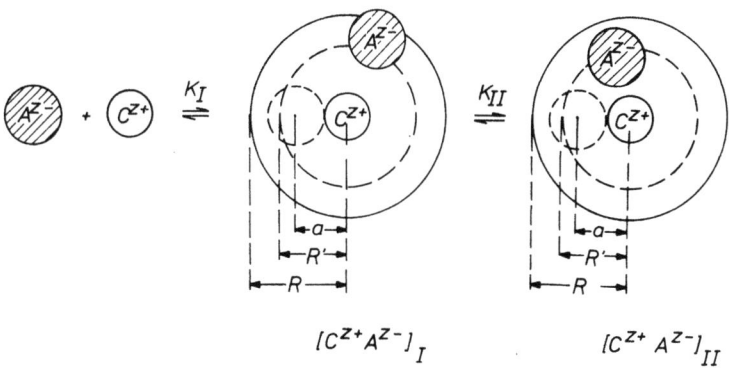

Figure 7. Scheme of a two-step ion-pair formation $C^{z_+} + A^{z_-} \rightleftarrows [C^{z_+}A^{z_-}]_I \rightleftarrows [C^{z_+}A^{z_-}]_{II}$: R and R' association limits of species I; R' and a of species II.

species given by $c'_A = (1-\alpha)c$. Then Eq. (38) can be written

$$\frac{1-\alpha}{\alpha^2 c}\frac{1}{y'^2_{\pm}} = K_I K_{II} + K_I \tag{39}$$

and K_A, Eq. (1b), is an overall constant:

$$K_A = K_I(1+K_{II}) \tag{40}$$

The ion pair $[C^{z_+}A^{z_-}]_I$ is generally determined only by long-range forces,

$$K_I = 4\pi N_A(10^n) \int_{R'}^{R} r^2 \exp\frac{2q}{r} dr \tag{41}$$

and hence K_{II} can be obtained from the measured constant K_A and the constant K_I based on appropriately estimated values of R' and R.

Association constants are obtained from Λ–c measurements over a concentration range of the electrolyte compound which permits the use of activity coefficients based on Debye–Hückel theory, Eq. (1c). The activity coefficient of the ion pairs, Eq. (1d), in this range is that of an ideally dissolved polar molecule (cf. Ref. 148), i.e., $y_A = 1$. Debye–Hückel theory is based on a potential $\psi_i(r)$ satisfying the boundary conditions[8]

$$\lim_{r\to\infty}\psi_i(r) = 0 \tag{42a}$$

$$-\varepsilon_0 \iint \varepsilon_r \text{ grad } \psi_i(r)\, d\mathfrak{f} = e_0 z_i \tag{42b}$$

The first of these conditions requires a vanishing potential if $r \to \infty$; the second one expresses the simple fact that the total flow of the dielectric displacement vector through a closed surface equals the sum of enclosed charges. Thus R_y, Eq. (1c), can be defined as the upper limit R of association. The activity coefficient

$$-\ln y'_{\pm} = \frac{\kappa q}{1+\kappa R} \tag{43}$$

satisfies Eq. (23) in connection with the association constant, Eq. (25). Bjerrum's choice, $R_y = R = q$, is a consistent assumption. Prue et al.[76,152,153] determined the value of $R_y = R$ from experimental data.

In a thorough analysis of activity in solvents of lower dielectric constant, Justice and Justice[161] rederived activity coefficients from the distribution functions of Meeron[166] and Debye–Hückel by using the compressibility and pressure equations of Rasaiah and Friedman[96] and the virial equation. The activity coefficient results as a series expansion versus concentration in the form[161]

$$\ln y_\pm = \kappa q + \kappa q \kappa R[1 - \delta + \delta'] - K_A c$$
$$- \tfrac{1}{2}(K_{++} + K_{--})c + 0(c^{3/2}) \quad (44)$$

K_A is the association constant, Eq. (35), with $\phi(r) = 1$ and $U^*(r) = \text{constant}$. K_{++} and K_{--} are association constants of paired states of ions of like sign, which can be neglected in dilute solutions and have not been considered up to now in our discussion. The quantities δ and δ' are functions of q/R^{161}; $\delta - \delta' = 0$ at $R \approx 1.1q$.

3. Ionic Conductance

Ionic conductances λ_i are available from the determination of the relevant transference numbers t_i. Progress in the theory of transference numbers is related to the enlargement of our knowledge of conductance. Investigations of concentration dependence of transference numbers, based on the limiting law of conductance,[167] have been carried out as well as on the complete conductance equations that are available today.[168,169]

Limiting ionic conductances $\lambda_+^\infty = \lambda_{C^{z+}}^\infty$ and $\lambda_-^\infty = \lambda_{A^{z-}}^\infty$ are additive, giving

$$\lambda_+^\infty + \lambda_-^\infty = \Lambda^\infty \quad (45)$$

Equation (45) can provide approximate values of $\lambda_+^\infty = \lambda_-^\infty$ from Λ^∞ if electrolytes having ions of similar shapes and sizes can be found and the interaction forces of cation and anion with the solvent are equal, e.g., as with $(i\text{-}C_5H_{11})_4N\text{-}B(i\text{-}C_5H_{11})_4$ (cf. Ref. 8).

Interpretation of ionic conductance is based on a hydrodynamic model (cf. Ref. 73), a kinetic model,[171] or on proton-jump or electron-jump mechanisms.[172,173]

In the hydrodynamic model, it is assumed that the ion is a sphere of radius r_i moving through a continuous incompressible fluid of viscosity η; this leads to

$$\lambda_i^\infty = \frac{1}{\Xi} \frac{N_A e^2 |z_i|}{k_i^\infty} \quad (46)$$

with a coefficient k_i^∞, which varies from $6\pi\eta r_i$ for a sphere dragging the adjacent solvent along with it, to $4\pi\eta r_i$ for a sphere moving through the solvent without drag (cf. Refs. 8 and 173).

Equation (46) is used to calculate approximate ionic limiting conductance in a solvent II from known values in a solvent I by the relationship

$$\frac{(\lambda_i^\infty)_{II}}{(\lambda_i^\infty)_I} = \frac{\eta_I}{\eta_{II}} \quad (47)$$

if $(r_i)_I$ and $(r_i)_{II}$ can be assumed to be equal. Together with $k_i = 6\pi\eta r_i$, it is used to estimate solvation numbers of small ions (cf. Ref. 73) and it yields Walden's rule

$$\lambda_i^\infty \eta = \text{constant} \quad (48)$$

Krumgalz' investigation on λ_i^∞ is based on these properties.[174] Fuoss[151,175] tried to take into account the orientation of solvent molecules in the vicinity of the moving ion by introducing a permittivity-dependent radius into Eq. (46). The statistical theory of this effect was developed by Boyd[176] and Zwanzig.[177] Equation (49), based on a revised theory,[178] takes into account the idea that solvent molecules immediately adjacent to the ions probably do not completely participate in the relaxation process[178,179]:

$$\lambda_i^\infty = \frac{1}{\Xi} \frac{N_A e^2 |z_i|}{k_i^\infty + (3/4s)(1/4\pi\varepsilon_0)[(e_0 z_i)^2/r_i^3][(\varepsilon_s - \varepsilon_\infty)\tau/\varepsilon_s(2\varepsilon_s + 1)]} \quad (49)$$

In Eq. (49), ε_s is the static permittivity of the solvent, ε_∞ the high-frequency permittivity, τ the relaxation time relating to the relaxation region between ε_∞ an ε_s; $s = 2$ if $k_i = 6\pi\eta r_i$ and $s = 1$ if $k_i = 4\pi\eta r_i$.

The kinetic model is based on transition state theory and treats λ_i in terms of a process of interchange of sites separated by a mean distance L. The ionic migration is related to an activation energy $\Delta G^{\ddagger}$:

$$\lambda_i^{\infty} = \frac{1}{\Xi} \frac{N_A e^2 |z_i| L^2}{6h} e^{-\Delta G^{\ddagger}/RT}$$

$$\Delta G^{\ddagger} = \Delta H^{\ddagger} - T\Delta S^{\ddagger} \qquad (50)$$

Proton-jump and electron-jump mechanisms are not considered here. For more comprehensive information, see Ref. 173 and, especially for proton transfer, see Ref. 172.

4. Analysis of Experimental Data

The use of the set of Eqs. (1) for evaluation of experimental data* requires knowledge of $\Lambda(c, T)$, $\varepsilon_r(T)$, $\eta(T)$, and, depending on the experimental method, $\rho(c, T)$.[32] Figure 8 shows a Λ–c–T diagram obtained by the method of stepwise concentrations and isologous sections (cf. Section II.2) from measurements at temperatures $-40 \le \theta \le +25°C$. The curves through the measured points in Fig. 8 are plotted by a computer on the basis of a four-parametric evaluation (for explanations, see the later text). Figure 9 illustrates reproducibility and comparison with literature data at 25°C. The dashed line represents the limiting law behavior $(E - J_1 - J_2 - 0)$.

The first step in the analysis of conductance data is the choice of the most useful conductance equation, Eq. (1), and the related procedure of evaluation. Both depend essentially on concentration range and accuracy of data. The most appropriate procedures today are least-squares methods based on a series expansion, e.g., $\Lambda(\Lambda^{\infty} + \Delta\Lambda^{\infty}, J_1 + \Delta J_1, \ldots)$, as described by Kay[181] and others.[12,182–185] For graphical methods, see Refs. 12, 92, 138, 139, 186. In the framework of numerical evaluation of data two trends can be observed: Λ is taken as depending on Λ^{∞} and the distance parameter a, e.g., $\Lambda = \Lambda(\Lambda^{\infty}, a)$, or in the form given above where J_1 and J_2 are quantities containing the distance parameter. In the following discussion we use the latter representation of Λ which enables the essential features to be made evident.

---

* When not indicated otherwise, dimensions are always Ω^{-1} cm^2 mol^{-1} for Λ and Λ^{∞}, mol^{-1} liter for K_A, Å for R_1, R_2, R, R_y, and q.

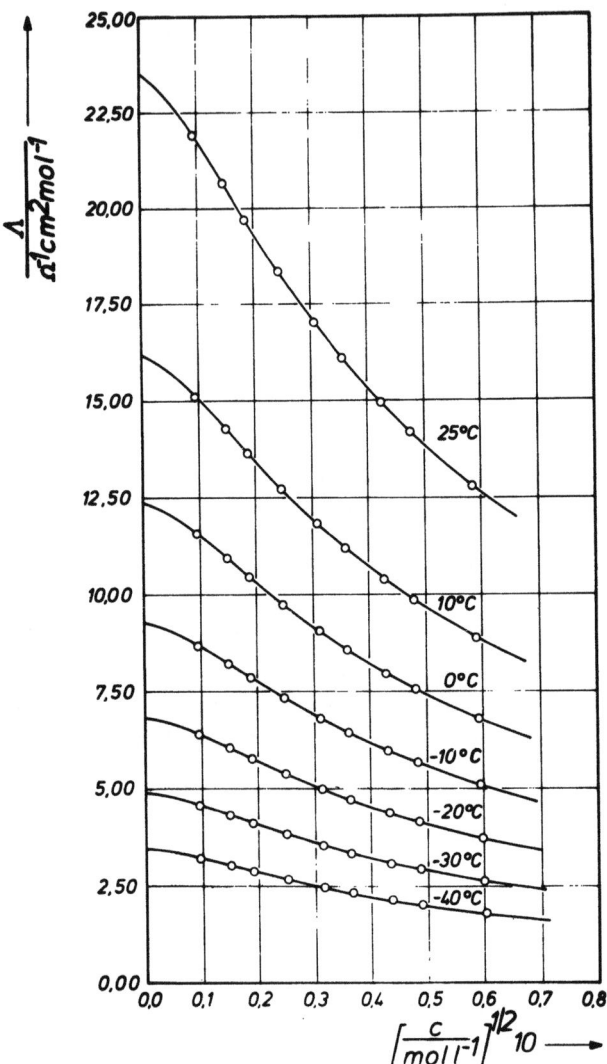

Figure 8. $\Lambda-c-T$ diagram as obtained by the method of stepwise concentration and isologous sections: $(n\text{-}Am)_4NI$ in n-propanol, $-40 \le \theta \le 25°C$.

Any conductance equation quoted in Section III.1 contains the coefficient S in the form $S = S_1 \Lambda^\infty + S_2$ of the limiting law. Three versions for E are used in the literature: $E = E_1 \Lambda^\infty - E_2$ without taking into account the "Chen effect"[117,118] and $E = E_1 \Lambda^\infty - 2E_2$ by doing so; thirdly, a recent contribution of Ebeling et al.[113,147,187] contains this coefficient as $E = E_1 \Lambda^\infty - 3E_2/2$. Numerical values of the coefficients in Eqs. (1) are, however, thereby affected only to a small extent (cf. Tables 5 and 7). In contrast to S and E, which do not contain an ion-distance parameter, J_1 and J_2 do. For the purpose of data analysis we regard the distance parameters derived from J_1 and J_2 as not necessarily equal, i.e., $J_1 = J_1(R_1)$ and $J_2 = J_2(R_2)$. This permits treatment of a conductance equation which is complete up to the term in $c^{3/2}$ as a function $\Lambda = \Lambda(\Lambda^\infty, J_1, J_2, K_A)$ and thus takes into account that all currently known conductance equations still contain the approximate expressions J_1 and J_2, S and E being represented by their theoretical values.

Among the "complete" conductance equations based on the model of rigid charged spheres, those of Pitts in the forms given by Fernández-Prini and Prue (PFPP equation[131]), Fuoss and Hsia/Fernández-Prini (FHFP equation[119]) and an equation of

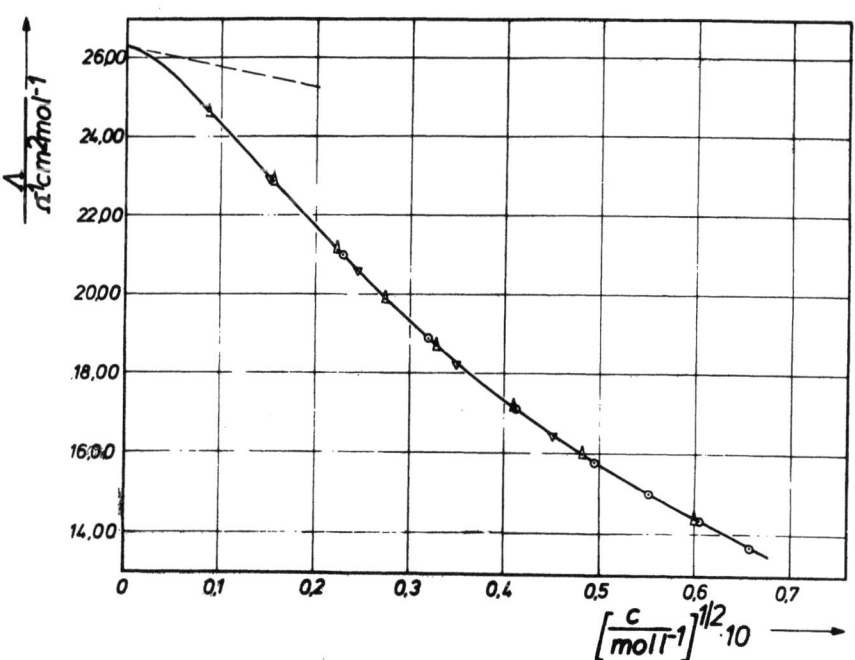

Figure 9. Reproducibility of measuring data (○, series I; △, series II) and comparison with literature (▽, measurements of Evans and Gardam[284]) at 25°C. The tangent at $c = 0$ represents the limiting law.

Justice (compare coefficients with Refs. 188 and 189) which covers the main features of the Fuoss–Hsia theory (FJ equation), are almost equally good approaches. The Fuoss–Hsia equation and hence the expression of Fuoss–Justice are based on consideration of the relaxation field as treated by Fuoss and Hsia;[69] they yield a more complete J_2 term than does the Pitts equation. Measurements must cover a sufficiently large concentration range in order to evaluate the four parameters of the conductance equations. A survey of various data shows that $\kappa q = 0.3$ [cf. Eq. (2)] must be chosen as a minimum value for the highest concentration of a run.[188] The range of validity of the equations is estimated as $\kappa q < 0.5$.[142]

Table 5* summarizes the results obtained from a data analysis carried out by determining Λ_0, J_1, J_2, and K_A by a least-squares method. The values of R_1 and R_2 were then calculated separately from the appropriate coefficients. This procedure is indicated in the following by a symbol 4 after the specification of the equation, e.g., FHFP 4. The procedures FHFP 4 and PFPP 4, of course, yield identical coefficients Λ_0, J_1, J_2, and K_A. Any deviation of FJ 4 is due to the E term of this equation which accounts for the Chen effect. All evaluations in Table 5 were performed by setting $R_y = q$ in Eq. (1c). The first and most important result which can be deduced from Table 5 is the compatibility of R_1 and R_2 within their limits of approximation. Furthermore, the distance parameter $R_1 = R_2$ of the conductance equation depends mainly on the solvent properties (and the net charges of the ions) and is almost independent of ionic radius and ionic size. This statement is verified for many electrolyte solutions and at various temperatures.[190] A further striking example in this context is the evaluation according to FHFP 4 of data for tetraalkylammonium salts in methanol as the solvent [141] from the experimental data of Kay, Zawoyski, and Evans[191] at 25°C, which also satisfy the condition $R_1 = R_2$. Four-parameter evaluations, however, are rare on account of the very high accuracy required for the data.

---

* Data are representative for the measures from our laboratory at $-45 \leq \theta/°C \leq 25$ in steps of 10 K (Ref. 190; cf. also Refs. 31 and 198). Publication of the completed data is in preparation.

Table 5
Conductance Data from Four-Parameter Evaluations

Electrolyte	Eq.	Λ^∞	K_A	R_1	R_2	$\frac{\sigma_\Lambda}{10^{-3}}$	Λ^∞	K_A	R_1	R_2	$\frac{\sigma_\Lambda}{10^{-3}}$	Λ^∞	K_A	R_1	R_2	$\frac{\sigma_\Lambda}{10^{-3}}$
		$\theta = 25°C; q = 11.5$ Å					Ethanol $\theta = -15°C; q = 10.3$ Å					$\theta = -45°C; q = 9.6$ Å				
Pr_4NBr	FHFP 4	46.83	125	10.8	10.4	7	20.35	119	12.3	11.5	3	9.347	115	10.6	10.5	3
	FJ	46.82	134	12.1	11.4	6	20.35	127	13.9	12.9	3	9.34	122	12.1	11.7	2
Pr_4NI	FHFP 4	49.93	169	10.6	10.3	4	21.92	172	12.3	12.0	3	10.15	187	12.1	12.2	1
	FJ	49.92	176	11.5	11.1	4	21.92	178	13.5	13.1	3	10.15	193	13.2	13.4	1
		$\theta = 25°C; q = 13.7$ Å					n-Propanol $\theta = -10°C; q = 12.3$ Å					$\theta = -40°C; q = 11.5$ Å				
Pr_4NI	FHFP 4	26.28	515	14.0	13.2	5	10.33	430	10.7	10.3	1	3.780	454	8.7	8.9	0.5
	FJ	26.28	524	14.6	13.8	5	10.33	436	11.0	10.6	1	3.780	459	8.8	8.9	0.5
$MeBu_3NI$	FHFP 4	25.45	594	12.0	11.2	3	10.02	512	10.7	10.6	0.3	3.678	555	10.8	12.0	0.3
	FJ	25.44	602	12.3	11.4	3	10.02	518	10.9	10.8	0.3	3.678	561	11.0	12.5	0.3
$(n-Am)_4NI$	FHFP	23.51	499	10.9	10.5	2	9.280	457	10.8	11.0	0.4	3.412	486	8.7	9.7	0.5
	FJ	23.51	508	11.2	10.6	2	9.279	464	11.1	11.4	0.4	3.412	492	8.7	9.9	0.4
LiCl	FHFP 4	20.01	264	10.0	10.0	0.5	7.657	95	8.5	8.8	0.3	2.754	45	7.5	8.1	0.1
	FJ	20.00	276	10.5	10.2	0.5	7.655	107	9.3	9.3	0.2	2.754	56	8.5	8.8	0.1
KSCN	FHFP	26.60	328	12.0	10.9	1	10.44	158	11.3	10.4	1	3.819	98	11.5	10.7	0.3
	FJ	26.59	336	12.5	11.1	1	10.44	165	12.1	10.9	1	3.818	106	12.7	11.5	0.3

In conclusion, it should be mentioned that the investigation of conductance of electrolyte solutions cannot provide information about ion size parameters and related properties of the dissolved electrolyte compound from J_1 and J_2 parameters, if association occurs in the solution.

On the other hand, for solutions without association, Eq. (10) is applied in which $J_1 = J_1(a)$ and $J_2 = J_2(a)$ are functions of ionic radii. Transition from Eq. (1a) to Eq. (10) with vanishing association has been repeatedly subject to theoretical investigation.[44,102,103,135,157]

Four-parameter fits of data for very weakly associating electrolytes are not feasible. Even precise Λ–c data do not allow separation of the K_A and J_1 contributions.[144] Solutions in acetonitrile or propylene carbonate are examples. This problem will be taken up in connection with three-parameter evaluations. Table 5 shows also that R_1 and R_2 are almost temperature-independent quantities. Their values are found to be comparable to those of Bjerrum's limit of association $R = q$. This coincidence, however, is rather a consequence of initially setting $R_y = q$ than of any intrinsic feature of the theory, as can be seen from Table 6.

Table 6 contains evaluations based on FJ 4 in which the assumption $R_y = q$ is dropped and R_y fixed arbitrarily at $0.7 < R_y/q < 1.3$. The σ values listed at the head of each column are valid for all R_y values at a given temperature. Variations of the quantities Λ^∞, K_A, J_1, and J_2 are small, and R_1 and R_2 vary less than does R_y. This tendency can be understood from the activity formula, Eq. (44). With regard to association, Table 6 provides the information that $R_y = q$ cannot be deduced as a condition from the conductance theory itself. Data analysis must presuppose the use of well-founded activity coefficients (cf. Ref. 192).

Where the set of Eqs. (1) is used for a four-parameter evaluation in order to determine K_A and Λ^∞, or for compilation of conductance data, the special theory implicit in Eq. (1a) need not be considered. If such an evaluation is not possible, e.g., where the association constants are very small, data analysis is performed by a three-parameter fit on the basis of a procedure first used by Justice,[141] who set $J_1(R_1)$ equal to a known value, as in the case of the coefficients S and E. This kind of procedure is

Table 6
Effect of R_y on Conductance Data of KSCN in n-Propanol at Various Temperatures

Conditions	R_y	Λ^∞	J_1	J_2	K_A	R_1	R_2
$\theta = 25°C$		(± 0.005)	(± 40)	(± 400)	(± 3)	(± 0.3)	(± 0.3)
$\sigma_\Lambda = 8 \times 10^{-4}$	17.78	26.592	2890	−9720	347	14.3	11.7
$q = 13.68$	15.05	26.591	2720	−9020	339	13.1	11.3
	q	26.590	2630	−8690	336	12.5	11.1
	12.31	26.589	2540	−8360	332	11.9	10.9
	10.94	26.588	2450	−8050	328	11.3	10.7
$\theta = -10°C$		(± 0.003)	(± 25)	(± 250)	(± 4)	(± 0.6)	(± 0.5)
$\sigma_\Lambda = 6 \times 10^{-4}$	15.05	10.439	800	−2720	171	13.1	11.2
$q = 12.28$	13.68	10.439	780	−2640	168	12.6	11.0
	q	10.438	760	−2590	165	12.1	10.9
	10.94	10.438	730	−2510	163	11.6	10.8
	8.21	10.437	680	−2390	157	10.5	10.5
$\theta = -40°C$		(± 0.001)	(± 15)	(± 120)	(± 6)	(± 1.1)	(± 0.7)
$\sigma_\Lambda = 3 \times 10^{-4}$	15.05	3.817	260	−960	113	14.0	11.8
$q = 11.45$	13.68	3.817	250	−940	110	13.5	11.7
	q	3.817	240	−900	108	12.7	11.5
	10.94	3.816	240	−900	106	12.5	11.5
	8.21	3.816	220	−870	101	11.6	11.3

indicated by a symbol 3 after the specification of the applied equation. With $R_1 = R_y = q$, the FHFP 3, PFPP 3, or FJ 3 equations are established, all of which are sufficient approximations for dealing with the conductance problem. For the purpose of data analysis, we use the intrinsic requirement of the conductance equation, $R_1 = R_2$, as a measure of the compatibility of the equations. Table 7 contains the results of three-parameter evaluations of conductance data obtained by this procedure.

In Table 7* all the quantities necessary for the compilation are summarized: Λ^∞, K_A, q, specification of the equation, and the value of R_2 for control of compatibility. Agreement with the data from four-parameter fits, contained in Table 5 is satisfying. The advantage of this method is best demonstrated for solutions of

---

* Data are representative for the measures from our laboratory at $-45 \leq \theta/°C \leq 25$ in steps of 10K (Ref. 190; cf. also Refs. 31 and 198). Publication of the complete data is in preparation.

Table 7
Conductance Data from Three-Parameter Evaluations ($R_1 = R_s = q$)

Electrolyte	Eq.	Λ^∞	K_A	R_2	$\frac{\sigma_\Lambda}{10^{-3}}$	Λ^∞	K_A	R_2	$\frac{\sigma_\Lambda}{10^{-3}}$
		\multicolumn{4}{c}{$\theta = 25°C$; $q = 11.5$ Å}	\multicolumn{4}{c}{Ethanol $\theta = -15°C$; $q = 10.3$ Å}
Pr$_4$NBr	FHFP 3	46.82	129	11.0	13	20.35	110	9.8	6
	FJ 3	46.82	131	10.9	6	20.35	112	9.6	3
	PFPP 3	46.84	140	8.1	7	20.35	118	7.5	3
Pr$_4$NI	FHFP 3	49.94	174	11.2	4	21.92	163	10.1	4
	FJ 3	49.92	176	11.1	4	21.91	165	10.0	4
	PFPP 3	49.95	186	8.3	4	21.92	171	7.9	4
		\multicolumn{4}{c}{$\theta = 25°C$; $q = 13.7$ Å}	\multicolumn{4}{c}{Propanol $\theta = -10°C$; $q = 12.3$ Å}
Pr$_4$NI	FHFP 3	26.28	513	13.0	5	10.34	443	12.2	2
	FJ 3	26.27	516	12.9	5	10.33	446	12.1	1
	PFPP 3	26.30	536	9.5	7	10.34	459	9.3	3
MeBu$_3$NI	FHFP 3	25.46	612	13.2	4	10.02	524	12.5	0.7
	FJ 3	25.46	616	13.2	3	10.02	528	12.6	0.6
	PFPP 3	25.48	636	9.9	6	10.03	541	9.8	1
(n-Am)$_4$NI	FHFP 3	23.53	528	13.5	6	9.283	469	12.7	1
	FJ 3	23.53	532	13.5	5	9.281	473	12.8	0.8
	PFPP 3	23.55	552	10.1	10	9.288	486	9.9	2
		\multicolumn{4}{c}{}	\multicolumn{4}{c}{$\theta = -45°C$; $q = 9.6$ Å}
	FHFP 3					9.344	112	9.4	3
	FJ 3					9.344	114	9.2	2
	PFPP 3					9.346	118	7.4	3
	FHFP 3					10.15	178	9.9	1
	FJ 3					10.15	180	9.9	1
	PFPP 3					10.15	185	8.0	1
		\multicolumn{4}{c}{}	\multicolumn{4}{c}{$\theta = -40°C$; $q = 11.5$ Å}
	FHFP 3					3.782	472	12.3	0.8
	FJ 3					3.782	476	12.5	0.8
	PFPP 3					3.784	485	9.8	1
	FHFP 3					3.679	560	12.8	0.3
	FJ 3					3.678	564	13.1	0.3
	PFPP 3					3.680	573	10.5	0.4
	FHFP 3					3.414	505	13.0	0.7
	FJ 3					3.413	509	13.4	0.6
	PFPP 3					3.415	519	10.6	1.0

Salt	Method												
		θ = 25°C; q = 7.8 Å				θ = −5°C; q = 7.6 Å				θ = −35°C; q = 7.5 Å			
LiCl	FHFP 3	20.06	307	13.0	6	7.670	128	11.9	2	2.758	72	11.2	0.7
	FJ 3	20.04	309	13.1	5	7.664	130	12.0	1	2.756	74	11.3	0.4
	PFPP 3	20.08	334	9.4	1	7.676	145	8.7	3	2.760	86	8.3	0.9
KSCN	FHFP 3	26.63	347	12.3	3	10.45	166	11.3	1	3.818	98	10.7	0.2
	FJ 3	26.62	348	12.2	2	10.44	167	11.2	0.6	3.816	99	10.6	0.3
	PFPP 3	26.68	374	8.5	7	10.46	183	7.9	2	3.820	110	7.5	0.4

Acetonitrile

Salt	Method												
		θ = 25°C; q = 7.8 Å				θ = −5°C; q = 7.6 Å				θ = −35°C; q = 7.5 Å			
Me$_4$NI	FHFP 3	196.53	33.0	8.6	13	142.82	31.1	8.5	10	94.05	31.4	8.2	4
	FJ 3	196.50	34.2	8.3	11	142.80	32.2	8.2	8	94.04	32.6	7.8	4
	PFPP 3	196.53	35.8	6.7	14	142.83	33.6	6.7	10	94.05	33.8	6.5	4
KI	FHFP 3	186.69	16.2	8.4	23	134.69	11.7	8.3	20	87.81	9.0	8.1	12
	FJ 3	186.65	17.4	8.3	20	134.66	12.9	8.2	16	87.80	10.2	8.0	10
	PFPP 3	186.71	19.5	6.5	27	134.70	14.8	6.4	22	87.82	11.9	6.3	14

Propylene carbonate

Salt	Method												
		θ = 25°C; q = 4.3 Å				θ = −5°C; q = 4.3 Å				θ = −35°C; q = 4.3 Å			
LiClO$_4$	FHFP 3	26.75	0.7	5.7	4	12.95	0.4	6.2	2	3.882	(0)	a	1
	FJ 3	26.75	1.2	5.7	4	12.95	0.9	6.2	2	3.882	0.1	7.4	1
	PFPP 3	26.75	1.3	5.0	4	12.95	1.0	5.5	2	3.882	0.2	6.5	1
KPF$_6$	FHFP 3	28.98	0.1	6.2	2	13.91	(0)	a	2	4.123	(0)	a	0.5
	FJ 3	28.97	1.2	6.1	2	13.90	0.7	6.7	2	4.123	0.4	7.2	0.5
	PFPP 3	28.98	1.3	5.3	2	13.91	0.8	5.8	2	4.123	0.5	6.2	0.6

a If $K_A = (0)$, the program jumps to a three-parameter evaluation (λ^∞, J_1, J_2).

weakly associating electrolytes, e.g., in acetonitrile or propylene carbonate as solvent.

A further three-parameter evaluation, based on a molecular model of association, is encouraged by the results in Table 6 (cf. also Refs. 142–144, 283). In consistency with thermodynamics of the association process (Section III.2), paired states of ions can be considered as ion pairs if the ions have approached to within a distance smaller than the dimensions s of a solvent molecule, i.e., $R = a + s$ in connection with Eq. (12) or $R' = a + s$ for an application of Eq. (36). Table 8 shows data obtained from an evaluation in which R is set $R = a + s$ and hence $R_1 = R_y = a + s$. Comparison of results shown in Tables 5, 7, and 8 justifies the approximation underlying this procedure. On the other hand, if $a + s < q$, data of Table 7 can be used also in connection with a two-step equilibrium[144] to comply with the above model: $R' = a + s$, $R = q$

Table 8
Conductance Data from Three-Parameter Evaluations on the Basis of the FJ Equation ($R_1=R_y=a+s$)

Electrolyte	a (Å)	Λ^∞	K_A	R_2	$\frac{\sigma_\Lambda}{10^{-3}}$	Λ^∞	K_A	R_2	$\frac{\sigma_\Lambda}{10^{-3}}$	Λ^∞	K_A	R_2	$\frac{\sigma_\Lambda}{10^{-3}}$
						Ethanol ($s = 5.30$ Å)							
			$\theta = 25°C$				$\theta = -15°C$				$\theta = -45°C$		
Pr$_4$NBr	6.47	46.82	133	11.6	5	20.35	118	10.7	3	9.345	121	10.8	2
Pr$_4$NI	6.68	49.93	178	11.9	4	21.92	171	11.2	3	10.150	188	11.6	1
						Propanol ($s = 6.90$ Å)							
			$\theta = 25°C$				$\theta = -10°C$				$\theta = -40°C$		
Pr$_4$NI	6.68	26.27	515	12.8	5	10.34	453	13.0	2	3.782	486	13.9	1
MeBu$_3$NI	5.64	25.43	602	12.5	6	10.02	529	12.7	0.5	3.678	568	13.8	0.2
						Acetonitrile ($s = 5.12$ Å)							
			$\theta = 25°C$				$\theta = -5°C$				$\theta = -35°C$		
Me$_4$NI	5.64	196.51	38	9.7	36	142.81	37	9.7	11	94.04	37	9.5	3
KI	3.52	186.68	19.4	8.8	22	134.68	15.2	8.8	18	87.81	12.5	8.7	11
						Propylene carbonate ($s = 5.40$ Å)							
			$\theta = 25°C$				$\theta = -5°C$				$\theta = -35°C$		
LiClO$_4$	3.14	26.75	4.4	8.2	4	12.95	4.1	8.6	2	3.882	3.2	9.5	1
KPF$_6$	3.87	28.98	3.9	8.2	2	13.91	3.4	8.6	2	4.123	3.1	9.0	0.6

in Eq. (41). Further variants are possible, e.g., setting the limit of association at $R_y = 1.1q$ in accordance with the result of Justice and Justice[161] or introducing the Gurney cosphere as a measure of the size of free ions.[109,142,145,157]

Van Evercooren, Merken and Thun[193] have used Justice's three-parametric procedure[141] in their investigation on hydrochloric acid in N-methylpropionamide in another way. In this solvent of high permittivity ($\varepsilon_r = 175$ at 25°C) they obtained association constants, $K_A < 1$, by varying $R_y = R_1$ until J_2 (expt.) equals J_2 (theor.).

Finally, three-parameter evaluations are always required if the conductance equation is used in a nondeveloped form, e.g., $\Lambda = \Lambda(\Lambda^\infty, R, K_A)$ or similar expressions. It is reduced to a two-parameter procedure for nonassociating electrolytes.[183]

Hitherto, the limiting law ($E = J_1 = J_2 = 0$) has been used for analysis of data despite the impossibility of attaining its region of validity. Figure 9 illustrates this fact by the plotted dashed line representing the limiting law for an electrolyte for which $K_A \approx 500 \text{ mol}^{-1}$ liter. A further inconsistency is the use of the limiting law for the activity coefficient, which is also an insufficient approximation. Rough evaluations can be performed to determine the limiting conductance, Λ^∞, and association constants which are high. To give an approximate limit of applicability, $K_A > 10^3 \text{ mol}^{-1}$ liter.

5. Results and Discussion

A survey of recent publications dealing with temperature-dependent conductance data for dilute and concentrated (for discussion, see Section IV) electrolyte solutions is given in Table 9. Some older contributions are listed where the field covered has not been explored subsequently.

(i) Limiting Conductance

The discussion of limiting conductances is based on the features given in Section III.3. Single-ion conductivities generally are not available at temperatures other than 25°C (cf. Table 9).

Table 9
Survey of Publications and Their Information

Solvent[a]	Electrolyte	Region of temperature (°C)	Region of concentration	Eq.[b]	Parameters	Ref.
Methanol	NaI, KI	5–45	10^{-5}–10^{-3}	LL	$K_A, \Delta G, \Delta H, \Delta S$	194
	TaaX, TaaPi	10, 25	10^{-4}–10^{-2}	FO	$\Lambda^\infty, \lambda^\infty, \Lambda^\infty \eta, a, K_A$	191
	Me$_3$SI, Et$_3$SI, Pr$_3$SI	10, 25	10^{-4}–10^{-2}	FO	$\Lambda^\infty, \lambda^\infty, \lambda^\infty \eta, a, K_A$	196
	NaCl	10, 25	10^{-4}–10^{-2}	FO	t_+^∞	197
	KSCN, LiClO$_4$, LiBr	-50–20	0.05–5	Emp.	E_m, E_κ	240
Ethanol	Pr$_4$NBr, Pr$_4$NI, Pr$_4$NClO$_4$	-45–25	10^{-4}–10^{-2}	Compl.	$\Lambda^\infty, K_A, \Delta G, \Delta H, \Delta S$	198
	NaI, KI	5–45	10^{-5}–10^{-3}	LL	$K_A, \Delta G, \Delta H, \Delta S$	194
	Na salts	-60–60	10^{-5}–sat.	Emp.	$\Lambda^\infty, \Lambda^\infty \eta$	199
Propanol	TaaI, KI, LiCl	-45–25	10^{-4}–10^{-2}	Compl.	$\Lambda^\infty, K_A, R, \Delta G, \Delta H, \Delta S$	31, 32, 144
	KSCN	-45–25	10^{-4}–10^{-2}	Compl.	Λ^∞, K_A	32
	CsOPr	5–35	10^{-4}–10^{-2}	Compl.	Λ^∞, K_A	188
	NaI, KI	5–45	10^{-5}–10^{-3}	LL	$K_A, \Delta G, \Delta H, \Delta S$	194
Higher aliphatic alcohols	NaI, KI	5–45	10^{-5}–10^{-3}	LL	$\Lambda^\infty, K_A, \Delta G, \Delta H, \Delta S$	194
Ethylene glycol	Et$_4$NX	25–45	10^{-3}–10^{-1}	LL, FOS	$\Lambda^\infty, \Lambda^\infty \eta, a$	200
Diethylene glycol	AX, TaaX	0–140	10^{-3}–10^{-2}	LL	$\Lambda^\infty, \Lambda^\infty \eta, \Delta H, \Delta S$	201
	HCl, HBr	25–45	10^{-3}–10^{-2}	FOS	$\Lambda^\infty, \Lambda^\infty \eta$	241
Glycerin	HCl	20–50	10^{-3}–10^{-2}	Emp. LL	$\Lambda(c)$	202
Formamide	TaaX, AX	10, 25	10^{-4}–10^{-2}	FO	$\Lambda^\infty, \lambda^\infty \eta, K_A$	203
	NaI	25–80	Conc.	Emp.	Λ, η, glass transition point	204
N-Methylformamide	AX, Et$_4$NBr, Et$_4$NPi	15, 25	10^{-3}–10^{-1}	Emp. LL	$\Lambda^\infty, \Lambda^\infty \eta$	205
	KBr	15–45	0–0.3		T_+, λ	206
	NaCl, CsCl	5, 25	10^{-3}–10^{-1}	FO	Λ, η, a	207
N,N-Dimethylformamide	KClO$_4$	-50–60	10^{-4}–10^{-2}	LL	Λ^∞	208
	TaaI, NaI, KI	20–55	10^{-3}–10^{-2}	LL	$\Lambda^\infty, \Lambda^\infty \eta$	209
	NaBr, KBr	15, 25	10^{-3}–10^{-1}	Emp. LL	$\Lambda^\infty, \Lambda^\infty \eta$	205
	Bu$_4$NI, KSCN, NH$_4$Br	-50–125	0.1–1.0	Emp.	$\Lambda, \eta, d, E_m, E_\eta$	210
	TaaI	15–35	10^{-3}–10^{-2}	Emp. LL	$\Lambda^\infty, \lambda_\pm^\infty \eta$	211
	NaBr, KBr	10–60	10^{-4}–10^{-2}	LL	Λ^∞, K_A	195

Solvent	Electrolyte	Temp. range	Conc. range	Method	Properties	Ref.
N-Methylacetamide	NaClO$_4$	30–50	10^{-4}–10^{-2}	LL	Λ^∞	212
	KBr	35–50	0–0.3		T^∞_+, λ	213
	AX	30–60	10^{-4}–2	LL	$\Lambda^\infty, \Lambda^\infty \eta$	214
	AX, Et$_4$NBr, Et$_4$NPi	35, 45	10^{-3}–10^{-1}	Emp. LL	$\Lambda^\infty, \Lambda^\infty \eta$	215
	AX	35–55	10^{-4}–10^{-2}	LL	$\Lambda^\infty, \lambda, \Lambda^\infty \eta$	216
	TaaI	35–55	10^{-4}–10^{-2}	Emp. LL	$\Lambda^\infty, \lambda, \lambda \eta$	217
N-Methylpropionamide	HCl	25–50	10^{-3}–10^{-1}	Compl.	$\Lambda^\infty, \Lambda^\infty \eta, K_\Lambda, a$	193
	AX	35–55	10^{-4}–10^{-2}	LL	$\Lambda^\infty, \lambda, \Lambda^\infty \eta$	216
	TaaI	30–50	10^{-3}–10^{-2}	Emp. LL	$\Lambda^\infty, \lambda^\infty_+\eta$	211
Acetone	NaI	−50–50	10^{-5}–0.5	LL	$\Lambda^\infty, \Lambda^\infty \eta, K_a, a, \Delta G, \Delta H, \Delta S$	218
	NaI, NaClO$_4$, ϕ_4AsI	25–45		LL	$\Lambda^\infty, K_\Lambda, \Lambda^\infty \eta, \Delta G, \Delta S$	219
	LiBr	25–40	~10^{-3}	FO	$\Lambda^\infty, \Lambda^\infty \eta, K_\Lambda, a$	239
Ethyl methyl ketone	TaaX	15–35	10^{-4}–10^{-3}	LL, FO	$\Lambda^\infty, K_\Lambda, a, \lambda, \Delta H, \Delta S$	220
Acetonitrile	Bu(i-Am)$_3$NB(C$_6$H$_5$)$_4$	−40–25	10^{-4}–10^{-2}	Compl.	$\Lambda^\infty, K_\Lambda, R, \Delta G, \Delta H, \Delta S$	144, 221
	Me$_4$NI, KI, KClO$_4$	−40–25	10^{-4}–10^{-2}	Compl.	$\Lambda^\infty, K_\Lambda, R, \lambda^\infty \eta, \Delta G, \Delta H, \Delta S$	190, 222
	TaaClO$_4$	10, 25	10^{-4}–10^{-2}	FO	$\Lambda^\infty, \lambda^\infty, \lambda^\infty \eta, a, K_\Lambda$	196
Phenylacetonitrile	Me$_3$SI, Et$_3$SI, Pr$_3$SI	25–200	10^{-4}–10^{-3}	FO	$\Lambda^\infty, \lambda^\infty, K_\Lambda, a$	265
Propylene carbonate	Bu$_4$NBϕ_4	−45–25	10^{-4}–10^{-2}	Compl.	$\Lambda^\infty, \Lambda^\infty \eta, K_\Lambda, R$	144, 221
	LiClO$_4$, KPF$_6$	25–40	Conc.	—	κ, η	223
	KPF$_6$, KSCN, LiClO$_4$	−45–25	Conc.	Emp.	$\kappa, \kappa_{max}, E_\kappa$	Section IV
	KPF$_6$, KSCN, LiClO$_4$	−50–125	0.05–0.5		$\Lambda, \eta, d, E_\eta, E_\lambda$	224
	LiPF$_6$, TaaPF$_6$	−25–50			κ, κ_{max}	225
	TaaX	−40–40	1 M		E_κ	239
γ-Butyrolactone	LiClO$_4$	25–45		LL	$\Lambda^\infty, K_\Lambda, a, \Delta G, \Delta S$	219
Nitromethane	LiClO$_4$	20–70	10^{-4}–10^{-3}	LL	High-pressure data	226
Nitrobenzene	ϕ_4AsI	20–45	10^{-4}–10^{-1}	FO	$\Lambda^\infty, \lambda^\infty \eta, K_\Lambda$	227
Dimethylsulfoxide	TaaPi	20–55	10^{-5}–10^{-2}	FO, compl.	$\Lambda^\infty, \Lambda^\infty \eta, a$	228
	HCl	25, 30	10^{-3}–10^{-1}	LL, FOS	Λ^∞, a	229
	HBr	25–60	10^{-4}–1	None	Ion-solvent complexes	230
	NH$_4$Br	−61–45	10^{-6}–10^{-4}	LL	$\Lambda^\infty, \Lambda^\infty \eta, K_\Lambda, \Delta H$	231
Tetrahydrofurane	NaAlBu$_4$, Bu$_4$NAlBu$_4$	−75–25	10^{-6}–10^{-4}	LL	$\Lambda^\infty, \Lambda^\infty \eta, K_\Lambda, \Delta H, \Delta S$	232
	Bu(i-Am)$_3$NBϕ_4, ABϕ_4					
	Na salts of aromatic radical ions					
	LiClO$_4$	−25–50	Conc.		κ, κ_{max}	225, 272, 273
	ABϕ_4, Iso-Am$_3$BuNBϕ_4	−70–25			$\Lambda^\infty, \lambda_i^\infty, K_\Lambda$	264

continued overleaf

Table 9 (continued)

Solvent[a]	Electrolyte	Region of temperature (°C)	Region of concentration	Eq.[b]	Evaluation information Parameters	Ref.
Tetrahydropyrane	NaBϕ_4, Bu$_4$NBϕ_4	−40–25	10^{-6}–10^{-3}	Compl.	Λ^∞, $\Lambda^\infty\eta$, K_A, triple ions ΔH, ΔS	233
Dimethoxyethane	fluorenyl salts (Na, K, Cs)	−75–25	10^{-6}–10^{-4}	LL	Λ^∞, $\Lambda^\infty\eta$, K_A, ΔH, ΔS	232
Methylene dichloride	Na salts of aromatic radical ions	−70–25			Λ^∞, λ_i^∞, K_A	264
	ABϕ_4, i-Am$_3$BuNBϕ_4				Triple ions	234
Ethylene chloride and ethyliden chloride	Et$_3$HNCl	−71–35	10^{-2}–10^{-1}	LL	Λ^∞, K_A, ΔH, ΔS	235
	(X-ϕ)Me$_3$NClO$_4$, X = m-, o-, p-chloro, methoxy					
Propylene chloride	TaaPi	6–35	10^{-5}–10^{-3}	LL	Λ^∞, K_A, ΔH, ΔS	236
	TaaPi	6–35	10^{-5}–10^{-3}	LL	Λ^∞, K_A, ΔH, ΔS	236
o-Dichlorbenzene	TaaPi	25–65	10^{-5}–10^{-4}	LL	Λ^∞, K_A	237
Anisole	Bu$_4$NNO$_3$	~33–95	10^{-4}–10^{-2}		Triple ions	238

[a] Abbreviations: Taa = tetraalkylammonium, A = alkali, X = halide, Pi = picrate, ϕ = phenyl.
[b] Equations: LL = limiting law or its extensions; FO = Fuoss–Onsager ($J_2 = 0$); FOS = Fuoss–Onsager–Skinner; compl. = complete Eq. (1a), including J_2; emp. = empirical.

Table 10
Limiting Values t_+^∞ of Transference Numbers in Acetonitrile at Temperatures $-35°C \le \theta \le +25°C^a$

Ion	$-35°C$	$-25°C$	$-15°C$	$-5°C$	$+5°C$	$+15°C$	$+25°C$
Me_4N^+	0.4753	0.4756	0.4759	0.4762	0.4765	0.4768	0.4771
Et_4N^+	0.4481	0.4488	0.4495	0.4501	0.4507	0.4513	0.4518

^{a} From Ref. 242.

Values based on a moving-boundary experiment for acetonitrile as the solvent[8,242] are given in Table 10. A program to measure transference numbers at -40 to $+25°C$ in other solvents is underway. Approximate values can be obtained with the hypothesis $\lambda_+^\infty = \lambda_-^\infty$ from appropriate electrolytes (e.g., Table 11) or be based on the constancy of the Walden product. A further approximation can be obtained from the assumption of temperature-independent transference numbers t_+^∞ of tetraalkylammonium ions (cf. Table 10). Alkali ions show a more pronounced increase of t_+^∞ with temperature, e.g., KBr in N-methylamides.[206,213] This method does not presuppose temperature-independence of the Walden product. It requires knowledge of t_i^∞ at only one temperature and Λ^∞ as a function of temperature. On the basis of $t_{+,298}^\infty$ (Table 10) limiting conductances λ_i^∞ in the temperature range $-35 \le \theta \le +25°C$ are obtained with an accuracy of 0.5%. Uncertainty of the two other methods is about 2–5%, depending on the solvent.

Limiting ionic conductances yield ionic radii by Stokes law (cf. Section III.3). This result can be obtained by measurements at a single temperature and is not considered in this context. For more comprehensive information on solvation numbers see Ref. 244. Temperature dependence of the limiting conductance, however, provides additional information about structural effects arising from the solvent. The conductivity–viscosity product $\lambda^\infty \eta$ in aqueous solutions shows a positive or a negative temperature variation in accordance with structure-promoting or -breaking properties of an ion. The Me_4N^+ ion, as well as the larger alkali metal ions, disorganize water structure in their immediate vicinity and yield a negative temperature coefficient of $\lambda^\infty \eta$, whereas

Table 11
Limiting Conductance of $(iso\text{-}Am)_3BuNB(C_6H_5)_4$ and Limiting Ionic Conductances in Propanol at $-40 \leq \theta \leq +25°C$

$\theta/C°$	-40	-30	-20	-10	0	$+10$	$+25$
$\Lambda^\infty[(i\text{-}Am)_3BuNB(C_6H_5)_4]^a$	2.99 ± 0.01	4.33 ± 0.01	6.07 ± 0.02	8.30 ± 0.02	11.08 ± 0.04	14.76 ± 0.04	21.01 ± 0.07
$\lambda^\infty[(i\text{-}Am)_3BuN^+]$	1.50	2.17	3.04	4.15	5.54	7.38	10.51
$\lambda^\infty[(i\text{-}Am)_4N^+]$	1.46	2.11	2.95	4.03	5.38	7.16	10.19
$\lambda^\infty[Bu_4N^+]$	1.58	2.29	3.20	4.38	5.84	7.77	11.07
$\lambda^\infty[Br_4N^+]$	1.81	2.62	3.67	5.01	6.69	8.88	12.68
$\lambda^\infty[Et_4N^+]$	2.28	3.19	4.45	6.18	8.20	10.70	14.38
$\lambda^\infty[K^+]$	1.76	2.54	3.56	4.87	6.51	8.66	12.43
$\lambda^\infty[(C_6H_5)_4B^-]$	1.50	2.17	3.04	4.15	5.54	7.38	10.51
$\lambda^\infty[I^-]$	1.97	2.82	3.92	5.32	7.10	9.18	13.59

^{a} $\Lambda^\infty[(iso\text{-}Am)_3BuNB(C_6H_5)_4]$ determined by FHFP 4 method.[243]

Pr_4N^+ and higher homologs behave in just the opposite manner.[196] This observation is related to a decrease of water structure by increasing temperature, in competition with the ionic effect. Water is considered to consist of a three-dimensional hydrogen-bonded network of water molecules. This picture suggests that examination of less structured solvents such as amides with a two-dimensional network,[203] N-methylamides and alcohols with linear association of their molecules, or inert solvents would be profitable. The latter interact with ions by means of their dipole moments only. Ion–solvent interactions of Lewis acid–Lewis base types have also been considered in this context.

As far as the existing but scarce material permits the deduction of a rule, no significant temperature variation of $\lambda_+^\infty \eta$ or $\Lambda^\infty \eta$ can be detected in dipolar aprotic solvents and in inert solvents for both tetraalkylammonium as well as alkali metal salts (with the possible exception of Li^+ salts[201,231,245]) viz., propylene carbonate (cf. Table 14),[221] acetonitrile,[221,222] N-N-dimethylformamide,[209] tetrahydrofurane,[231] and tetrahydropyrane.[233] Values of $d(\Lambda^\infty \eta)/dT$ for HBr in dimethylsulfoxide are greater than zero,[228] but for HCl, however, a constant Walden product is found.[227] In acetone, only $\lambda_{Na^+}^\infty \eta$ values exhibiting a negative temperature coefficient[219] are found. In N-methylamides all tetraalkylammonium salts yield negative temperature coefficients, $d(\lambda_+^\infty \eta)/dT < 0$[211,217] or $d(\Lambda \eta)/dT < 0$,[215] and so do the alkali metal salts[207,214,215,216] and HCl,[193] this decrease being less pronounced for small alkali metal ions.[216] Sodium salts in ethanol show constant Walden products over a wide temperature range[199] and also no significant variation could be found for tetrapropylammonium salts.[198] In propanol, $d(\Lambda^\infty \eta)/dT = 0$ for tetraalkylammonium,[31,243] sodium, and potassium salts,[31,32,243] while $d(\Lambda^\infty \eta)/dT < 0$ for LiCl[32,243] and $d(\Lambda^\infty \eta)/dT > 0$ for $CsOC_3H_7$.[188] Solutions in methanol show behavior comparable to that of aqueous solutions: Me_4N^+ and Me_3S^+ have negative temperature coefficients, while the other R_4N^+ and R_3S^+ ions yield zero or slightly positive slopes.[196] A marked decrease of $\Lambda^\infty \eta$ with increasing temperature is observed for tetraethylammonium halides in ethylene glycol.[200] For solutions in formamide, reliable values are not available.[203]

The temperature dependence of Λ^∞ can be approximated in the framework of the kinetic conductance theory (cf. Section III.3) by the equation

$$\ln \Lambda^\infty + \frac{2}{3} \ln \rho = -\frac{\Delta H^\ddagger}{RT} + B \qquad (51)$$

which is obtained from Eqs. (50) and (45) with the additional assumption of temperature-independent transference numbers[246,247]; ρ is the density of the solvent.

Figure 10 provides the information that $\Delta H^\ddagger$ [Eq. (51)] is mainly a function of the solvent properties; e.g., $\Delta H^\ddagger \approx$ 13 kJ mol^{-1} for solutions of 1.1 electrolytes in ethanol, $\Delta H^\ddagger \approx$ 18 kJ mol^{-1} for those in propanol, and $\Delta H^\ddagger \approx 8.5$ kJ mol^{-1} in acetonitrile as solvent. For propylene carbonate, see Section IV.3 (iv).

The contribution of Barreira and Hills[226] is a comprehensive

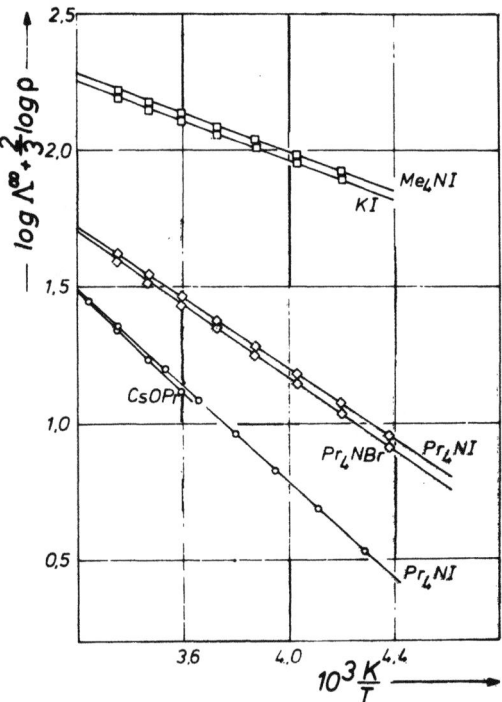

Figure 10. Plot of $\log \Lambda^\infty + \frac{2}{3} \log \rho = f(1/T)$ according to Eq. (51) for various salts in ethanol ($\Diamond$), propanol ($\bigcirc$), and acetonitrile($\square$).

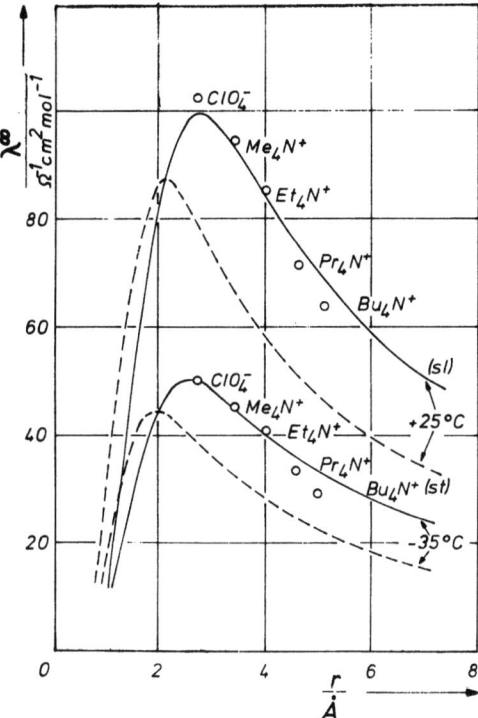

Figure 11. Representation of ionic limiting conductance λ_i^∞ according to Eq. (49) in acetonitrile at 25°C and −40°C.[222,242] sl: slipping movement [$s = 1$, $k_i = 4\pi\eta r_i$ in Eq. (49)]; st: sticking movement [$s = 2$; $k_i = 6\pi\eta r_i$ in Eq. (49)].

application of the kinetic model to ionic mobility. Activation energies for ionic mobility in acetonitrile have been investigated in our laboratory.[222,242]

The Boyd–Zwanzig relation, Eq. (49), gives insight into the nature of ionic movement. As an example, Fig. (11) shows that tetraalkylammonium ions and large anions between 25 and −40°C move essentially without dragging solvent molecules with them.[222,242]

(ii) Ion-Size Parameters

When conductance data are evaluated by a complete conductance equation, the relation $R_1 = R_2 = R_y$ is implied or found.

We therefore avoid discussion of the distance parameters obtained by the older theories established with a set of coefficients excluding J_2. Distance parameters deduced from J_1 are then adaptation parameters and provide only implicit information.

(iii) Association Constants

The Gibbs energy ΔG_A^0 of the association process is obtained from the experimentally determined association constant K_A, Eq. (1), by means of Eq. (21), and hence ΔH_A^0 and ΔS_A^0 are also known when measurements are conducted at different temperatures.

Shkodin and Kurova[194] investigated the temperature dependence of K_A in a series of aliphatic alcohols and found for NaI and KI positive ΔH_A^0 and ΔS_A^0 values which increase with increasing chain length of the alcohol with the exception of NaI in methanol and ethanol for which $\Delta H_A < 0$ in both solvents. Increasing values of K_A as a function of temperature were also obtained in propanol for various other alkali salts,[31,144,243] (cf. Fig. 12). Tetraalkylammonium salts in alcoholic solutions show

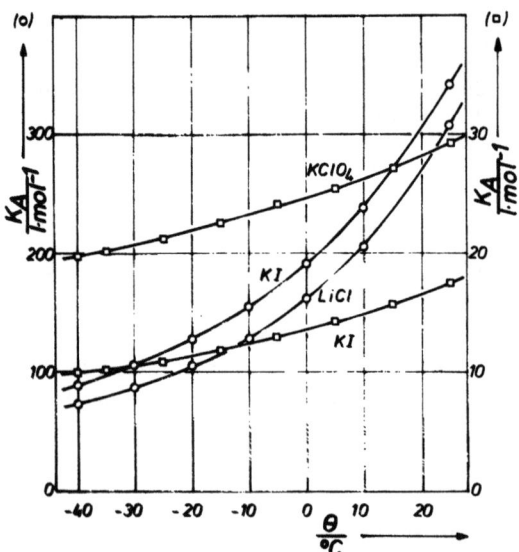

Figure 12. Association constants K_A of alkali metal salts in n-propanol (O) and acetonitrile (□) as a function of temperature, $-40 \leq \theta \leq 25°C$.

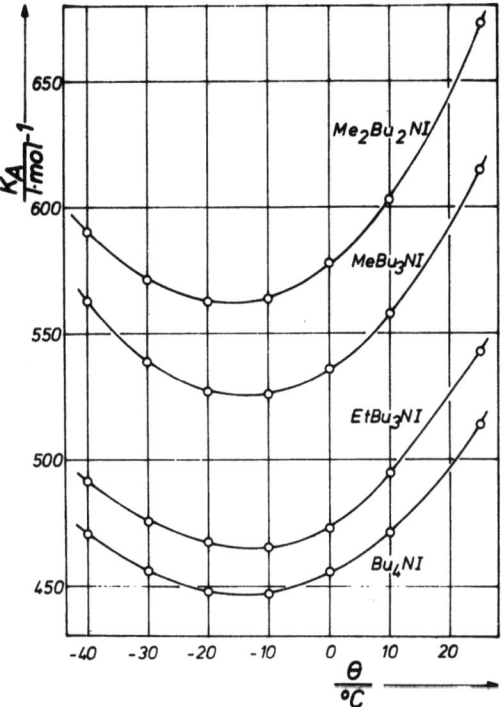

Figure 13. Association constants K_A of tetraalkylammonium salts $X_{4-n}Bu_nNI$ in n-propanol as a function of temperature, $-40 \leq \theta \leq 25°C$.

an unusual temperature-dependence function, Fig. 13, which hardly can be understood with the assumption of purely long-range forces. Figure 13 shows plots of K_A vs. θ of some tetraalkylammonium iodides, $Bu_{4-n}R_nNI$, in propanol which all go through a minimum at a temperature which is independent of the cation. This type of temperature function is generally found for tetraalkylammonium salts in ethanol[198] and propanol,[31,144,243] and can be expected for theoretical reasons also for other solvents. It has been verified for acetonitrile.[144,243] Variation of the anion displaces the minimum significantly.[198]

Formamide and N-methylamides as solvents yield complete dissociation of 1.1-electrolytes (cf. examples in Table 9). Small association constants were determined for HCl in N-methylpropionamide.[193]

Alkali salts in aprotic dipolar solvents generally show increasing association constants with increasing temperature as in the case of acetonitrile,[144] propylene carbonate,[144] dimethylformamide,[195] acetone,[218,219] and dimethylsulfoxide[229] as solvents. HCl in dimethylsulfoxide shows decreasing dissociation constants,[227] while HBr appears almost completely dissociated.[228] Tetraalkylammonium salts were investigated in acetonitrile,[144,222,243] where K_A functions with a minimum were obtained.[144,243] In ethylmethyl ketone, $dK_A/dT > 0$ between 15 and 35°C.[220] Tetraphenylarsonium iodide shows increasing association constants with increasing temperature in acetone (20–40°C) and nitromethane (25–45°C).[219]

A multistep association process is proposed for tetraalkylammonium picrates in solvents of low dielectric constant[236] and for tetraborates and fluorenyl salts in tetrahydropyran.[233] Triple-ion formation is found in the latter systems,[233] in methylene dichloride,[234] and in anisole.[238]

A consideration similar to that in Section III.5 (ii) leads us to avoid discussing association constants by means of Eq. (27). Setting the upper limit of the integral at a fixed value, e.g., $R = q$, compels us to use the lower limit $a = a_K^B$ as an adaptation parameter for the short-range interactions of the association process. This generally occurs if Bjerrum's association constant—or any other of the K_A^C constants in connection with long-range forces—is chosen as the basis of interpretation. Abnormally low a_K^B values for tetraalkylammonium salts in various solvents provide a striking example for this procedure (cf. Ref. 144). Inversion of the temperature coefficient of K_A or minima of the temperature functions $\Delta G_A^0(T)$ cannot be understood. Short-range forces can be better taken into account by Eq. (35) or need even more developed models.

Figure 14 shows the result of applying Eq. (35) to experimental K_A values as obtained in Table 7. Evaluation of the short-range interaction function $B(T)$ (Fig. 14) is based on the assumptions $\phi(r) = 1$ and $U^*(r) = $ constant. The lower limit a of the integral is set equal to the calculated distance of closest approach and not used for adaptation; the upper limit is $R = q$. As $N_A U^* = \Delta G_A^*$ is the molar Gibbs energy attributed to the short-range interactions [cf. Section III.3], $\partial B(T)/\partial(T^{-1}) = -\Delta H_A^*$.

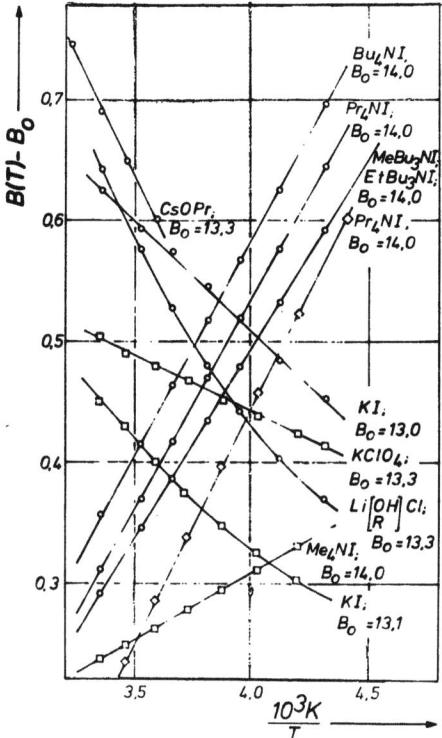

Figure 14. Representation of the noncoulombic Gibbs' energy of the association process, Eq. (35) for various salts in ethanol (◇), propanol (○), and acetonitrile (□) as a function of $10^3/T$; $-40 \leq \theta \leq 25°C$. $B(T) - B_0 = \ln(K_A^{expt.}/K_A^B) + \ln[4\pi(10^n)Ne^6/(4\pi\varepsilon_0 k)^{-3}]$ (cf. Ref. 144).

The plot of $B(T)$ vs. T^{-1} yields straight lines with a positive slope for all tetraalkylammonium salts in ethanol, propanol, and acetonitrile, indicating $\Delta H_A^* < 0$. The same plot for the alkali metal salts does not always yield linear functions but it indicates always $\Delta H_A^* > 0$.

Table 12 summarizes the results. ΔG_A^0 values are obtained with the help of Eq. (21) from the measured association constants at 25°C. The absolute values of entropies, ΔS_A^*, are not very significant; only their variations from one electrolyte to the other in a given solvent provide information. The quantity a has been identified with the center-to-center distance of hard spheres if the

Table 12
Thermodynamic Data of Ion-Pair Formation ($R=q$)

Electrolyte	a (Å)	ΔH_A^* (J mol^{-1})	ΔS_A^* (J mol^{-1} K^{-1})	$\Delta G_A^*(298)$ (J mol^{-1})	$\Delta G_A^0(298)$ (J mol^{-1})
			Ethanol		
Pr$_4$NBr	6.47	−6000	−10.6	−2800	−12100
Pr$_4$NI	6.68	−7500	−12.7	−3700	−12800
			Propanol		
Pr$_4$NI	6.68	−6800	−8.0	−4400	−15500
Bu$_4$NI	7.10	−6900	−8.0	−4500	−15500
i-Am$_4$NI	7.10	−7500	−8.4	−4900	−15700
i-Am$_3$BuNI	7.10	−7300	−8.4	−4900	−15700
MeBu$_3$NI	5.64	−6200	−6.7	−4100	−15900
EtBu$_3$NI	6.17	−6200	−6.7	−4100	−15600
Me$_2$Bu$_2$NI	5.64	−6000	−5.0	−4400	−16200
KI	3.52	+3500	+13.4	−460	−14500
LiCl^a	5.30	+7400	+32	−2100	−14200
CsOPr	3.20	+8300	+36	−2800	−17100
			Acetonitrile		
Me$_4$NI	5.64	−2300	+5.4	−4000	−8700
KI^a	3.52	+4100	+13.4	+80	−7100
KClO$_4$	3.70	+1800	+10.5	−1400	−8400

^{a} ΔH_A^* and ΔS_A^* determined from the initial slope.

ions are spherically symmetrical, e.g., KI or R$_4$NI. For nonsymmetrical ions such as MeBu$_3$N$^+$ or C$_3$H$_7$O$^-$, the shortest possible distance between the localized positive and negative charges is chosen. These values are listed as the a parameters in Table 12. They are identified with crystallographic radii as far as these can be determined or are calculated from bond lengths or from the molar volumes of the corresponding isosteric alkanes.[73] An appropriate choice of the a parameter may be used to distinguish solvent-separated and contact ion pairs, e.g., LiCl in propanol.[144]

The assumption $R = a + s$ instead of $R = q$ or a two-step equilibrium with $R' = a + s$, $R = q$ ($a + s < q$) does not provide other information.[144] The same classification of the salts is observed: $\Delta H_A^* > 0$ for tetraalkylammonium and $\Delta H_H^* > 0$ for alkali salts. Entropy values are found to be $\Delta S_A^* = 0$ for tetraalkylammonium and $\Delta S_A^* > 0$ for alkali salts.[144,283]

All methods show that an important part of the energy of

formation, ΔG_A^0, of an ion pair from the initially infinitely separated ions is due to non-Coulombic interaction. For alkali metal salts $\Delta H_A^* > 0$, but this heat required for ion pair formation is compensated for by an increase of entropy in this process. The ionic shell loosens oriented solvent molecules. $\Delta H_A^* < 0$ and $\Delta S_A^* \approx 0$ or slightly negative indicates that interaction of tetraalkylammonium ions and solvent molecules in the vicinity is of a different nature: the interaction effects arise mainly from dispersion forces. Solutions of LiCl in propanol show special properties. Their ΔH_A^* values as determined from crystallographic radii ($a = 2.50$ Å), are positive but small. If, however, a solvent-separated ion pair is postulated by setting $a = 5.30$ Å, values are obtained which are comparable to those of the other alkali salts. The distance parameter $a = 5.30$ Å is calculated from Pauling's radii of Li^+ and Cl^- and the van der Waals volume of —OH.[248]

As a conclusion, the knowledge of K_A as a function of temperature over a sufficiently large temperature range allows an unambiguous differentiation to be made between types of ion-solvent interactions on the basis of models which take into account non-Coulombic interaction potentials.

Empirical extensions of the treatment of the association constant have been used in solutions with o-dichlorobenzene[237] as the solvent. A Born cycle for ion-pair formation has been studied by Denison and Ramsey.[235]

(iv) **Mixed Solvents**

Investigations in aqueous–organic-solvent mixtures must be mentioned as a further group: water–methanol,[249] water–ethanol,[250–256] water–butanol,[259] water–methoxyethanol,[257] water–dioxane,[258] water–acetone,[245,260] water–N-methylacetamide,[261] and water–methyl-acetate.[262] Acetonitrile–benzene has been used for NaI,[263] methanol–acetone for KSCN, $LiClO_4$, and LiBr,[240] and propylene-carbonate–tetrahydrofurane for $LiClO_4$.[225] The main interest in mixed solvents is in the field of concentrated solutions and hence this problem will be treated when that topic is considered (Section IV). As far as dilute solutions are concerned, use is made of electrolyte theory as outlined in Sections III.2 and III.3.

IV. CONCENTRATED ELECTROLYTE SOLUTIONS

1. Survey

As a supplement to the brief survey in Section I concerning concentrated electrolyte solutions it must be said that the present situation in this field is characterized by the fact that the literature contains no conductance data $\kappa = \kappa(m, T, \xi)$ permitting a comprehensive discussion of conductance-determining factors as a function of electrolyte concentration m (mol kg^{-1}), temperature T or θ, and solvent composition ξ (wt.%) of binary solvent mixtures. The existing review articles[26,27,29] contain only few dependences of the type $\kappa = \kappa(m, T)$ or $\kappa = \kappa(\xi)$ together with data for single concentrations at 25°C, and this situation has scarcely changed since then.*

Some contributions concerning aqueous solutions are included in the following discussion where solvent structure effects are considered and data from organic solvents were not available. This discussion is built up around investigations providing extensive information on the complete conductance dependences $\kappa = \kappa(m, T, \xi)$ for various electrolytes in mixtures of propylene carbonate (PC) and dimethoxyethane (DME) over the temperature range $-45 \leq \theta \leq +25°C$.[48] Table 13 contains a survey of the solutions investigated and may serve as a basis of orientation in the framework of this section. Figures 15–19 show the specific conductances of these solutions as a family of functions $\kappa = \kappa(m, \xi, \theta)_{\xi,\theta}$ at four selected temperatures.

Some qualitative items of information can be deduced from Figs. 15–19. The maxima $\kappa_{max}(\theta, \xi)$ of the specific conductances are attained at concentrations $\mu(\theta)_\xi$ if the solubility of the electrolyte permits a sufficiently high concentration to be attained. The displacement of μ as a function of temperature in any solvent mixture is positive, i.e., $(\partial \mu/\partial \theta)_{\xi,\theta} > 0$. This behavior is generally observed in organic solvents. Comparison of maximum specific conductances in solvent mixtures of different compositions shows a shift of both κ_{max} and μ to higher values with

---

* After going to press, a publication of Tikhonov, Ivanova, and Ravdel'[225] concerning conductance measurements of LiClO$_4$ in THF (tetrahydrofurane), PC (propylene carbonate), and their mixtures over the range $-20 \leq \theta \leq +50°C$ came to our knowledge which contains functions $\kappa_{LiClO_4/PC}(c, \theta)$, $\kappa_{LiClO_4/THF}(c, \theta)$, and $\kappa_{LiClO_4/PC/THF}(\xi, \theta)_{c=1M}$.

Table 13
Maximum of Specific Conductances as a Function of Temperature and Solvent Composition. Solvent: PC and PC–DME Mixtures (ξ = wt.% PC)

Electrolyte	ξ	$\theta = 25°C$					$\theta = -5°C$					$\theta = -45°C$				
		$\kappa_{max}/10^{-3}$	$\sigma/10^{-5}$	μ	$\sigma/10^{-3}$	$\Delta_{max}/10^{-5}$	$\kappa_{max}/10^{-3}$	$\sigma/10^{-5}$	μ	$\sigma/10^{-3}$	$\Delta_{max}/10^{-5}$	$\kappa_{max}/10^{-3}$	$\sigma/10^{-5}$	μ	$\sigma/10^{-3}$	$\Delta_{max}/10^{-5}$
LiClO$_4$ (Fig. 15)	100	5.42	1	0.662	8	2	2.36	1	0.540	5	1	0.275	0.1	0.341	5	0.1
	76.12	11.57	10	1.09	20	11	5.71	1	0.841	4	1	1.044	0.6	0.520	5	0.6
	51.32	14.16	7	1.24	30	12	8.20	3	0.953	11	2	2.322*	—	0.793*	—	—
	42.13	14.59	1	1.386	4	2	8.65	1	1.04	4	2	2.66	5	0.68	20	5
	28.14	14.38	3	1.652	7	5	9.03	2	1.186	5	2	3.314	0.1	0.735	1	0.1
	12.86	12.82	12	2.00	70	2	8.51	4	1.51	60	4	—	—	—	—	—
LiPF$_6$ (Fig. 19a)	100	5.41	2	0.86	30	2	2.21	1	0.686	7	1	0.234*	—	0.426*	—	—
KPF$_6$ (Fig. 16)	100	7.31	2	0.97	30	2	3.09	1	0.725	4	1	0.347	0.1	0.436	4	0.1
	76.12	10.32	1	1.066	9	1	5.17	1	0.813	2	1	0.978	0.1	0.507	1	0.1
	42.13	12.75	1	1.314	10	1	7.67	1	1.005	2	1	2.487*	—	0.676*	—	—
KSCN (Fig. 18)	100	7.00	<1	1.097	5	1	2.885	0.1	0.767	1	0.2	0.335	0.1	0.426	0.1	0.1
	76.12	9.01	1	1.106	8	1	4.446	0.5	0.797	5	1	0.870	0.2	0.478	1	0.2
	51.32	9.86	<1	1.319	1	0.2	5.485	0.1	0.949	<1	0.1	1.509	0.1	0.588	<1	0.1
	42.13	9.70	<1	1.490	7	2	5.566	0.1	1.050	<1	0.2	1.685	0.1	0.644	1	0.1
	28.14	9.04	<1	1.874	17	1	5.283	0.7	1.290	5	1	1.786	0.7	0.760	10	1
Et$_4$NPF$_6$ (Fig. 17)	100	12.23	5	1.80	90	0.4	5.102	0.3	1.272	10	0.2	0.581	1	0.76	12	2
	76.12	15.63*	—	2.24*	—	—	7.86	2	1.41	40	2	1.516	1	0.85	30	0.1
	42.13	Only data at -35°C are evaluated										4.63	6	0.96	80	1
n-Pr$_4$NPF$_6$ (Fig. 19b)	100	8.47	14	1.17	190	3	3.44	1	0.86	30	2	0.377	0.1	0.503	6	0.1
n-Bu$_4$NPF$_6$ (Fig. 19c)	100	6.31	2	0.840	6	1	2.58	2	0.666	15	3	0.298	0.3	0.408	14	0.3

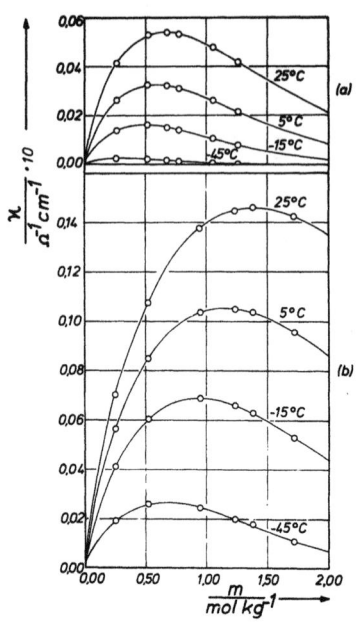

Figure 15. Specific conductance $\kappa(m, \theta, \xi)$ of $LiClO_4$ in mixtures (ξ = wt.% PC) of propylene carbonate and dimethoxyethane. $\xi = 100$ (a); 42.13 (b).

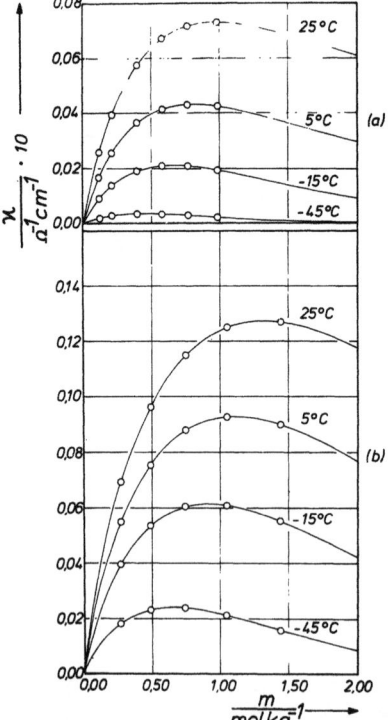

Figure 16. Specific conductance $\kappa(m, \theta, \xi)$ of KPF_6 in mixtures (ξ = wt.% PC) of propylene carbonate and dimethoxyethane. $\xi = 100$ (a); 42.13 (b).

decreasing ξ when ξ = wt.% of PC (cf. Figs. 15, 16, 17, 18). Obviously κ_{max} at constant temperature decreases with increasing radii of R_4N^+ cations (cf. Figs. 17a, 19b, and 19c) and so does κ at any other concentration. Furthermore, conductance in a given solvent is predominantly determined by the cation properties, the influence of anions being small (see Figs. 15a and 19a in comparison with 16a and 17a); for aqueous solutions, see Refs. 267 and 268.

2. Analysis of Experimental Data

A more quantitative insight than the preceding observations allow is provided by the application of Eq. (4) to our measurements. This procedure yields more precise values of κ_{max} and μ and enables their standard deviations $\sigma(\kappa_{max})$ and $\sigma(\mu)$ to be accounted for. Comparison of $\kappa(m, \theta, \xi)$ for various electrolyte solutions at any state (m, θ, ξ) is now also possible. Reliable extrapolated values κ_{max} and μ are obtained if solubility prevents these values from being attained experimentally and the region for extrapolating is not too large. If the extrapolation region was too large or measurements in highly concentrated solutions could not be made at low temperatures because of decreasing solubility, values of κ_{max} and μ were obtained with the help of appropriate temperature functions, e.g., Eq. (55). These evaluations are indicated in Table 13 by an asterisk (*).

Table 13 contains values of κ_{max}, μ, and their standard deviations, as well as the maximum deviation $\Delta_{max} = \max |\kappa_{calc} - \kappa_{expt}|$ for selected temperatures. For information about the complete temperature program, $-45 \leq \theta \leq 25°C$ in steps of 10 K, including the parameters a and b of Eq. (4), see Ref. 48. Precision of measuring data (better than 0.1%) and good aptitude for adaptation of Eq. (4) to conductance data permit data analysis with a minimum of 5 to 6 measuring points.

3. Electrolytes in Pure Organic Solvents

(i) Maximum Specific Conductance

The maximum in specific conductance κ_{max} and its interpretation has been the main interest for many authors.[26-29,48,49,239,268,269,272] In his work on aqueous solutions,

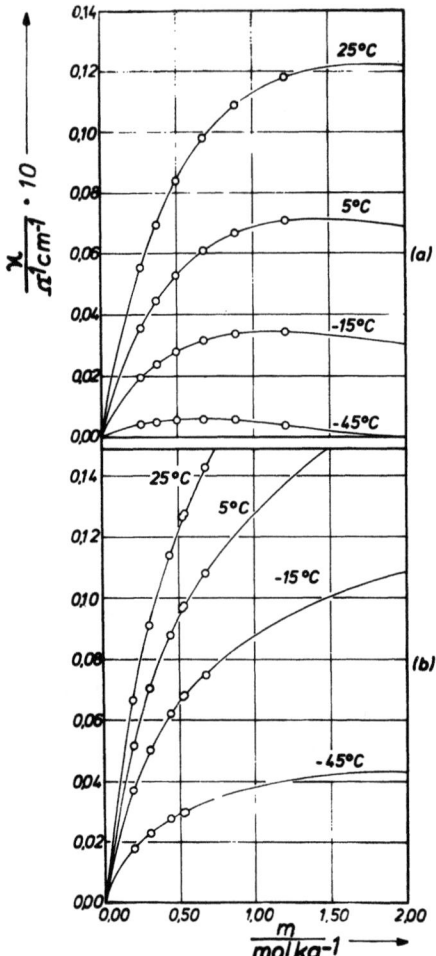

Figure 17. Specific conductance $\kappa(m, \theta, \xi)$ of Et_4NPF_6 in mixtures ($\xi = $ wt.% PC) of propylene carbonate and dimethoxyethane. $\xi = 100$ (a); 42.13 (b).

Molenat[269] obtained from the definition of specific conductance, which we write as

$$\kappa = \gamma \Lambda c$$

with γ as the conversion factor appropriate for the chosen system of units, a relation

$$d\kappa = \gamma[\Lambda \, dc + c \, d\Lambda] \qquad (52)$$

Experimental evidence allows the assumption $d\Lambda < 0$ if $dc > 0$ and if triple-ion formation can be excluded. As a consequence of

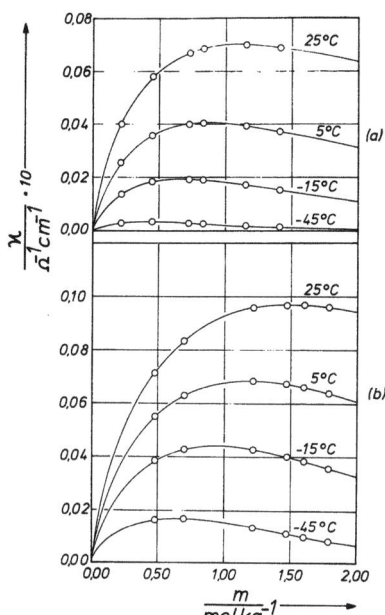

Figure 18. Specific conductance $\kappa(m, \theta, \xi)$ of KSCN in mixtures ($\xi =$ wt.% PC) of propylene carbonate and dimethoxyethane. $\xi = 100$ (a); 42.13 (b).

this assumption and Eq. (52), a maximum in specific conductance is obtained if $\Lambda\, dc = |c\, d\Lambda|$. $d\kappa > 0$ if $\Lambda\, dc > |c\, d\Lambda|$. Hence the existence of a κ_{max} follows from the competition of two factors, a conductance-increasing term caused by an increase of the density of free ions and a κ-decreasing term resulting from a lowering of ionic mobilities as electrolyte concentration increases. According to Molenat, the existence of a maximum in the specific conductance does not need any hypothesis about changing solvent structure.[269] In contrast to the latter statement, Valyashko and Ivanov[268] on the basis of Samoilov's theory[271] have stressed the role of ion–solvent interaction and hydration in competition with ion–ion interaction. They used this concept for explaining the shift of μ for alkali metal salts to higher values of m in a series of aqueous alkali sulfate solutions from Li to Cs and the observation that $(\partial\mu/\partial\theta) > 0$. Jasinski[26] favors ion association as the most important feature for explaining the maximum in the specific conductance. According to this author, the different behavior of $LiClO_4$ in PC yielding a maximum κ_{max} in contrast to solutions of the same electrolyte

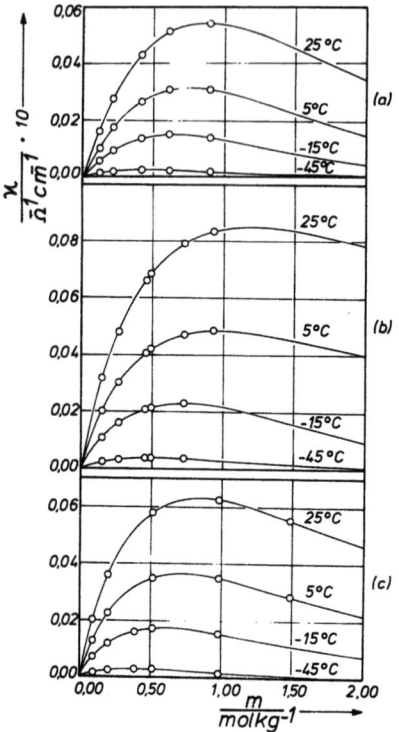

Figure 19. Specific conductance $\kappa(m, \theta)$ of $LiPF_6$ (a), $n\text{-}Pr_4NPF_6$ (b), and $n\text{-}Bu_4NPF_6$ (c) in propylene carbonate solvent.

in THF, is related to a different "shielding" of the ions by solvent molecules.

Molenat's formulation is a clear formal statement but leaves undecided the question about the factors governing variation of ionic mobility. Increasing electrolyte concentration requires, in addition to changes of ion–ion interaction, a variation of ion-solvent interaction. At the concentration μ, the ratio of the electrolyte and the solvent molecules is of the order of 1:10 if 1.1 electrolytes are considered in both aqueous[268] as well as nonaqueous solutions (cf. Figs. 16a and 18a), i.e., almost all solvent molecules are more or less oriented within the ionic fields. Changing solvent structure with increasing electrolyte concentration changes the dielectric properties of the solution (cf. Ref. 8). Solvent structure around the ions is one of the mobility-determining effects. Ion solvation and ion association are in competition, as known from dilute solutions when short-range forces beyond coulombic interaction forces are taken into account (cf. Section III). However, at present, a range of validity of the ion association concept is difficult to estimate. At low concentrations, association has been introduced as a component of the electrolyte activity and a consequence of ion distribution functions, and thus should be understood at higher concentrations only as an approximate measure for estimation of the same types of interaction. For example, Jasinski's example[26] would require $K_A(\text{LiClO}_4)$ in THF to be much smaller than $K_A(\text{LiClO}_4)$ in PC, which can be hardly understood from measurements at low concentrations [cf. $K_A(\text{LiB}\phi_4/\text{THF}) \sim 10^4$;[231] $K_A(\text{LiClO}_4/\text{PC}) = 1.3$].[48] This example should be understood rather on the basis of the distinctly different viscosities of these solvents. The maximum of specific conductance in THF does not appear as a consequence of insufficient solubility just as can be extrapolated from Figs. 15a and 15b for DME.[48]

(ii) *Specific Conductance and Ionic Radii*

Inspection of Table 13 encourages an investigation of the relation between $\kappa(\mu)$ and ionic radii for which tetraalkylammonium salts are the most appropriate examples (cf. Section IV.1). Additivity of ionic conductances in connection with Stokes law (cf. Section III.3) suggests the use of a relationship

$$\frac{\kappa(m)}{m} = C\left(\frac{1}{r_-} + \frac{1}{r_+}\right) \qquad (53)$$

the coefficient C being a function of η^{-1} and other nonspecified parameters. A series of electrolytes with a common anion allows Eq. (53) to be written as

$$\frac{\kappa(m)}{m} = C_1 + \frac{C_2}{r_+}, \qquad C_1 = \lim_{r_+ \to \infty} \frac{\kappa(m)}{m} \qquad (54)$$

At moderate concentrations ($m = 0.1$ mol kg^{-1}; Fig. 20) Eq. (54) is satisfied by the experimental data for $R_4\text{NPF}_6$ insofar as κ/m is proportional to r_+^{-1} and $[\partial(\kappa/m)/\partial(r_+^{-1})]_\theta < 0$. As in dilute solutions, C_1 should not be used to discuss an anion contribution. It is interesting to note that $(\kappa/m)_{m,\theta}$ values for K^+ and $n-\text{Pr}_4\text{N}^+$ are equal at any temperature. As $n-\text{Pr}_4\text{N}^+$ is not solvated by PC,[266]

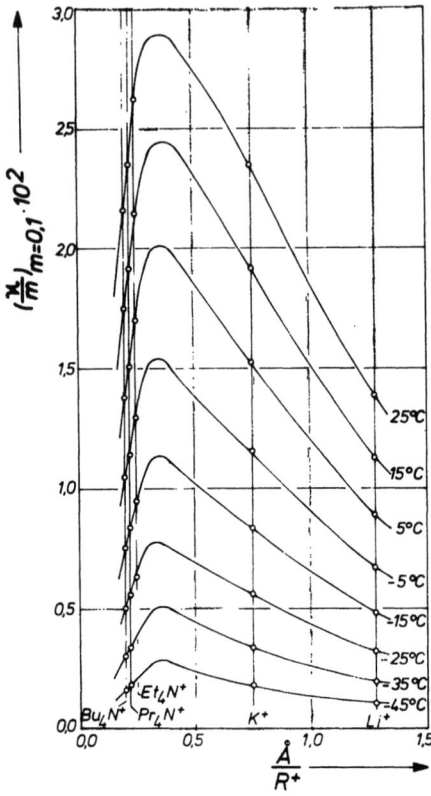

Figure 20. Representation of $\kappa(m)/m$, Eq. (53) as a function of ionic radii at $m = 0.1 \text{ mol kg}^{-1}$ (plotted curves through the measuring points have no theoretical significance) for various salts in propylene carbonate at temperatures $-45 \leq \theta \leq 25°C$.

the Stokes radius $R(K^+)$ should be equal to the ionic radius $r(n\text{-Pr}_4N^+)$ and independent of temperature in accordance with the fact that the Walden product of KPF_6 is temperature independent (cf. Table 14). $R(Li^+) > R(K^+)$, but a more quantitative statement is not possible in consequence of the low purity of $LiPF_6$.[48] It suffices to note that $R(Li^+) \approx r(n\text{-BuN}^+)$ and is independent of temperature, a conclusion which is based on comparison of κ/m of $LiClO_4$ and $n\text{-BuNPF}_6$ by means of Eq. (53), where $R(PF_6^-)$ must be set $\approx R(ClO_4^-)$.

Application of Eq. (54) to $m = \mu$, $\kappa = \kappa_{max}$ for all the salts investigated yields $(\kappa_{max}/\mu)_\theta \approx$ constant, and hence is almost independent of ionic radius in contrast to the behavior of those salts at low and moderate concentrations, e.g., $\kappa_{max} = (8 \pm 7) \times 10^{-4} + (6.3 \pm 0.6) \times 10^{-3} \mu$ at 25°C.[48] Figure 21 illustrates this linear relationship. If μ is attained only at high values, the decrease of ionic mobility caused by ion–solvent and ion–ion interactions overwhelms the increase of charge density only at this high concentration. The corresponding high charge density yields high conductance. The reasons for high μ values can be quite varied: weak ion–solvent interactions, small ionic radii, or low solvent viscosity. Thus lithium salts as well as tetrabutylammonium salts allow the expectation of small values of μ in contrast to tetraethylammonium salts. The influence of viscosity can be estimated by separating η in the coefficients of Eq. (53).[48]

The different behavior of dilute or moderately concentrated to concentrated solutions can be shown from the examples in Table 14.

Comparison of Eq. (53) and the Walden product, Eq. (48), yields an expression $A(m) = \Lambda^\infty m/\kappa(m)$ which should be constant if the Stokes radii, as determined from the properties of infinitely

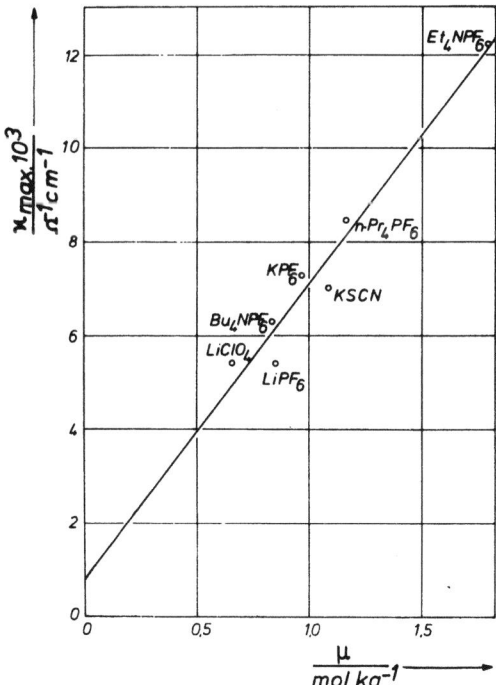

Figure 21. Linear dependence $\kappa_{max} = \kappa_{max}(\mu)$ for various salts in propylene carbonate at 25°C.

Table 14
Walden Product and Values of $A(m) = m\Lambda_0/\kappa(m)$ of $LiClO_4$ and KPF_6 in Propylene Carbonate at Various Temperatures

Temperature (°C)	$LiClO_4$				KPF_6			
	$\eta\Lambda_0$	$A(0.1)$	$A(1.0)$	$A(\mu)$	$\eta\Lambda_0$	$A(0.1)$	$A(1.0)$	$A(\mu)$
+25	0.676	1233	5418	3265	0.732	1234	3966	3848
+15	0.680	1224	5767	3163	0.734	1222	4132	3616
+5	0.684	1217	6267	3063	0.742	1205	4376	3427
−5	0.687	1215	7023	2965	0.738	1202	4722	3263
−15	0.691	1224	8202	2879	0.739	1198	5245	3131
−25	0.697	1217	10157	2779	0.742	1198	6084	3004
−35	0.706	1238	13719	2690	0.748	1202	7500	2889
−45	0.707	1263	20945	2603	0.747	1216	10164	2783

^{a} η is used with the unit "Poise."

dilute solutions and at concentration m, are equal, and the coefficient C in Eq. (53) can be represented in the form $C = \eta^{-1}C'$. From Table 14 it can be deduced that these conditions are fulfilled only at moderate concentrations ($m = 0.1$ mol kg^{-1}), yielding a constant value of A (0.1), independent of temperature for both electrolytes. The latter fact results from referring each ionic mobility to its value at infinite dilution. From this statement, an important feature can be seen. At concentrations up to $m = 0.1$ mol kg^{-1}, specific conductance is governed by almost the same factors as conductance in highly dilute solutions when the difference of ionic mobilities is eliminated. At high concentrations, however (e.g., when $m = 1.0$ mol kg^{-1} in Table 14), A_{LiClO_4} (1.0) > A_{KPF_6} (1.0) at any temperature. Temperature dependence of A (1.0) is significant for both electrolytes and more pronounced for $LiClO_4$ than for KPF_6; thus $(dA/dT)_{m=1.0} < 0$. In contrast to the behavior of moderately concentrated solutions, ionic mobilities depend on factors which are not eliminated by referring to infinite dilution. Loss of mobility by these additional effects is bigger and depends more strongly on temperature for $LiClO_4$ than for KPF_6. An explanation in terms of solvate clusters, as has been considered in aqueous solutions,[267,268] can be taken into account. As the position of μ itself on the concentration axis is defined by these effects (cf. Section IV.3 (i)), it is found that $A_{LiClO_4}(\mu) \approx A_{KPF_6}(\mu)$.

(iii) Temperature Dependence of Specific Conductance

Temperature dependence of $\kappa(m, \theta)$ is accounted for with appropriate accuracy by an equation of the form

$$\ln \kappa(m, \theta) = a_0 + a_1 T^{-1} + a_2 T^{-2} + \cdots \quad (55)$$

This type of equation has been applied to moderately concentrated[210,266] and concentrated[239,240,268] solutions.

Application of Eq. (55) truncated at the T^{-2} term, cf. also the Girifalco viscosity equation,[274] yields temperature coefficients $B(T)$,

$$\frac{d \ln \kappa(m, \theta)}{d(1/T)} = B(T) = a_1 + 2a_2 T^{-1} \quad (56)$$

which are equal for all electrolytes investigated at $m = \mu$ and which provide activation energies if concentration dependence of this quantity is neglected.

Activation energies including their concentration dependence have been obtained[48] for various concentrations by means of Eq. (56) applied to appropriate regions of the $\kappa(m, \theta)$ diagram. This yields the functions $E_\theta(m)$ and $E_m(\theta)$, which are summarized together with their corresponding coefficients and the maximum deviations of E, Δ_{max}, in Tables 15 and 16 for some values of m and θ. Δ_{max} values are maximum deviations of the single values from the appropriate equations (a) and (b) of Tables 15 and 16, respectively, but are no measure of the accuracy of the E values themselves.

The kinetic model of conductance provides the same information as the hydrodynamic model (Section IV.3 (ii)). Activation energies $E_\theta(m)$ for all electrolytes become equal at low concentrations (cf. $a_0^{(\theta)}$, Table 16), indicating that these values are controlled mainly by the properties of the solvent. This result agrees well with the statement that the ratio of activation energies of conductance and viscosity is ≈ 1 and does not depend much on concentration, temperature, and electrolyte.[266,275] At high concentrations ($m = 1.0$ mol kg^{-1}) the sequence of mobility activation energies for a series of electrolytes is $LiClO_4 > Bu_4NPF_6 > KPF_6 \approx KSCN > Et_4NPF_6$, which is in accordance with the sequence of these electrolytes in Fig. 21. The electrolyte in which the

Table 15
Activation Energies $E_m/\text{kJ mol}^{-1}$ as a Function of Temperature for Various Salts in Propylene Carbonate:
$$E_m(\theta) = a_0^{(m)} + a_1^{(m)}\theta + a_2^{(m)}\theta^2 \quad \text{(a)}$$

Electrolyte	$m = 0.1$			$m = 0.5$				$m = 1.0$			
	$a_0^{(m)}$	$a_1^{(m)}$	$10^4 \times a_2^{(m)}$	Δ_{max} (%)	$a_0^{(m)}$	$a_1^{(m)}$	$10^4 \times a_2^{(m)}$	Δ_{max} (%)	$a_0^{(m)}$	$a_1^{(m)}$	$10^4 \times a_2^{(m)}$	Δ_{max} (%)
LiClO$_4$	17.17	−0.2194	8.95	0.3	19.72	−0.2990	11.54	0.7	24.48	−0.4255	17.60	0.4
KPF$_6$	17.08	−0.2182	9.04	0.3	19.03	−0.2790	11.54	0.3	22.35	−0.3642	15.04	0.4
KSCN	17.05	−0.2040	8.42	0.3	19.21	−0.2584	10.58	0.4	22.24	−0.3317	13.69	0.4
Et$_4$NPF$_6$	16.84	−0.2272	9.40	0.3	18.58	−0.2112	8.75	0.2	19.78	−0.2715	10.60	0.3
n-Bu$_4$NPF$_6$	17.41	−0.2148	8.93	0.3	20.04	−0.2508	10.38	0.3	22.83	−0.4534	18.53	0.4

Table 16
Activation Energies $E_\theta/\text{kJ mol}^{-1}$ as a Function of Concentration for Various Salts in Propylene Carbonate: $E_\theta(m) = a_m^{(\theta)} + a_1^{(\theta)} m + a_2^{(\theta)} m^2$ (b)

Electrolyte	$\theta = 15°C$				$\theta = -5°C$				$\theta = -35°C$			
	$a_0^{(\theta)}$	$a_1^{(\theta)}$	$a_2^{(\theta)}$	Δ_{max} (%)	$a_0^{(\theta)}$	$a_1^{(\theta)}$	$a_2^{(\theta)}$	Δ_{max} (%)	$a_0^{(\theta)}$	$a_1^{(\theta)}$	$a_2^{(\theta)}$	Δ_{max} (%)
LiClO$_4$	13.82	2.16	2.53	0.2	17.67	5.29	3.66	0.4	23.7	17.0	0.07	2.1
KPF$_6$	13.81	1.94	1.50	0.2	17.71	4.75	1.95	0.1	24.3	9.4	1.85	1.0
KSCN	13.82	3.48	0.29	0.3	17.54	5.58	0.85	0.1	24.85	9.24	2.81	0.1
Et$_4$NPF$_6$	13.39	3.87	−1.39	0.4	17.70	3.23	0.21	0.1	25.5	3.34	1.82	0.4
Bu$_4$NPF$_6$	13.52	8.99	−6.01	0.3	17.86	6.78	0.40	0.2	25.8	2.0	13.0	0.8

strongest interactions arise, LiClO$_4$, has to cross the highest activation barrier. At a concentration $m = 1.0$ mol kg^{-1}, the maximum specific conductance of Et$_4$NPF$_6$ has not yet been attained ($\mu = 1.35$ mol kg^{-1} at 0°C), whereas it has been surpassed for KSCN ($\mu = 0.83$) and is still further for LiClO$_4$ ($\mu = 0.56$). Decreasing concentration reduces the differences of activation energies. At $m = \mu$, the activation energies become equal for all electrolytes at any temperature (cf. Table 17). In this table, the temperatures at the limits of the temperature program (25°C and −45°C) are omitted because of the ill-defined tangents at these limits. It should be remembered also that the μ values for Et$_4$NPF$_6$ at temperatures $\theta > -25$°C are affected by greater uncertainty than that for the others. Constancy of $E(\mu, \theta)_\theta$ can be chosen to be an alternative criterion of the position μ of maximum specific conductance: maximum specific conductance for an electrolyte arises when the conductance-determining effects have established a critical energy barrier which depends only on solvent and temperature.

(iv) The Coefficients a and b

The dimensionless a parameter of Eq. (4) turns out to be the same for all electrolytes investigated when the relevant standard deviations are taken into account. The b parameter generally decreases with decreasing temperature. In contrast to the a parameter, this parameter depends strongly on the electrolyte compound, e.g., $b = -0.08$ for LiClO$_4$ at 25°C and $b = +0.12$ at 25°C for Bu$_4$NPF$_6$ in PC. A simple type of relation of these parameters to dielectric constant or viscosity, as proposed in the theoretical significance to Eq. (4). For a comprehensive discussion see Ref. 48.

4. Organic Solvent Mixtures

(i) Introduction

Decrease of the parameter ξ, i.e., an increase of DME concentration in e.g., PC–DME mixtures, affects conductance in a complex way, both by molecular as well as by macroscopic

Table 17
Activation Energies $E_\theta(\mu)$/kJ mol^{-1} at Various Temperatures in Electrolyte Solutions with Propylene Carbonate as Solvent

Electrolyte	$\theta = 15°C$		$\theta = 5°C$		$\theta = -5°C$		$\theta = -15°C$		$\theta = -25°C$		$\theta = -35°C$	
	μ	$E_\theta(\mu)$	μ	$E_\theta(\mu)$	μ	$E_\theta(\mu)$	μ	$E_\theta(\mu)$	μ	$E_\theta(\mu)$	μ	$E_\theta(\mu)$
LiClO$_4$	0.624	16.2	0.583	18.9	0.540	21.6	0.495	24.5	0.445	27.1	0.394	30.0
KPF$_6$	0.881	16.7	0.799	19.3	0.725	22.6	0.655	24.8	0.582	27.5	0.509	30.3
KSCN	0.976	17.5	0.867	19.9	0.767	22.3	0.674	24.7	0.586	27.1	0.505	29.6
Et$_4$NPF$_6$	1.590	16.1	1.423	19.5	1.272	22.2	1.131	24.7	0.997	27.2	0.885	29.9
Bu$_4$NPF$_6$	0.778	16.9	0.715	19.8	0.666	22.6	0.601	25.0	0.536	27.4	0.475	29.7

properties of the solvent: viscosity $\eta(PC) > \eta(DME)$, e.g., $\eta_{298}(PC) = 2.526$ cP[48] and $\eta_{298}(DME) = 0.455$ cP,[264] and $d\eta(PC)/dT > d\eta(DME)/dT$, e.g., $\eta_{228}(PC) = 33.9$ cP[48] and $\eta_{228}(DME) = 1.25$ cP.[264] Decrease of viscosity usually leads to an increase of conductance[210,240,266] (cf. also Ref. 276 and Section IV.3 (iii)) and a displacement of $\kappa(\mu)$. A consequence, having technological interest, for solvent mixtures of PC and DME is that increase of conductance as a function of solvent composition ξ is much higher at low than at room temperatures (cf. Figs. 15–18).

Parallel to the decrease of viscosity, a decrease of relative permittivity ε_r is also observed, e.g., $\varepsilon_r(PC) = 64.9$[277] and $\varepsilon_r(DME) = 7.30$[278] at 25°C. The theory of dilute solutions predicts an increase of the association constant with decreasing permittivity in most solutions (cf. Ref. 8). When applied with necessary caution to moderate and concentrated solutions as a statement concerning ion interaction, this should produce a diminution of conductivity. This effect becomes more pronounced with increasing temperature. Possible exceptions are the tetraalkylammonium salts (cf. Section III.4).

Molecular effects, such as selective solvation in mixed solvents, cannot be taken into account through macroscopic variables; dipole moments μ or donor numbers D_{SbCl_5},[3,279] are more appropriate properties. PC and DME show quite different dipole moments: $\mu(PC) = 4.94$ D,[280] $\mu(DME) = 1.75$ D,[278] but the D_{SbCl_5} values of PC and an ether similar to DME, diethyl ether, are, however, comparable.[3,281] Donor numbers, which are the more relevant properties, indicate only small differences between PC and DME. The molar masses of these molecules are approximately equal and hence their difference does not have to be taken into account.

(ii) Maximum Specific Conductance

The discussion in this section must be restricted essentially to $LiClO_4$ and KPF_6 in PC–DME mixtures, these systems being the only ones investigated up to now with sufficient detail concerning both temperature and solvent composition. Figures 22a–c show the relevant representations of $\kappa_{max} = \kappa_{max}(\xi, \theta)_\theta$. Values

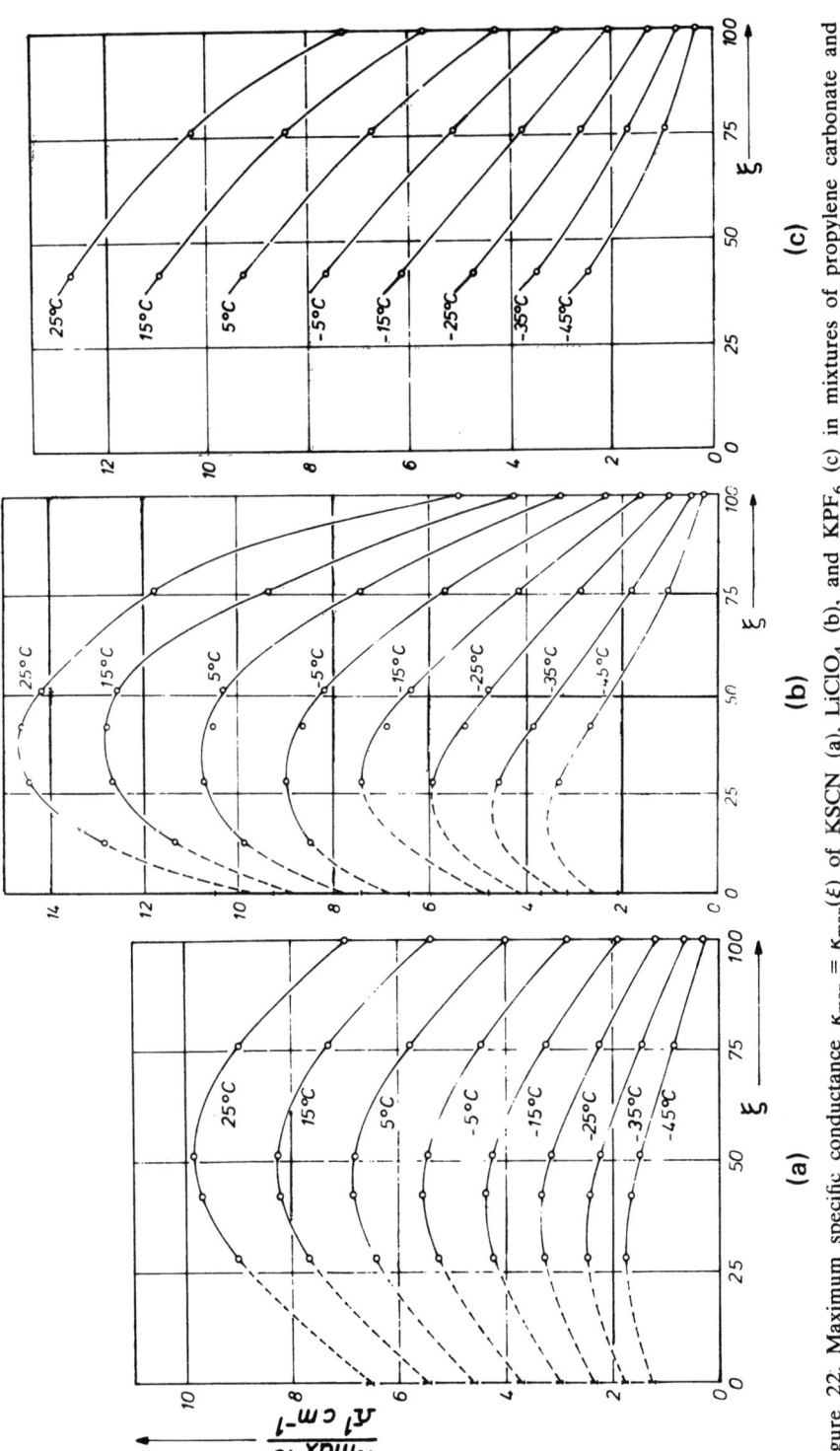

Figure 22. Maximum specific conductance $\kappa_{max} = \kappa_{max}(\xi)$ of KSCN (a), LiClO$_4$ (b), and KPF$_6$ (c) in mixtures of propylene carbonate and dimethoxyethane: ξ = wt.% PC. Standard deviations of the extrapolated values at $\xi = 0$ are plotted as a measure of uncertainty.

$\kappa_{max}(\xi, \theta)_{\xi=0}$ have been obtained by extrapolation according to $\log \kappa_{max}(\theta) = b_0 + b_1\xi + b_2\xi^2$. This function fits $\kappa_{max}(\xi, \theta)_\theta$ for $12.86 \leq \xi \leq 100$ with sufficient accuracy and the standard deviations $\sigma(\kappa_{max}(\theta)_{\xi=0})$ are plotted in Figs. 22a and 22b for representative examples. From these figures some information is already available: at any temperature $\kappa_{max}(\xi, \theta)_\theta$ possesses a maximum value $\kappa^*_{max}(\xi, \theta)_{\xi=\xi^*}$ which shifts according to $\partial\xi^*/\partial T > 0$. κ^*_{max} is the highest specific conductance which can be attained in the mixture of solvents at a given temperature: $\kappa^*_{max}(\xi^*, \theta) = \kappa_{max}(\mu, \xi^*, \theta)$. Further, the increase of maximum specific conductance $\kappa(\mu, \xi, \theta)_{\xi=100}$, i.e., $S = -(\partial\kappa(\mu, \xi, \theta)/\partial\xi)_{\theta,\xi=100}$, is steeper for LiClO$_4$ than for KSCN at any temperature. These observations are consistent with the assumptions of Section IV.4 (i).

The viscosity $\eta_0(\xi)$ of the pure mixed solvent decreases with decreasing value of the ξ parameter. Association is small in pure PC and increases when DME is added. κ^*_{max} is the result of the competing effects. The observed shift $(\partial\xi^*/\partial T) > 0$ is consistent with this statement. As $(\partial K_A/\partial T)_\xi > 0$ the influence of viscosity overwhelms the association effect at low temperatures already at low ξ values. The effect of interaction forces is illustrated by the following example: $\kappa^*_{max}(KSCN) < \kappa^*_{max}(LiClO_4)$, whereas $\kappa_{max}(KSCN)_{\xi=100} > \kappa_{max}(LiClO_4)_{\xi=100}$. The association constant of LiClO$_4$ ($K_A = 1.3 \pm 0.1$;[48] $K_A = -1.5$[282]) is smaller than that of KSCN ($K_A = 3.1$[282]) in PC as a result of non-Coulombic interaction forces and consequently leads to the observed steeper variation of κ_{max} at $\xi = 100$ of LiClO$_4$ compared to KSCN. The lower association constant of LiClO$_4$ leads to $[100 - \xi^*(LiClO_4)] > [100 - \xi^*(KSCN)]$.

When KSCN is compared with KPF$_6$ ($K_A = 1.3 \pm 0.1$[48]), the slope S at $\xi = 100$ of KSCN should be also less steep than that for KPF$_6$. Figures 22a and 22c confirm this expectation. On the other hand, comparing LiClO$_4$ and KPF$_6$, $S_{LiClO_4} > S_{KPF_6}$.

To the effects caused by ion association and solvent viscosity at increasing DME concentrations must be added yet a further contribution to the specific conductance κ and its appropriate maximum value κ_{max}. Increase in DME content is accompanied by an enhanced conductance which can in part be attributed to a change in ion solvation via the adjustment of the Stokes radii [cf. Eq. (53)]. The maximum change can be expected at $\xi \approx 100$, i.e., in PC-rich mixtures. The ratio of the mobility functions $q = (\kappa/m)_{\xi=76.12}/(\kappa/m)_{\xi=100}$ at $m = 0.1$ is chosen to illustrate this effect.[48] For all investigated electrolyte solutions $q > 1$; this results simply from changes in viscosity. Ion-specific effects related to solvation are observed as follows: $q_{\text{LiClO}_4} > q_{\text{KPF}_6} = q_{\text{Et}_4\text{NPF}_6} > q_{\text{KSCN}}$; these are independent of temperature. The value for KSCN must be excluded from discussion, because the association constant of KSCN is distinctly greater than those of the other three electrolytes, which show equal association in PC $[K_A(\text{Et}_4\text{NPF}_6) \approx K_A(\text{Et}_4\text{NClO}_4) = 1.2^{282}]$. Since Et_4NPF_6 is solvated neither by PC nor by DME, the distinctly high value $q_{\text{LiClO}_4} = 1.54$ compared to $q_{\text{KPF}_6} = q_{\text{Et}_4\text{NPF}_6} = 1.33$ is a clear indication of a specific solvation effect caused by DME.

ACKNOWLEDGMENT

The authors thank Dr. G. Schmeer for establishing the computer programs and carrying out data analysis for some of the systems discussed in this review.

REFERENCES

[1] M. L. McGlashan, *Manual of Symbols and Terminology for Physicochemical Quantities and Units*, Butterworths, London, 1970.
[2] *The Chemistry of Non-Aqueous Solvents*, Ed. by J. J. Lagowski, Academic Press, New York, 1966, Vol. 1; 1967, Vol. 2; 1970, Vol. 3; 1976, Vol. 4.
[3] V. Gutmann, *Coordination Chemistry in Non-Aqueous Solutions*, Springer Verlag, Wien, 1968; cf. *Chemische Funktionslehre*, Springer Verlag, Wien, 1971.
[4] *Non-Aqueous Electrochemistry*, Ed. by J. C. Marchon, Butterworths, London, 1971.
[5] G. J. Janz and R. P. T. Tomkins, *Nonaqueous Electrolytes Handbook*, Academic Press, New York, 1972, Vol. 1; 1973, Vol. 2.
[6] *Ions and Ion Pairs in Organic Reactions*, Ed. by M. Szwarc, Wiley, New York, 1972, Vol. 1; 1974, Vol. 2.
[7] *Physical Chemistry of Organic Solvent Systems*, Ed. by A. K. Covington and T. Dickinson, Plenum Press, New York, 1973.
[8] J. Barthel, *Ionen in nichtwäßrigen Lösungen*, Dr. Dietrich Steinkopff Verlag, Darmstadt, 1976.
[9] J. E. Gordon, *The Organic Chemistry of Electrolyte Solutions*, Wiley, New York, 1975.
[10] *Nonaqueous Solutions—4*, Ed. by V. Gutmann, Butterworths, London, 1975.

[11] *Nonaqueous Solutions—5*, Ed. by J. B. Gill, Butterworths, London, 1977.
[12] J. Barthel, *Angew. Chem.* **80** (1968) 253; *Int. Ed.* **7** (1968) 260.
[13] A. J. Parker, *Chem. Rev.* **69** (1969) 1.
[14] B. Kratochvil and H. L. Yeager, *Fortschr. Chem. Forsch.* **27** (1972) 1.
[15] G. J. Hills, in *Electrochemistry*, Ed. by G. J. Hills (Spec. Period. Reports), The Chemical Society, London, 1970, Vol. 1, p. 73.
[16] A. K. Covington and T. H. Lilley, in *Electrochemistry*, Ed. by G. J. Hills (Spec. Period. Reports), The Chemical Society, London, 1970, Vol. 1, p. 1.
[17] R. H. Wood and P. J. Reilly, *Ann. Rev. Phys. Chem.* **21** (1970) 387.
[18] J. I. Padova, in *Modern Aspects of Electrochemistry*, No. 7, Ed. by B. E. Conway and J. O'M. Bockris, Butterworths, London, 1972, p. 1.
[19] I. M. Kolthoff, *Pure Appl. Chem.* **25** (1971) 305.
[20] J. Padova, in *Water and Aqueous Solutions: Structure, Thermodynamics and Transport Processes*, Ed. by R. A. Horne, Wiley, New York, 1972, p. 109.
[21] K. P. Mishchenko, *Zh. Fiz. Khim.* **46** (1972) 2987.
[22] B. G. Cox, *Ann. Rep. Prog. Chem. Sect. A* **70** (1973) 249.
[23] M. J. Wootten, in *Electrochemistry*, Ed. by G. J. Hills (Spec. Period. Reports), The Chemical Society, London, 1973, Vol. 3, p. 20.
[24] T. H. Lilley, in *Electrochemistry*, Ed. by G. J. Hills (Spec. Period. Reports), The Chemical Society, London, 1973, Vol. 3, p. 187; 1975, Vol. 5, p. 1.
[25] J.-C. Lestrade, J.-P. Badiali, and H. Cachet, in *Dielectric and Related Molecular Processes*, Ed. by M. Davies (Spec. Period. Reports), The Chemical Society, London, 1975, Vol. 2, p. 106.
[26] R. Jasinski, in *Advances of Electrochemistry and Electrochemical Engineering*, Ed. by P. Delahay and Ch. W. Tobias, Wiley, New York, 1971, Vol. 8, p. 253.
[27] J. M. Freund and W. C. Spindler, in *The Primary Battery*, Ed. by G. W. Heise and N. C. Cahoon, Wiley, New York, 1971, Vol. 1, p. 341.
[28] H. N. Seiger, A. E. Lyall, and R. C. Shair, in *Power Sources 2, Proceedings of the Sixth International Power Sources Symposium*, Brighton (England), September 1968, Ed. by D. H. Collins, Pergamon Press, Oxford, 1970.
[29] R. Jasinski, *High-Energy Batteries*, Plenum Press, New York, 1967.
[30] K. F. Kordesch, in *Modern Aspects of Electrochemistry*, No. 10, Ed. by J. O'M. Bockris and B. E. Conway, Plenum Press, New York, 1975, p. 339.
[31] R. Wachter, Habilitationsschrift, Regensburg, 1973.
[32] R. Wachter and J. Barthel, *Ber. Bunsenges. Phys. Chem.* **83** (1979) 634.
[33] P. Bruno and M. Della Monica, *J. Phys. Chem.* **76** (1972) 3034.
[34] P. Bruno, C. Gatti, and M. Della Monica, *Electrochim. Acta* **20** (1975) 533.
[35] P. Bruno and M. Della Monica, *Electrochim. Acta* **20** (1975) 179.
[36] C. A. Angell, *J. Phys. Chem.* **70** (1966) 3988; C. A. Angell and E. J. Sare, *J. Chem. Phys.* **52** (1970) 1058.
[37] C. T. Moynihan, in *Ionic Interactions*, Ed. by S. Petrucci, Academic Press, New York, 1971, Vol. 1, p. 261.
[38] B. F. Wishaw and R. H. Stokes, *J. Am. Chem. Soc.* **76** (1954) 2065.
[39] M. Leist, *Z. Phys. Chem.* (Leipzig) **205** (1955) 16.
[40] G. J. Janz, A. E. Marcinkowsky, and I. Ahmad, *J. Electrochem. Soc.* **112** (1965) 104.
[41] M. Postler, *Collect. Czech. Chem. Commun.* **35** (1970) 535.
[42] W. Ebeling, D. Geisler, D. Kraeft, and R. Sändig, *Wiss. Z. Univ. Rostock, Math.-Naturwiss. Reihe* **23** (1974) 903.
[43] H. Falkenhagen, *Theorie der Elektrolyte*, Hirzel Verlag, Leipzig, 1971.
[44] H. Falkenhagen, W. Ebeling, and W. D. Kraeft, in *Ionic Interactions*, Ed. by S. Petrucci, Academic Press, New York, 1971, Vol. 1, p. 61.

[45] D. Kremp, *Ann. Phys. (Leipzig)* **17** (1966) 278; D. Kremp, W. D. Kraeft, and W. Ebeling, *Ann. Phys. (Leipzig)* **18** (1966) 246.
[46] W. Ebeling, R. Feistel, and D. Geisler, *Z. Phys. Chem. (Leipzig)* **257** (1976) 337.
[47] D. Geisler, R. Feistel, and R. Sändig, *Wiss. Z. Univ. Rostock Math.-Naturwiss. Reihe* **24** (1975) 687.
[48] J. Barthel, H.-J. Gores, and G. Schmeer, *Ber. Bunsenges. Phys. Chem.* **83** (1979).
[49] J. F. Casteel and E. S. Amis, *J. Chem. Eng. Data* **17** (1972) 55.
[50] D. Eagland and F. Franks, *Chem. Ind. (London)* (1965) 1601.
[51] J. L. Hawes and R. L. Kay, *J. Phys. Chem.* **69** (1965) 2420.
[52] R. L. Kay, B. J. Hales, and G. P. Cunningham, *J. Phys. Chem.* **71** (1967) 3925.
[53] R. Wachter and J. Barthel, *Electrochim. Acta* **16** (1971) 713.
[54] J. Barthel and G. Schwitzgebel, *Z. Phys. Chem. N.F.* **54** (1967) 173.
[55] G. Jones and R. C. Josephs, *J. Am. Chem. Soc.* **50** (1928) 1049.
[56] T. Shedlovsky, *J. Am. Chem. Soc.* **52** (1930) 1793.
[57] T. Shedlovsky, in *Techniques of Organic Chemistry*, Ed. by A. Weissberger, Wiley, New York, 1959, Vol. 1, Part H, p. 3011.
[58] W. Walisch and J. Barthel, *Z. Phys. Chem. N.F.* **34** (1962) 38; J. Barthel, F. Schmithals, and H. Behret, *Z. Phys. Chem. N.F.* **71** (1970) 115.
[59] J. Braunstein and G. D. Robbins, *J. Chem. Ed.* **48** (1971) 52.
[60] D. F. Evans and M. A. Matesich, in *Techniques of Electrochemistry*, Ed. by E. Yeager and A. J. Salkind, Wiley, New York, 1973, Vol. 2, p. 1.
[61] M. Palma and Y. Baratoux, *Bull. Soc. Sci. Nat. Phys. (Maroc)* **52** (1972) 7.
[62] K. Takahashi, S. Katayama, and R. Tamamushi, *Rikagaku Kenkyusho Hokoku* **49** (1973) 280673.
[63] R. Tamamushi and K. Takahashi, *J. Electroanal. Chem. Interfac. Electrochem.* **50** (1974) 277.
[64] S. J. Khang and T. J. Fitzgerald, *Ind. Eng. Chem. Fundam.* **14** (1975) 208.
[65] G. Jones and C. Bradshaw, *J. Am. Chem. Soc.* **55** (1933) 1780.
[66] G. Jones and J. Prendergast, *J. Am. Chem. Soc.* **59** (1937) 731.
[67] J. F. Lind, Jr., J. J. Zwolenik, and R. M. Fuoss, *J. Am. Chem. Soc.* **81** (1959) 1557.
[68] J.-C. Justice, *J. Chim. Phys.* **65** (1968) 353.
[69] R. M. Fuoss and K. L. Hsia, *Proc. Natl. Acad. Sci. (USA)* **57** (1967) 1550.
[70] R. Sändig, R. Feistel, A. Grosch, and J. Einfeldt, quoted in Ref. 71.
[71] G. J. Janz and R. P. T. Tomkins, *J. Electrochem. Soc.* **124** (1977) 55C.
[72] D'Ans-Lax, *Taschenbuch für Chemiker und Physiker*, 2nd ed., Springer Verlag, Berlin, 1949.
[73] R. A. Robinson and R. H. Stokes, *Electrolyte Solutions*, 2nd rev. ed., Butterworths, London, 1970.
[74] J. P. Buttler, H. I. Schiff, and A. R. Gordon, *J. Chem. Phys.* **19** (1951) 752.
[75] J. Barthel and G. Schwitzgebel, *Z. Phys. Chem. N.F.* **54** (1967) 181.
[76] J. E. Prue and P. J. Sherrington, *Trans. Faraday Soc.* **57** (1961) 1795.
[77] A. K. R. Unni, L. Elias, and H. I. Schiff, *J. Phys. Chem.* **67** (1963) 1216.
[78] J. H. Exner and E. C. Steiner, *J. Am. Chem. Soc.* **96** (1974) 1782.
[79] J. F. Coetzee, G. P. Cunningham, D. K. McGuire, and G. R. Padmanabhan, *Anal. Chem.* **34** (1962) 1139.
[80] R. De Lisi and M. Goffredi, *J. Chem. Soc. Faraday Trans. 1* **70** (1974) 787.
[81] J. A. Riddick and W. B. Bunger, *Organic Solvents*, 3rd. ed. (Vol. 2 of *Techniques of Chemistry*, Ed. by A. Weissberger), Wiley, New York, 1970.

[82] R. E. Gibson and O. H. Loeffler, *J. Am. Chem. Soc.* **61** (1939) 2515.
[83] J. S. Burlew, *J. Am. Chem. Soc.* **62** (1940) 690.
[84] J. Barthel et al., publication in preparation.
[85] W. Gildseht, A. Habenschuss, and F. H. Spedding, *J. Chem. Eng. Data* **17** (1972) 402.
[86] A. A. Mariott and M. Buckley, *Natl. Bur. Stand. Circ.* (1953) 537.
[87] J. F. Swindells, J. R. Coe, and R. B. Godfrey, *J. Res. Natl. Bur. Stand.* **48** (1952) 1; **56** (1956) 1.
[88] J. Barthel, *Ber Bunsenges. Phys. Chem.* **83** (1979) 252; *Chem. Ing. Tech.* **50** (1978) 259.
[89] L. Onsager, *Phys. Z.* **28** (1927) 277.
[90] P. Debye and H. Falkenhagen, *Phys. Z.* **29** (1928) 402.
[91] H. S. Harned and B. B. Owen, *The Physical Chemistry of Electrolytic Solutions*, 3rd. ed., Reinhold, London, 1970.
[92] R. M. Fuoss and F. Accascina, *Electrolytic Conductance*, Interscience, New York, 1959.
[93] H. Falkenhagen and G. Kelbg, in *Modern Aspects of Electrochemistry*, No. 2, Ed. by J. O'M. Bockris, Butterworths, London, 1959, p. 1.
[94] G. Kelbg, *Z. Phys. Chem. (Leipzig)* **214** (1960) 8.
[95] H. L. Friedman, *Ionic Solution Theory*, Interscience, New York, 1962.
[96] J. C. Rasaiah and H. L. Friedman, *J. Chem. Phys.* **48** (1968) 2742; **50** (1969) 3965.
[97] J. C. Rasaiah, *J. Solution Chem.* **2** (1973) 301.
[98] R. M. Fuoss and L. Onsager, *J. Phys. Chem.* **68** (1964) 1.
[99] L. Onsager, *Phys. Z.* **27** (1926) 388.
[100] H. L. Friedman, *Physica (Utrecht)* **30** (1964) 537; *J. Chem. Phys.* **42** (1965) 450.
[101] R. M. Fuoss and L. Onsager, *J. Phys. Chem.* **61** (1957) 668; **62** (1958) 1339; *Proc. Natl. Acad. Sci. (USA)* **41** (1955) 274, 1010; R. M. Fuoss, *J. Am. Chem. Soc.* **80** (1958) 3163; **81** (1959) 2659; *J. Phys. Chem.* **63** (1959) 633.
[102] R. M. Fuoss and L. Onsager, *J. Phys. Chem.* **66** (1962) 1722; **67** (1963) 621, 628; **68** (1964) 1.
[103] R. M. Fuoss, L. Onsager, and J. F. Skinner, *J. Phys. Chem.* **69** (1965) 2581.
[104] E. Pitts, *Proc. R. Soc. (London)* **217A** (1953) 43.
[105] H. Falkenhagen, M. Leist, and G. Kelbg, *Ann. Phys. (Leipzig)* **11** (1952) 51.
[106] G. Kelbg, *Wiss. Z. Univ. Rostock, Math.-Naturwiss. Reihe* **9** (1959/60) 41.
[107] J. Quint and A. Viallard, *J. Chim. Phys. Physicochim. Biol.* **69** (1972) 1095, 1100; **72** (1975) 335.
[108] J.-C. Justice, *J. Phys. Chem.* **79** (1975) 454.
[109] R. M. Fuoss, *J. Phys. Chem.* **79** (1975) 525, 1983.
[110] M. J. Pikal, *J. Phys. Chem.* **75** (971) 663.
[111] W. Ebeling and R. Feistel, *Chem. Phys. Lett.* **36** (1975) 404.
[112] D. Kremp, H. Ulbricht, and G. Kelbg, *Z. Phys. Chem. (Leipzig)* **240** (1969) 65, 80.
[113] E. Ebeling, R. Feistel, G. Kelbg, and R. Sändig, *J. Nonequil. Thermodyn.* **3** (1978) 11.
[114] J. P. Valleau, *J. Phys. Chem.* **69** (1965) 1745.
[115] D. S. Berns and R. M. Fuoss, *J. Am. Chem. Soc.* **82** (1960) 5585,
[116] E. Pitts, B. E. Tabor, and J. Daly, *Trans. Faraday Soc.* **65** (1969) 849.
[117] M. S. Chen, Ph.D. thesis, Yale University, 1969.
[118] P. C. Carmann, *J. Phys. Chem.* **74** (1970) 1653.
[119] R. Fernández-Prini, *Trans. Faraday Soc.* **65** (1969) 3311.

[120] R. Fernández-Prini, in *Physical Chemistry of Organic Solvent Systems*, Ed. by A. K. Covington and T. Dickinson, Plenum Press, New York, 1973, p. 525.
[121] T. J. Murphy and E. G. D. Cohen, *J. Chem. Phys.* **63** (1970) 2177.
[122] W. Ebeling, H. Falkenhagen, and W. D. Kraeft, *Ann. Phys. (Leipzig)* **18** (1966) 15.
[123] G. Kelbg and H. Ulbricht, *Z. Phys. Chem. (Leipzig)* **244** (1970) 125.
[124] H. Ulbricht and H. Falkenhagen, *Z. Phys. Chem. (Leipzig)* **243** (1970) 305, 313.
[125] C. Treiner and J.-C. Justice, *J. Chim. Phys. Physicochim. Biol.* **68** (1971) 56.
[126] R. Bury, M.-C. Justice, and J.-C. Justice, *J. Chim. Phys. Physicochim. Biol.* **67** (1970) 2045.
[127] P. C. Carmann, *J. South African Chem. Inst.* **28** (1975) 80.
[128] R. M. Fuoss, *J. Phys. Chem.* **78** (1974) 1383; **79** (1975) 1038.
[129] J.-C. Justice, *J. Phys. Chem.* **79** (1975) 454, 1039.
[130] R. M. Fuoss, *J. Phys. Chem.* **81** (1977) 1529.
[131] R. Fernández-Prini and J. E. Prue, *Z. Phys. Chem. (Leipzig)* **228** (1965) 373.
[132] D. Kremp, W. D. Kraeft, and W. Ebeling, *Ann. Phys. (Leipzig)* **18** (1966) 246.
[133] W. Ebeling, W. D. Kraeft, and D. Kremp, *J. Phys. Chem.* **70** (1966) 3338.
[134] W. D. Kraeft, *Z. Phys. Chem. (Leipzig)* **237** (1968) 289.
[135] W. D. Kraeft and W. Ebeling, *Z. Phys. Chem. (Leipzig)* **240** (1969) 141.
[136] W. D. Kraeft and R. Sändig, *J. Chim. Phys.* **67** (1970) 1265.
[137] J. E. Prue, *Ionic Equilibria*, Pergamon Press, Oxford, 1966.
[138] R. M. Fuoss and C. A. Kraus, *J. Am. Chem. Soc.* **55** (1933) 476.
[139] T. Shedlovsky, *J. Franklin Inst.* **225** (1938) 739.
[140] R. M. Fuoss, *J. Am. Chem. Soc.* **79** (1957) 3301; **80** (1958) 3163; **81** (1959) 2659.
[141] J.-C. Justice, *Electrochim. Acta* **16** (1971) 701.
[142] J.-C. Justice and M.-C. Justice, in *Ion–Ion and Ion–Solvent Interactions*, The Chemical Society Faraday Division, General Discussion 64 (Oxford 1977), p. 265.
[143] J. Barthel, R. Wachter and H.-J. Gores, 5th International Conference on Non-Aqueous Solutions, (Leeds, 1976).
[144] J. Barthel, R. Wachter, and H.-J. Gores, in *Ion–Ion and Ion–Solvent Interactions*. The Chemical Society Faraday Division, General Discussion 64 (Oxford 1977), p. 285.
[145] R. M. Fuoss, *Proc. Nat. Acad. Sci. USA* **75** (1978) 16; *J. Solution Chem.* **7** (1978) 771; *J. Phys. Chem.* **82** (1978) 2477.
[146] R. M. Fuoss, in *Ion–Ion and Ion–Solvent Interactions*, The Chemical Society Faraday Division, General Discussion 64 (Oxford 1977), pp. 324, 327.
[147] W. Ebeling, in *Ion–Ion and Ion–Solvent Interactions*, The Chemical Society Faraday Division, General Discussion 64 (Oxford 1977), p. 335.
[148] J. G. Kirkwood and I. Oppenheim, *Chemical Thermodynamics*, McGraw-Hill, New York, 1961.
[149] W. R. Gilkerson, *J. Chem. Phys.* **25** (1956) 1199.
[150] R. M. Fuoss, *J. Am. Chem. Soc.* **80** (1958) 5059.
[151] H. Sadek and R. M. Fuoss, *J. Am. Chem. Soc.* **81** (1959) 4507.
[152] J. E. Prue, in *Chemical Physics of Ionic Solutions*, Ed. by B. E. Conway and R. G. Barradas, Wiley, New York, 1966, p. 163.
[153] R. Fernández-Prini and J. E. Prue, *Trans. Faraday Soc.* **62** (1966) 1257.
[154] N. Bjerrum, *Danske Vidensk. Selsk., Mat.-Fysiske Medd.* **7** No. 9 (1926).
[155] J. E. Prue, *J. Chem. Ed.* **46** (1969) 12.

[156] H. Falkenhagen and W. Ebeling, in *Ionic Interactions*, Ed. by S. Petrucci, Academic Press, New York, 1971, Vol. 1, p. 1; W. Ebeling, *Z. Phys. Chem. (Leipzig)* **238** (1968) 400; W. D. Kraeft and W. Ebeling, *Z. Phys. Chem. (Leipzig)* **240** (1969) 141.

[157] M.-C. Justice and J.-C. Justice, *Colloq. Int. C.N.R.S.* **246** (1975) 241.

[158] G. Kelbg, *Z. Phys. Chem. (Leipzig)* **214** (1960) 8.

[159] F. Booth, *J. Chem. Phys.* **19** (1951) 391.

[160] S. Petrucci, in *Ionic Interactions*, Ed. by S. Petrucci, Academic Press, New York, 1971, Vol. 1, p. 117.

[161] M.-C. Justice and J.-C. Justice, *J. Solution Chem.* **5** (1976) 543; **6** (1977) 819.

[162] W. Ebeling, *Z. Phys. Chem. (Leipzig)* **249** (1972) 140.

[163] E. A. Guggenheim, in *Chemistry at the Centenary (1931) Meeting of the British Association for the Advancement of Science*, Heffer and Sons, Ltd., Cambridge, 1932, p. 58.

[164] E. A. Moelwyn-Hughes, *A Short Course of Physical Chemistry*, Longmans, London, 1966.

[165] L. D. Pettit and S. Bruckenstein, *J. Am. Chem. Soc.* **88** (1966) 4783.

[166] E. Meeron, *J. Chem. Phys.* **26** (1957) 804.

[167] R. H. Stokes, *J. Am. Chem. Soc.* **76** (1954) 1988.

[168] R. L. Kay and J. L. Dye, *Proc. Natl. Acad. Sci. USA* **49** (1963) 5.

[169] D. P. Sidebottom and M. Spiro, *J. Chem. Soc. Faraday Trans. 1* **69** (1973) 1287.

[170] J. Einfeldt and A. Grosch, *Exp. Tech. Phys.* **26** (1978) 477.

[171] F. Barreira and G. J. Hills, *Trans. Faraday Soc.* **64** (1968) 1359; S. B. Brummer, *J. Chem. Phys.* **42** (1965) 1636; S. B. Brummer and G. J. Hills, *Trans. Faraday Soc.* **57** (1961) 1816, 1823; G. J. Hills, in *Chemical Physics in Ionic Solutions*, Ed. by B. E. Conway and R. G. Barradas, Wiley, New York, 1966, p. 521.

[172] B. E. Conway, in *Modern Aspects of Electrochemistry*, Ed. by J. O'M. Bockris and B. E. Conway, Butterworths, London, 1964, Vol. 3, p. 43.

[173] M. Spiro in *Physical Chemistry of Organic Solvent Systems*, Ed. by A. K. Covington and T. Dickinson, Plenum Press, New York, 1973, p. 635.

[174] B. S. Krumgal'z, *Electrokhimiya* **8** (1972) 1320; *Zh. Fiz. Khim.* **46** (1972) 1498; *Russ. J. Phys. Chem.* **47** (1973) 528.

[175] R. M. Fuoss, *Proc. Natl. Acad. Sci. USA* **45** (1959) 807.

[176] R. H. Boyd, *J. Chem. Phys.* **35** (1961) 1281.

[177] R. Zwanzig, *J. Chem. Phys.* **38** (1963) 1603.

[178] R. Zwanzig, *J. Chem. Phys.* **52** (1970) 3625.

[179] R. Fernández-Prini and G. Atkinson, *J. Phys. Chem.* **75** (1971) 239.

[180] J. Thomas and D. F. Evans, *J. Phys. Chem.* **74** (1970) 3812.

[181] R. L. Kay, *J. Am. Chem. Soc.* **82** (1960) 2099.

[182] V. P. Barabanov, V. M. Tsentovskii, and V. S. Tsentovskaya, *Elektrokhimiya* **10** (1974) 432.

[183] J. Einfeldt, R. Feistel, A. Grosch, and R. Sändig, *Wiss. Z. Univ. Rostock, Math.-Naturwiss. Reihe* **24** (1975) 681.

[184] E. Schollmeyer and W. Seidel, *Z. Phys. Chem. (Leipzig)* **257** (1976) 1103.

[185] P. Beronius, G. Wikander, and A.-M. Nilsson, *Z. Phys. Chem. N.F.* **70** (1970) 52.

[186] R. M. Fuoss and F. Accascina, *Proc. Natl. Acad. Sci. USA* **45** (1959) 1383.

[187] R. Feistel, *Z. Phys. Chem. (Leipzig)* **259** (1978) 369.

[188] J. Barthel, J.-C. Justice, and R. Wachter, *Z. Phys. Chem. N.F.* **84** (1973) 100.

[189] C. De Rossi, B. Sesta, M. Battistini, and S. Petrucci, *J. Am. Chem. Soc.* **94** (1972) 2961.
[190] B. Kaukal, Diplomarbeit, Regensburg (1976); F. Strasser, Diplomarbeit, Regensburg (1977).
[191] R. L. Kay, C. Zawoyski, and D. F. Evans, *J. Phys. Chem.* **69** (1965) 4208.
[192] H. Friedman, in *Ion–Ion and Ion–Solvent Interactions*, The Chemical Society, Faraday Division, General Discussion 64, Oxford (1977), p. 332.
[193] J. E. Van Evercooren, G. V. Merken, and H. P. Thun, *Bull. Soc. Chim. Belg.* **84** (1975) 533.
[194] A. M. Shkodin and T. I. Kurova, *Elektrokhimiya* **10** (1974) 340.
[195] B. S. Krumgal'z, Y. I. Gerzhberg, and V. M. Ryabikova, *Elektrokhimiya* **12** (1976) 1178.
[196] D. F. Evans and T. L. Broadwater, *J. Phys. Chem.* **72** (1968) 1037.
[197] G. A. Vidulich, G. P. Cunningham, and R. L. Kay, *J. Solution Chem.* **2** (1973) 23.
[198] J. Barthel, G. Schmeer, F. Strasser, and R. Wachler, *Rev. Chim. Minérale* **15** (1978) 99.
[199] I. P. Nikitina, B. S. Krumgal'z, and D. G. Traber, *Elektrokhimiya* **8** (1972) 644.
[200] R. L. Blokhra and Y. P. Sehgal, *J. E.ectroanal. Chem. Interfac. Electrochem.* **62** (1975) 381.
[201] Y. N. Eichis and A. N. Zhitomirskii, *Zh. Fiz. Khim.* **50** (1976) 2169.
[202] V. I. Vigdorovic, I. T. Pchel'nikov, and L. E. Tsygankova, *Electrokhimiya* **8** (1972) 1024.
[203] J. Thomas and D. F. Evans, *J. Phys. Chem.* **74** (1970) 3812.
[204] P. Bruno, C. Gatti, and M. Della Monica, *Electrochim. Acta*, **20** (1975) 533.
[205] C. M. French and K. H. Glover, *Trans. Faraday Soc.* **51** (1955) 1418.
[206] R. Gopal and O. N. Bhatnagar, *J. Phys. Chem.* **70** (1966) 3007.
[207] R. I. Mostkova, Yu. M. Kesslei, and V. N. Semenova, *Elektrokhimiya* **7** (1971) 642.
[208] B. S. Krumgal'z and V. M. Ryabikova, *Elektrokhimiya* **8** (1972) 1162.
[209] R. D. Singh and R. Gopal, *Z. Phys. Chem. N.F.* **83** (1973) 25.
[210] T. V. Rebagay, J. F. Casteel, and P. G. Sears, *J. Electrochem. Soc. Electrochem. Sci. Technol.* **121** (1974) 977.
[211] R. D. Singh and R. Gopal, *Bull. Chem. Soc. Japan* **45** (1972) 2088.
[212] J. F. Casteel and E. S. Amis, *J. Chem. Eng. Data* **19** (1974) 121.
[213] R. Gopal and O. N. Bhatnagar, *J. Phys. Chem.* **69** (1965) 2382.
[214] R. L. Dawson, P. G. Sears, and R. H. Graves, *J. Am. Chem. Soc.* **77** (1955) 1986.
[215] C. M. French and K. H. Glover, *Trans. Faraday Soc.* **51** (1955) 1427.
[216] R. D. Singh, *Z. Phys. Chem. N.F.* **90** (1974) 34.
[217] R. D. Singh, P. P. Rastogi, and R. Gopal, *Can. J. Chem.* **46** (1968) 3525.
[218] B. S. Krumgal'z and Yu. I. Gherzberg, *Zh. Obshch. Khim.* **43** (1973) 462.
[219] V. M. Tsentovskii, V. P. Barabanov, R. B. Bairamov, and B. D. Chernokal'skii, *Zh. Obshch. Khim.* **44** (1974) 2379.
[220] S. R. C. Hughes and D. H. Price, *J. Chem. Soc. A* (1968) 1464.
[221] H.-J. Gores, Dissertation, Saarbrücken (1974).
[222] B. Kaukal, Diplomarbeit, Regensburg (1976).
[223] G. Pistoia, *J. Electrochem. Soc.* **118** (1971) 153.
[224] J. F. Casteel, J. R. Angel, H. B. McNeeley, and P. G. Sears, *J. Electrochem. Soc. Electrochem. Sci. Technol.* **122** (1975) 319.

[225] K. J. Tikhonov, V. A. Ivanova, and B. A. Ravdel', *Zh. Prikl. Khim.* (*Leningrad*) **50** (1977) 49.
[226] F. Barreira and G. J. Hills, *Trans. Faraday Soc.* **64** (1968) 1359.
[227] J. A. Bolzan and A. J. Arvía, *Electrochim. Acta* **15** (1970) 827.
[228] J. A. Bolzan and A. J. Arvía, *Electrochim. Acta* **16** (1971) 531.
[229] R. L. Blokhra and M. L. Parmar, *J. Electroanal. Chem. Interfac. Electrochem.* **57** (1974) 117.
[230] J. Imhof, T. D. Westmoreland, Jr., and M. C. Day, *J. Solution Chem.* **3** (1974) 83.
[231] J. Comyn, F. S. Dainton, and K. J. Ivin, *Electrochim. Acta* **13** (1968) 1851.
[232] P. Chang, R. V. Slates, and M. Szwarc, *J. Phys. Chem.* **70** (1966) 3180.
[233] S. Boileau and P. Hemery, *Electrochim. Acta* **21** (1976) 647.
[234] J. H. Beard and P. H. Plesch, *J. Chem. Soc. (London)* (1964) 4075.
[235] J. T. Denison and J. B. Ramsey, *J. Am. Chem. Soc.* **77** (1955) 2615.
[236] K. H. Stern and A. E. Martell, *J. Am. Chem. Soc.* **77** (1955) 1983.
[237] H. L. Curry and W. R. Gilkerson, *J. Am. Chem. Soc.* **79** (1957) 4021.
[238] G. S. Bien, C. A. Kraus, and R. M. Fuoss, *J. Am. Chem. Soc.* **56** (1934) 1860.
[239] M. D. Surova and S. I. Zhdanov, *Elektrokhimiya* **9** (1973) 350.
[240] P. G. Sears and R. L. Dawson, *J. Chem. Eng. Data* **13** (1968) 124.
[241] C. Kalidas and V. Srivinas-Rao, *Indian J. Chem.* **14A** (1976) 129.
[242] H. Hammer, Dissertation, Saarbrücken (1975); U. Ströder, Diplomarbeit, Regensburg (1975).
[243] J. Barthel and H. Hilbinger, unpublished.
[244] J. F. Hinton and E. S. Amis, *Chem. Rev.* **71** (1971) 627.
[245] D. Singh and A. Mishra, *Bull. Chem. Soc. Japan* **40** (1967) 2801.
[246] R. Wachter, Dissertation, Saarbrücken (1968).
[247] J. Barthel, R. Wachter, and M. Knerr, *Electrochim. Acta* **16** (1971) 723.
[248] A. Bondi, *J. Phys. Chem.* **68** (1964) 441.
[249] V. I. Korobkov and A. D. Mikhilev, *Elektrokhimiya* **6** (1970) 1002.
[250] A. Than and E. S. Amis, *Z. Phys. Chem. N.F.* **58** (1968) 196.
[251] H. O. Spivey and T. Shedlovsky, *J. Phys. Chem.* **71** (1967) 2165.
[252] E. S. Amis and J. F. Casteel, *J. Electrochem. Soc. Electrochem. Sci. and Technol.* **117** (1970) 213.
[253] C. Hibbs and E. S. Amis, *J. Inorg. Nucl. Chem.* **33** (1971) 1659.
[254] J. J. Lee, *Hwahak Konghak* **12** (1974) 203.
[255] R. L. Kay and T. L. Broadwater, *J. Solution Chem.* **5** (1976) 57.
[256] G. Pistoia, M. De Rossi, and B. Scrosati, *J. Electrochem. Soc.* **117** (1970) 500.
[257] G. V. Merken, H. P. Thun, and F. Verbeek, *Electrochim. Acta* **19** (1974) 947.
[258] D. Singh and A. Mishra, *Indian J. Chem.* **7** (1969) 86.
[259] N. Yui, Y. Kurokawa, and M. Nakayama, *Bull. Chem. Soc. (Japan)* **46** (1973) 1027.
[260] V. V. Shcherbakov, N. M. Silkina, and V. I. Ermakov, *Zh. Fiz. Khim.* **50** (1976) 2718.
[261] J. F. Casteel and E. S. Amis, *J. Chem. Eng. Data* **19** (1974) 121.
[262] R. L. Blokhra and P. C. Verma, *Indian J. Chem.* **14A** (1976) 696.
[263] I. N. V'yunnik, A. M. Zholnovach, and A. M. Shkodin, *Zh. Fiz. Khim.* **51** (1977) 485.
[264] C. Carvajal, K. J. Tölle, J. Smid, and M. Szwarc, *J. Am. Chem. Soc.* **87** (1965) 5548.
[265] E. J. del Rosario and J. E. Lind, Jr., *J. Phys. Chem.* **70** (1966) 2876.
[266] J. F. Casteel, J. R. Angel, H. B. McNeeley, and G. P. Sears, *J. Electrochem. Soc. Electrochem. Sci. Technol.* **122** (1975) 319.

[267] R. I. Poluden', V. V. Pavlova, and M. A. Tolstaya, *Zh. Prikl. Khim.* **49** (1976) 2585.
[268] V. M. Valyashko and A. A. Ivanov, *Russ. J. Inorg. Chem.* **19** (1974) 1628; *Zh. Neorg. Khim.* **19** (1974) 2978.
[269] J. Molenat, *J. Chim. Phys.* **66** (1969) 825.
[270] G. Pistoia, *J. Electrochem. Soc.* **118** (1971) 153.
[271] O. Ya. Samoilov, *Structure of Aqueous Electrolyte Solutions*, 3rd printing (transl. by D. J. G. Ives), Consultants Bureau Enterprises, New York, 1969.
[272] B. K. Makarenko, E. A. Mendheritskii, R. P. Sobolev, Y. M. Povarov, and P. A. Sereda, *Elektrokhimiya* **10** (1974) 355.
[273] B. K. Makarenko, Y. M. Povarov, P. A. Sereda, and A. S. Lileev, *Elektrokhimiya* **12** (1976) 518.
[274] L. A. Girifalco, *J. Chem. Phys.* **23** (1955) 2446.
[275] N. M. Baron and M. U. Shcherba, *Zh. Prikl. Khim.* **47** (1974) 1855.
[276] J. Johnston, *J. Am. Chem. Soc.* **31** (1909) 1010.
[277] R. Payne and I. E. Theodorou, *J. Phys. Chem.* **76** (1972) 2892.
[278] V. Viti and P. Zampetti, *Chem. Phys.* **2** (1973) 233,
[279] V. Gutmann, *Angew. Chem.* **82** (1970) 858; *Int. Ed.* **9** (1970) 843.
[280] R. Kempa and W. H. Lee, *J. Chem. Soc. (London)* (1958) 1936.
[281] U. Mayer, V. Gutmann, and K. Kösters, *Monatsh. Chem.* **107** (1976) 845.
[282] M. L. Jansen and H. L. Yeager, *J. Phys. Chem.* **77** (1973) 3089.
[283] J. Barthel, in: *Nonaqueous Solutions—6*, Ed. by W. E. A. McBride, Pergamon Press, New York, 1979.
[284] D. F. Evans and P. Gardam, *J. Phys. Chem.* **72** (1968) 3281.
[285] H. L. Friedman and B. Larson, *J. Chem. Phys.* **70** (1979) 92.

2

Solvent Adsorption and Double-Layer Potential Drop at Electrodes

Sergio Trasatti

University of Milan, Milan, Italy

I. INTRODUCTION

1. Scope of the Chapter

Although the primary importance of the solvent dipole contribution [g^S(dip)] to the inner-layer potential drop ($\Delta\phi$) was recognized long ago,[1] the problem appears to be still unsettled quantitatively. This is because g^S(dip) is not amenable to experimental measurement[2] and thus it can only be derived indirectly through assumptions involving a model. There has recently been a strong revival of interest in the topic and a number of attempts[3-15] have been made to quantify g^{H_2O}(dip) for Hg and other metals. However, results often are in quantitative disagreement with each other and qualitative differences may be found especially in the starting models.

It is not the purpose of this chapter to review the field of double-layer theories from the beginning. This has already been done a number of times,[16] and recently this topic has been covered very thoroughly by Reeves[17a] in this same series. However, Reeves' chapter was written just when most of the new developments started and the number, variety, and relevance of the later results are such as to deserve one more critical review to assess the new situation. Another extensive treatment of the state of water and the related question of its interaction with ions at charged interfaces, which dealt with some of the recent work, was given by Conway.[17b]*

A complete description of the boundary region between a metal and an electrolyte solution, and of the electric potential profile across it must take into account all of the possible factors by which electrical double layers may arise[2]: separation of free charges on either side of the interface, preferential orientation of solvent dipoles, specific adsorption of ions of the electrolyte, and partial charge transfer[18–22] from molecules or ions (additional dipoles). Theories usually proceed step by step by complicating successively a primitive naive picture of the system. Thus, in the present case, the presence of electroactive substances may be avoided. The possibility of additional surface dipoles (for example[23,24] Pt–H and Pt–O) is removed if metals not reacting chemically with the solvent are first considered. The use of suitable electrolytes may minimize the effect of specific adsorption or confine it to potential regions far from those being investigated. Limitingly, even the presence of an electrolyte may prove to be unnecessary conceptually if the condition of uncharged metal (potential of zero charge, pzc) is considered and the diffuse layer contribution to the potential is properly allowed for. It follows that the simplest model we may treat is that of a sp-band metal[3] in contact with a pure solvent.

The inner-layer potential drop at the pzc is determined[3,13] by the two dipolar layers on the metal and on the solution surface, respectively. The former is described[15] in theories of metals by physicists and its structure may reasonably be assumed as independent of the charge on the metal (regarded as a perfect conductor). This is tantamount to including any change in the surface electron distribution in the computation of the free charge on the metal side. The potential drop associated with the solvent side of the interface is, on the contrary, charge dependent[25] and an adequate description of this aspect is required. In fact, the kinetics of electrode processes and especially the double-layer capacity may depend on the solvent potential drop significantly, whereas effects of the double layer on the metal may be neglected to a first approximation. The first step to be taken in

* Reference 17b appeared after the present chapter had already been submitted.

double-layer theories is therefore the description of the state of solvent molecules at the electrode.

Two approaches to the problem of the solvent dipole contribution to the electrode potential may be distinguished. One starts[4,6,11] from a detailed molecular model for the solvent at the interface and derives g^S(dip) by investigating the way the model reacts as the charge on the metal is changed and suitable values are given to some essential parameters. An alternative possibility[7,12] is to start from a quite indefinite model for the solvent at a molecular level and to derive g^S(dip) by fitting the model to experimental data. This requires that comparison be made at a lower level of complexity of experimental data (for instance, charge–potential relationships[12] instead of capacity–charge relationships[4,6] which are linked to the former by an integration route). Molecular parameters and a more detailed picture of the interface may then be derived from the results obtained.

The two above approaches may be complementary to one another because they must necessarily correspond at some point if the choice of models is reasonable. Also, some parameters of the less sophisticated route may be used in the other approach for a best-fit procedure. In this chapter, a conceptual procedure based on the less sophisticated approach as suggested by the present author[7,12] will be followed first. Results will then be critically discussed in the light of predictions of other models,[4,6,11] and also of other existing views or independent results. This does not imply any *a priori* choice or preference. Simply, it is easier and more relevant for the author to present his own work and to discuss it in the light of the work of others than vice versa. It is hoped that the discussion that will emerge will be more consequential and concise, and that, in some cases, more definite conclusions than those presently available will be arrived at.

The chapter will be divided into sections concerned with the determination of g^S(dip) at the potential of zero charge, and with the effect of charge and of temperature on it, respectively. Emphasis will be placed on the role of the nature of the metal on g^S(dip). Semiconductor surfaces will not be considered, although some of the conclusions may easily be extended to that case. Results for polycrystalline surfaces will be exclusively considered since data for single crystals are still too scanty for a general treatment. Most of the material presented will be concerned with aqueous solutions for which there exists the greatest amount of experimental work but a short section will be devoted to metal–nonaqueous-solvent interfaces. Although thermodynamic aspects of double layers will not be treated because they have been exhaustively discussed elsewhere,[2,26,27] a section will, however, be devoted to the configurational entropy of the double-layer because this quantity is closely related to structural aspects.[28,29]

2. Historical Survey

In this section, the chronological development of some of the fundamental aspects which have enabled a well-established rational basis to be reached in double-layer theories will be outlined. The importance of the role of solvent molecules in contributing to double-layer potential drops has been recognized for a long time.[1,30] This aspect is implicit in the early electrocapillary work by Gouy.[31] However, systematic studies were begun by the late Professor Frumkin, whose work laid the basis of modern electrochemistry. It appears that without the experimental work on solid metals carried out by Frumkin's school, most of the present knowledge on double-layer structure at such surfaces would not have been possible.

Although since the beginning of his work[32,33] Frumkin realized qualitatively the problem of the role of the solvent, a first attempt at a quantitative treatment is not found until his paper of 1928. At that time Frumkin and Gorodetzkaya,[34] on the basis of electrocapillary results for Hg and Tl amalgams, suggested that a definite relation should exist between differences in pzc ($\Delta E_{\sigma=0}$) and differences in work function ($\Delta \Phi$) amongst various metals. This statement implied that the solvent orientation at two interfaces was to be regarded as the same or negligibly different. The problem of the dependence of solvent orientation on the nature of the solid phase had still to wait some 35 years before an acceptable solution was proposed.[35,36] However, the conclusions from the early paper by Frumkin and Gorodetzkaya may nowadays be regarded as substantially, even though somewhat fortuitously, correct. In fact, Tl and Hg, though different, possess similar hydrophilicities,[37,38] and account must also be taken of the inaccuracy in the values of the work functions known at that time.[39]

Another factor which retarded the development of double-layer theories was that measurements on Ga failed for many years[40-44] to show the now well-known striking difference between this metal and Hg. This is thought to be due to the extreme sensitivity of Ga to metallic impurities[45,46] which, in the liquid metal, are especially surface active.[47] Therefore, only Ga of special purity gives the correct experimental response.[36] Thus, it appears that the development of double-layer theories has been retarded not by conceptual inadequacies but rather by experimental difficulties related to the measurement of capacitance on metals other than Hg.

Until 1964 accurate experimental results were available only for Hg electrodes and theories were devoted to interpret, exclusively, capacity curves for this metal, with the implicit assumption that the double-layer structure should not differ substantially on other metals. Figure 1 shows that, in the absence of ionic specific adsorption, the capacity curve for Hg exhibits four distinct features: a steep rise at positive charges (σ), a hump at small positive charges, a shallow minimum at negative charges, and a moderate rise at strongly negative charges. Since the metal–water interaction is intrinsically weak for Hg, the primary role played by this factor was not recognized in the early theories. All models proposed before 1964[17] were successful in reproducing capacity curves in the region of the hump, but all of them failed to reproduce the capacity rise at positive charges. Such a rise was even suggested[48] as something to disregard in developing theories.

Some theories actually accounted for the capacity rise by incipient dissolution of the metal (adatom formation),[49] specific adsorption of ions,[25,50,51] or electrostriction.[52] However, these models can no longer be regarded as acceptable simply because they are able to reproduce some features of the capacity curve for Hg correctly. Models are now required to be able to give a reasonable answer also to the effect of a change in the nature of the metal. It is now well known that the different properties of metals manifest themselves particularly in the region of positive charges.[53,54] Unfortunately, some theories,[52,55] developed after

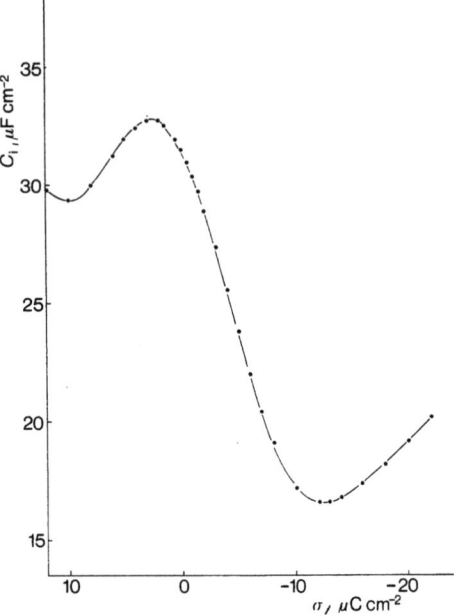

Figure 1. Inner-layer capacity as a function of charge for Hg electrodes[65] in contact with NaF aqueous solutions at 0°C.

some of the above matters had been pointed out,[35,36] failed to take into account their importance.

The first reliable results for Ga in aqueous solutions were published in 1964.[36] They should be regarded as a cornerstone for more modern double-layer theories in that it was shown[35] that differences observed between Ga and Hg could be rationalized in terms of different chemical interactions between solvent and metal surfaces. Differentiated solvent chemisorption was for the first time introduced as a new concept in double-layer studies, although water chemisorption on Hg had been spoken of previously[49] but from a quite different point of view and with different purposes in mind. However, almost at the same time, Frumkin[56] published a paper where, although the peculiarity of the results for Ga was stressed, the early idea[34] was again proposed on a more enlarged experimental basis (i.e., for many more metals) that $\Delta E_{\sigma=0}$ was probably equal or very close to $\Delta\Phi$. This was suggested on the basis of a plot of $E_{\sigma=0}$ vs. Φ where metals were shown to gather around a common line of unit slope. A $E_{\sigma=0}$ vs. Φ plot was, in fact, not proposed for the first time by Frumkin, but previous plots[2,57-59] also exhibited the same trend, thus suggesting similar conclusions. Actually, Vasenin[60] proposed a plot with a slope different from unity and he explained this in terms of different water orientation. However, the result was probably due to inadequate choice of Φ values in that it is now almost certain[3,10,14,15,61] that where the slope differs from one, it is >1 and not <1 as proposed by Vasenin. As a matter of fact, the plot was not substantiated theoretically in a satisfactory way and had no practical relevance later on. On the other hand, an attempt[57,62] was also made to show that the equality between $\Delta\Phi$ and $\Delta E_{\sigma=0}$ is, in fact, predictable on a sound thermodynamic basis. However, this conclusion was the consequence of some misunderstanding of the meaning of Volta potential[2] between a metal and a solution at the pzc, as remarked elsewhere.[61]

Although more results of capacity measurements with solid metals became available after 1965, the view held in Frumkin's paper was not modified explicitly until 1971 when the present author[3] published a paper where evidence for the nonequality of $\Delta\Phi$ and $\Delta E_{\sigma=0}$ was given and the behavior rationalized. For the first time transition metals were shown to be differentiated strongly from sp metals within which group it was possible to predict some aspect of behavior proved experimentally only later, as admitted by Frumkin himself.[53] It is thought that this paper may have catalyzed, to some extent, the transition from an electrostatic to a "chemical" theory of the double layer.

The first model based on a chemical view of the double layer was put forward by Damaskin and Frumkin[4] in 1974. It was later improved by Parsons[6] and retouched twice by Damaskin et al.[8,9] Although the model in itself is not without obscurities, the results obtainable from it are certainly striking and extremely attractive and one can hardly think that it does not lay the basis for a first real understanding of the constitution of an electrode double layer. Such a model emphasizes the role of the solvent and suggests that all of the change in capacity with charge and with the nature of the metal should be attributable to change in solvent orientation.

Bockris and Habib[11] have raised objections to the above model and proposed another model resting on early suggestions by Bockris et al.[25] The model is based on a less chemical and more electrostatic view of the double layer. It tends to deny that the solvent plays any role in determining the features of the capacity curve, although obviously a potential drop is admittedly associated with it and contributes to the electrode potential. More specifically, the rate of change of $g^S(\text{dip})$ with charge is predicted to be much lower than for the other models at the potential of zero charge.

The mode and extent of orientation of water molecules at the pzc on Hg were for a long time a matter of discussion.[25,48–50,63–70] Now it appears[7] that there is a general consensus of opinion about some preferential orientation with oxygen towards the metal. Frumkin[30] warned long ago against considering $g^{H_2O}(\text{dip}) = 0$ at the pzc. The early theories[48–50,63] assumed, for water, a preferential orientation with the positive ends of dipoles towards the metal. Such a conclusion was based on the conviction that the hump in capacity curves for Hg was associated with a dielectric maximum corresponding to random orientation of water molecules on the metal surface. Experimental facts suggesting that the opposite orientation was probably more correct led Levine et al.[55] to develop a model where the two aspects were made to coexist by introducing two different limiting orientations for water with intermolecular dielectric polarization. According to the model proposed by Damaskin and Frumkin,[4] the hump is at positive values of charge simply because it is shifted from its natural position by a combined effect due to water chemisorption.

No definite attempts[67] to estimate $g^{H_2O}(\text{dip})$ quantitatively at the pzc $[g^{H_2O}(\text{dip})_0]$ from experimental results may be found until 1960. Frumkin[68] showed that on the basis of the value[71] for the Volta potential[2] between Hg and water, $g^{H_2O}(\text{dip}) \simeq 0.32$ V at the pzc. The sign is consistent with the negative (O) ends of dipoles pointing to the metal. This value led to the suggestion that water molecules in contact with Hg are much more oriented than at the free surface (χ^{H_2O}). This was a quantitative confirmation of a qualitative conclusion[67] suggested earlier by Frumkin himself.

The present author,[3] however, suggested in 1971 that the above view originated from some neglect of the role of the surface potential of the metal. He proved on the basis of the temperature coefficient of $g^{H_2O}(dip)_0$ and χ^{H_2O} that probably the orientation of water at Hg is less than that at the free water surface. This conclusion was also reached on the basis of some different arguments by Trasatti[13] and independently by Frumkin et al.[61] The value for $g^{H_2O}(dip)$ on Hg proposed by the present author[72] in 1970 still seems reasonable but models[4,6,11] tend to predict somewhat lower values.

The mode of preferential orientation of the solvent, and the charge where it vanishes are closely interrelated.[25] Definite insight into the latter parameter has been gained only recently by means of measurements of surface excess entropy. The first experimental results were obtained by Hills and Payne.[73] Conway and Gordon[74] were the first to give a theoretical treatment in terms of configurational and field-dependent librational entropy. An extended analysis of this problem was carried out later by Bockris and Habib.[28] Both the experimental[29,73,74–77] and theoretical[9,28,78,79] results converge to identify a value of about $-4\ \mu C\ cm^{-2}$ as the charge where the entropy passes through a maximum for the Hg/water interface in aqueous solutions. If the change in entropy is simply related to configurational effects of a two-state system, then the maximum in entropy and the minimum in orientation of water are expected to coincide.[11,28] However, the librational component, as in hydration of ions, appears to be more important.[74]

Models based on a chemical view of the double layer, however, consistently predict[6,9] for the charges, relative to the two above effects, two different values. This is one point that has never been raised before. If the water layer consists of an ensemble of particles not admitting a configurational point of symmetry, then the excess entropy is possibly not related structurally to the orientation distribution of dipoles. Future work on the double layer should be addressed to clarify this point in order to put experimental parameters in the correct physical framework.

II. WATER DIPOLE CONTRIBUTION TO THE POTENTIAL OF ZERO CHARGE

Results from models for water at electrodes are usually compared to experimental data at the level of capacity–charge curves.[4,6,8,9] These are very sensitive relationships but they can hardly be directly and intuitively related to molecular parameters. Comparison of a less structured model with a less featured relationship, for example, such as the charge–potential curve,[12] may be a more manageable first step. Such experimental data enable one to argue straightforwardly in terms of a dipole contribution to the electrode potential.

1. Basic Double-Layer Model

The so-called Gouy–Stern–Grahame model of the double layer[66] will be adopted as the starting point. This model is now largely accepted in its general features. In the absence of specific adsorption, the metal–solution interface may be treated as a simple water capacitor. Diffuse-layer effects, unless otherwise stated, will always be considered as allowed for. The total electric potential drop across the inner layer is a function of two interrelated variables:[2,80] charge on the metal and dipole orientation. The possibility of reorientation of solvent molecules under the effect of the electric field is implied in this model. The two contributions may be separated by the conceptual method of keeping alternatively constant the two variables. Thus

$$\Delta_0^\sigma \Delta\phi = \Delta_0^\sigma g(\text{dip}) + g(\text{ion}) \tag{1}$$

where $\Delta_0^\sigma \Delta\phi$ is the change in inner-layer potential drop associated with the onset of a charge on the metal, $\Delta_0^\sigma g(\text{dip})$ is the contribution given by the dipolar layers on the phases, and $g(\text{ion})$ is the contribution given by free charges at constant dipole orientation. $g(\text{ion})$ is zero by definition at the pzc. Further,[2]

$$g(\text{dip}) = g^M(\text{dip}) - g^S(\text{dip}) \tag{2}$$

where superscripts M and S stand for metal and solvent, respectively. The minus sign arises as a consequence of the two surfaces facing one another. g^M(dip) may be regarded as charge independent, if the solid is a perfect conductor, in that no additional potential drop is created inside the metal by the onset of an excess charge and all of the change in surface electron distribution is accounted for in the value of the net charge. Thus

$$\Delta_0^\sigma g(\text{dip}) = -\Delta_0^\sigma g^S(\text{dip}) \tag{3}$$

It follows that

$$\Delta_0^\sigma \Delta\phi = g(\text{ion}) - \Delta_0^\sigma g^S(\text{dip}) \tag{4}$$

Actually, in early theories, penetration of the electric field into the metal[49,50,63] was taken to be plausible but this idea was later abandoned for the case of metals and retained only for semiconductors.[81] It is now known that this idea may prove useful[8,53] also for some semimetals like Bi and perhaps Sb. However, to a first approximation, it will be assumed that $\Delta_0^\sigma g^M(\text{dip}) = 0$. In fact, although the surface of a metal can hardly be regarded as a perfect conducting surface ($\varepsilon = \infty$), its apparent dielectric constant is yet some orders of magnitude higher[82] than that on the solution side of the interface. This implies that no contribution by the metal surface to the double-layer capacity will be considered as possible.

The separation implied in Eq. (4) parallels the splitting of the polarizability of a dielectric phase into molecular and orientation components.[83] The former is related to the displacement in the molecule of electrons and nuclei (induced dipole), and the latter is related to the possible reorientation of the permanent dipole.[84] The orientation polarizability is responsible for the first relaxation step in plots of dielectric constant (ε) vs. frequency for water.[85] Equation (1) actually involves a somewhat fictitious separation in that water molecules presumably rotate until some saturation orientation will be attained and *then* they will be polarized molecularly. Practically, some g(dip) is probably accounted for, or enclosed, in g(ion). However, this limitation is intrinsic in the model which, on the other hand, is universally accepted.[2,4,6,11]

Equation (1) implies only the existence of the outer Helmholtz plane (OHP) somewhere in the solution but it requires neither that the thickness of the inner layer between the metal and the OHP be fixed nor that a detailed model for the state of water molecules in between be specified. The sole assumption is that about the behavior of water molecules at constant orientation. The molecular polarizability may be assumed to be isotropic. No experimental data support this explicitly but calculations[86,87] suggest that the anisotropy may be as high as 20%. To a first approximation, especially if the reorientation of molecules is small as a whole, the molecular polarizability may be taken as constant. If so, Eq. (4) may be written as

$$\Delta_0^\sigma \Delta\phi = \sigma/K_{ion} - \Delta_0^\sigma g^S(dip) \quad (5)$$

where K_{ion} is the integral capacity associated with a capacitor at constant solvent orientation.

Equation (5) expresses the change in electrode potential with respect to the value of $\Delta\phi$ at the potential of zero charge. Thus, the relation

$$E_{pzc} = E - E_{\sigma=0} = \Delta_0^\sigma \Delta\phi \quad (6)$$

results, where E_{pzc} is the electrode potential in the rational scale[66,88] and may easily be derived from experimental results. Experimental charge–potential curves can thus prove useful for gaining insight into the separation involved in Eq. (5).

The shape of charge–potential curves at negative charges may easily be discussed theoretically in that over this region, conceptual complications are minimized. Figure 2 shows that the

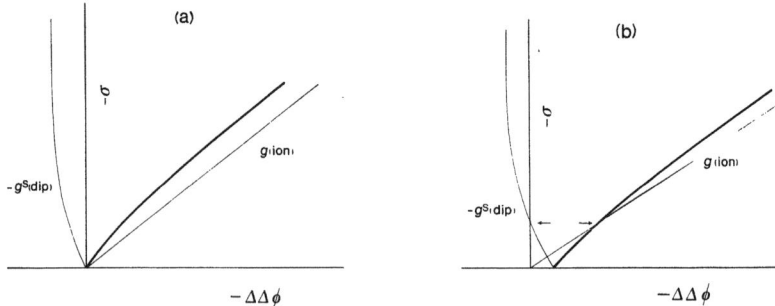

Figure 2. Sketch to show the components g(ion) and g^S(dip) giving the experimental charge–potential curve. (a) No preferential orientation of solvent dipoles at the pzc. (b) Preferential orientation with the negative end toward the electrode surface. Arrows indicate the point where the net orientation of dipoles is zero.

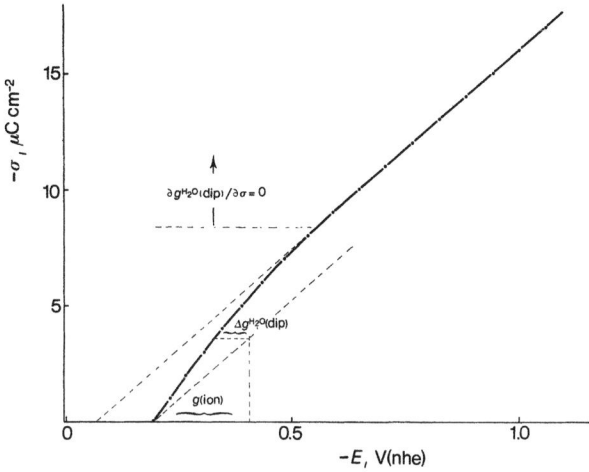

Figure 3. Charge–potential curve for Hg electrodes in contact with NaF aqueous solutions at 25°C. The resolution of g(ion) and g^S(dip) for a given change in charge is sketched. The limit beyond which the orientation polarization vanishes is also shown.

free-charge contribution to the potential is predicted by Eq. (5) to be a straight-line relation starting from the origin and having a slope equal to $1/K_{ion}$. In a very simple way, the dipole contribution may be envisaged as a curve reaching some saturation potential values at strongly negative charges. The curve may start either from the origin or from another potential value depending on whether or not the solvent layer possesses some definite preferential orientation at the pzc. The observable charge–potential relationship follows from the graphical combination of the two components.

Figure 3 shows the experimental charge–potential curve for Hg electrodes in fluoride solutions as obtained by integration of capacity curves.[44] Comparison of the behavior shown in Figs. 2 and 3 reveals that the experimental curve exhibits the theoretically expected shape. The relationship is in fact nonlinear at low charges and linear at strongly negative charges. If the slope of the linear portion is assumed to equal $1/K_{ion}$, then Fig. 3 provides a clue to a graphical resolution of free-charge and dipole contributions. However, only g^S(dip) relative to the value at the pzc can be derived in this way. If, at the pzc, water molecules possess a definite preferential orientation, then the straight line defining the molecular polarizability is expected to start not at the point where the experimental curve starts but at a potential more positive or more negative than the pzc by the amount $g^S(dip)_0$, where subscript 0 stands for $\sigma = 0$. Such a possibility corresponds to case b in Fig. 2. The figure suggests the way in which the different terms of Eq. (5) can be individually derived from Fig. 3, provided that reasonable values for K_{ion} and $g^S(dip)_0$ are either assumed or estimated independently.

Figure 4 displays charge–potential relationships for a number of metals in aqueous solutions.[12,37,89–92] All curves exhibit the same basic features, which implies that Eq. (5) is essentially satisfactory for representing the situation for a number of metals. Since linear parallel sections are observed for the various curves at strongly negative charges where the orientation of water molecules is expected to approach saturation, the limiting slope may be assumed, as a first approximation, to coincide with the value of K_{ion}, which thus proves to be metal invariant, being only a property of the solvent layer. Actually, some metals exhibit[91,92] higher apparent values for K_{ion}. The reasons for this are still obscure and this point will be discussed later. If this point is first

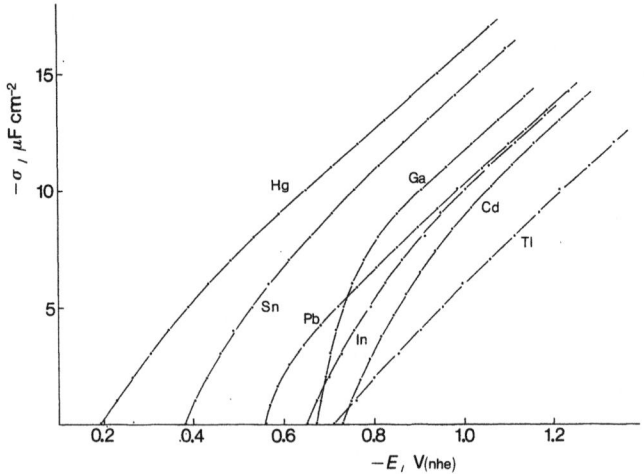

Figure 4. Experimental charge–potential curves for a number of metals in contact with aqueous solutions of not specifically adsorbed electrolytes at 25°C. Sources of data quoted in text.

neglected, then according to Eq. (5) the differences among the various curves in Fig. 4 can be attributed entirely to $g^S(\text{dip})$, which thus proves to be metal dependent at sufficiently small negative charges.

Routes to the estimation of a reasonable value for $g^S(\text{dip})_0$ are much more involved. This parameter cannot be obtained directly from considerations regarding a single charge–potential curve but requires a more elaborate approach. This will be discussed in detail below.

2. Relative Values of Surface Potential of Water at Various Metals

Quantitative comparison of potential values at constant charge for two metals cannot be made in terms of the equations presented above. An additional term, taking into account the different electronic energy of the solid, must appear in the expression of the potential.[13,15,93,94] Consider the simple cell

$$M \mid S \mid \text{Ref.} \mid M' \qquad (7)$$

where M' indicates a different electrical state of M and Ref. is the reference electrode. Let E be the cell potential difference. The work to transfer electrons from M to Ref. (electronic equilibrium is present at the Ref./M' contact) through the external circuit must equal the work to transfer electrons from M to Ref. through the internal circuit, viz., through the liquid phase. The resulting relation is

$$E = \Phi/e + \delta\chi^M - g^S(\text{dip}) + g(\text{ion}) + \text{const.} \qquad (8)$$

where the constant term includes the energy barrier at the solution–reference-electrode boundary. $\delta\chi^M$ accounts for any possible change in the electron distribution at the surface of M as vacuum is replaced by the liquid phase.[2,3,67,95–97]

Figure 4 shows that at strongly negative charges all curves become parallel. The orientation of water molecules may therefore be assumed to be metal independent. This was first suggested by Frumkin et al.[35,36] from a comparison of curves for Ga and Hg. It was later extended by the present author[3] to all other metals for which reliable charge–potential curves were available. Actually, constancy in the limiting slopes of the curves in Fig. 4 should mean that the water layer possesses the same apparent dielectric constant at strongly negative charges on all metals. This, in principle, does not necessarily imply the same orientation of solvent molecules in view of the assumed isotropy for the molecular polarizability. However, at strongly negative charges where water molecules encounter a dense sea of electrons, the limiting orientation is probably governed by the field rather than by the chemical properties of the shielded ion cores. It is important to recall that the energy of electrons in the metal at the same potential on the same scale is the same irrespective of the nature of the solid phase.[15,98]

Since K_{ion} is metal invariant, it follows that g(ion) is also metal independent. Actually, the capacity at the minimum has been found for some metals (In,[92] In–Ga,[99,100] Pb[91]) to be higher than that for Hg. Possible explanations for this have not been offered concretely. Corrections of the charge to recover a common value of K_{ion} have been made[61,101] in charge–potential curves. This is tantamount to assuming that the higher value of capacity is merely due to surface roughness but it is doubtful whether this is really so. It has been suggested[102] that such an effect might originate from the electron distribution at the surface of the metal. However, since this γ is a capacity in series with the solution double layer, an increase in K_{ion} on some metals with respect to Hg can hardly be envisaged if the latter is regarded as a perfect conductor. It would be necessary to postulate for Hg a very low capacity of the surface region in the solid. Evidence for this is, however, unavailable.

In principle, $\delta\chi^M$ cannot be regarded as metal invariant. The same holds for $g^S(dip)$. Thus, at the pzc,

$$E_{\sigma=0} = \Phi/e + \delta\chi^M - g^S(dip)_0 + \text{const.} \quad (9)$$

and the difference in pzc between two metals is given by

$$\Delta^{M_1}_{M_2} E_{\sigma=0} = \Delta^{M_1}_{M_2}\Phi/e + \delta\chi^{M_1} - \delta\chi^{M_2} - \Delta^{M_1}_{M_2} g^S(dip)_0 \quad (10)$$

Equation (10) shows that the Φ–$E_{\sigma=0}$ relationship which has been used several times[2,3,56,103] to gain insight into the behavior of solvent molecules at electrodes may be vitiated by the difference of the two interaction terms $\delta\chi^M$. In principle, $E_{\sigma=0}$ vs. Φ

plots may even be unsuitable for giving direct evidence for changes in $g^S(\text{dip})$. Although there may be acceptable reasons for assuming that $\delta\chi^{M_1} \simeq \delta\chi^{M_2}$, the following route for derivation of information on $g^S(\text{dip})$ seems undoubtedly much more rigorous. At strongly negative charges, for the arguments above,

$$\Delta^{M_1}_{M_2}E_{\sigma \ll 0} = \Delta^{M_1}_{M_2}\Phi/e + (\delta\chi^{M_1} - \delta\chi^{M_2}) \tag{11}$$

It has been suggested[3,15] that the reasonable equality between ΔE and $\Delta \Phi$ in Eq. (11), as observed experimentally for a number of metals, may be taken as an indication that actually $\delta\chi^{M_1} \simeq \delta\chi^{M_2}$. However, such an assumption may be avoided if the difference between Eqs. (10) and (11) is noted. In such a case

$$\Delta^{M_1}_{M_2}E_{\sigma=0} - \Delta^{M_1}_{M_2}E_{\sigma \ll 0} = -\Delta^{M_1}_{M_2}g^S(\text{dip}) \tag{12}$$

Equation (12) means that if curves for different metals are plotted with potential on the rational scale, the difference in potential at constant charge between two curves in regions where they are straight lines gives an experimental measure of the relative value of the solvent dipole contributions at the pzc.[104]

The above procedure rests on a definite postulate: that water molecules rotate under the action of the electric field and attain a metal-invariant saturation orientation[5,74] at some sufficiently strong negative charge. Table 1 summarizes values of $g^{H_2O}(\text{dip})_0$ for a number of metals as referred to Hg. The table gives two sets of data: one has been selected by the present author on the basis of published available charge-potential and capacity curves; the other one has been suggested by Frumkin et al.[61] Some of the values differ; one reason for this may be that Frumkin et al. selected values of $\Delta E_{\sigma \ll 0}$ at $-18 \, \mu\text{C cm}^{-2}$. At such a charge, the capacity at most of the metals is beyond the minimum and tends to rise again. Some differences may also be attributed to different sources of experimental capacity data. The present author of course used only published curves for which graphical integrations could be made. Reasons for the difference with Ga are unclear. The value reported by Frumkin et al.[61] refers to $1 \, N$ Na_2SO_4 solutions.[35,36] The present author integrated[12] capacity curves for $0.1 \, M$ and $0.01 \, M$ $NaClO_4$ solutions,[46] for which the deep minimum indicating the position of the potential of zero

Table 1
Relative Values of Surface Potential of Water at Metal/Aqueous-Solution Interfaces at the Potential of Zero Charge^a

Metal	$\Delta g^{H_2O}(dip)_0$ (V)	
	(a)^b	(b)^c
Au	−0.07 (231, 478)	0.00
Tl	−0.05 (37)	(0.11)^e
Pb	0.04^d (91)	0.01$^{d?}$
Bi	0.05 (188)	0.03
Sn	0.09 (89)	0.10
In	0.11^d (92)	0.12^d
Cd	0.15^d (90)	0.21$^{d?}$
Ga	0.25 (12, 46)	0.32

^{a} Reference metal, Hg.
^{b} As independently derived in this work from published data (references in parentheses).
^{c} From Ref. 61.
^{d} Charge corrected so as to have parallel curves in Fig. 4.
^{e} Calculated by using $\Delta E_{\sigma \ll 0}$ for Ga–Tl alloy.

charge is well developed, and obtained consistent results. However, such differences do not invalidate the conclusions which can be drawn from the results in Table 1. They simply suggest that some marginal questions must still be settled.

The above procedure can be applied only with metals for which reliable capacity data are available. In fact, this is possible for all sp metals, although some of them are still too difficult to be handled in this context because of the lack of easy-to-read experimental results. This is the case for Ag, whose position has not yet been assessed. Previous inclusion[3] of Ag in a $E_{\sigma=0}$ vs. Φ plot has been rightly criticized[53] because it was based on an unreliable $E_{\sigma=0}$ value.

The situation for transition metals is much more complicated. Capacity curves as reliable as those for sp metals are at present available only for Fe (for which satisfactory experimental results have become available only recently[105]) and partly for Pt[106]; for the latter metal, however, the nature of the potential drop at the interface is complicated by the presence of Pt–H surface dipoles[107-109] the sign of whose contribution changes about a point close to the potential of zero charge.[23,108]

For transition metals, recourse to $E_{\sigma=0}$ vs. Φ plots as described by Eq. (9) with the assumption of $\delta\chi^M = \text{const.}$ is thus necessary to derive information on the behavior of the solvent layer at the interface. $E_{\sigma=0}$ values are unfortunately not very reliable for most of the d metals. Although for the Pt group metals $E_{\sigma=0}$ may be measured reliably,[24] complications arise due to the unknown role played by Pt–O and Pt–H dipoles, and specific adsorption of anions, even though such interferences may be thought to be quantitatively at a low level only. However, conclusions derivable from a $E_{\sigma=0}$ vs. Φ plot for transition metals should be regarded as only tentative with the aim of obtaining a qualitative picture of metal–solvent interactions.

Figure 5 shows the $E_{\sigma=0}$ vs. Φ plot for some transition metals.[3] The point for Hg is also shown for comparison. If it is assumed that $\delta\chi^M = \text{const.}$, the plot suggests that at the pzc the contribution by solvent dipoles is negative for all the metals with respect to that for Hg. Further, a straight line with unit slope passes satisfactorily through the points. This implies that the orientation of solvent molecules is apparently metal invariant. The equation to the straight line is

$$E_{\sigma=0} = \Phi/e - 5.02 \tag{13}$$

as suggested previously.[3] Although Eq. (13) is to be considered

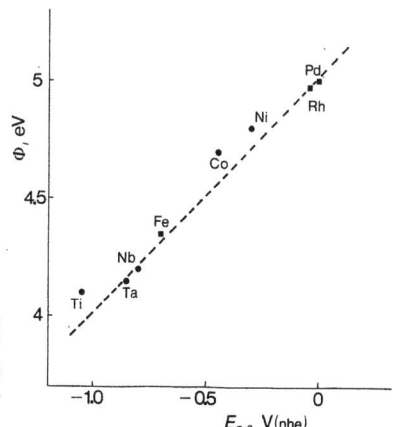

Figure 5. Plot of the potential of zero charge[3,105] against the work function[15] for transition metals. More (■) and less (●) reliable values of pzc are distinguished.

quantitatively as tentative, and some of the points undoubtedly require some revision, the data suggest for d metals a noticeably higher hydrophilicity than for Hg. The inclusion of the point for Fe[105] at a different position compared to that in the previous plot[3] would require some further discussion, which will be given in a future paper.

A plot of $E_{\sigma=0}$ vs. Φ has been suggested by the present author also for sp metals.[3,10,14,15] However, especially for low-Φ metals, the reliability of the work function data is doubtful.[39] All sp metals have been shown to fall in a different group from that for d metals. Some definite trend in the plot may, however, be discerned only if work function values derived from Eq. (11) on the assumption that $\delta\chi^M = $ const. are used. Such quantities have been defined as "electrochemical work functions."[3] This matter has been discussed recently elsewhere[15] and will not be dealt with here any further.

3. Surface Potential of Water at Mercury

For absolute values to be derived from Table 1, the value of $g^{H_2O}(\text{dip})_0$ for Hg should be estimated. No direct experimental measurements are possible[2], however, and recourse to routes making use of model assumptions are therefore necessary.

(i) Adsorption Potential Shifts

This route rests on the measurement of the electrode potential shift upon adsorption of an organic substance.[110,111] At full coverage ($\vartheta = 1$), it follows from Eq. (9) that

$$E^{org}_{\sigma=0} = \Phi/e + \delta\chi^M_{(org)} - g^{org}(\text{dip})_0 + \text{const.} \qquad (14)$$

To a first approximation, if $\delta\chi^M$ is assumed to be solvent invariant, $\delta\chi^M_{(org)} = \delta\chi^M_{(H_2O)}$. This is not unreasonable since the electronic distribution of molecular groups differs little from one substance to another. Thus, from Eqs. (9) and (14), the following relation may be derived for the pzc condition:

$$\Delta^{org}_{H_2O}E_{\sigma=0} = g^{H_2O}(\text{dip})_0 - g^{org}(\text{dip})_0 \qquad (15)$$

where the second term on the right-hand side is the potential drop associated with the adsorbate dipole layer which has replaced the solvent molecules on the metal surface. Since it is always very difficult to account quantitatively for $g^{org}(dip)_0$, attempts are usually made to derive $g^{H_2O}(dip)_0$ by using as probes organic substances thought to contribute a zero net permanent dipole perpendicular to the interface. This condition has been claimed[5,72,112] to have been met a number of times but it is now doubtful whether it is in fact so, although in the work of Ref. 5 the conformationally rigid, nondipolar molecule pyrazine was used.

Adsorption potential shifts can seldom be definitely measured up to $\vartheta = 1$. More often, adsorption potential shifts at $\vartheta = 1$ are extrapolated from some $\Delta E_{\sigma=0}$ vs. ϑ plots. Sometimes,[112] initial slopes in such plots are measured and taken as satisfying Eq. (15). However, the reliability of extrapolations depends on the reasonableness of the model adopted for the adsorption layer.[7,111,113] If the model of two capacitors in series is valid, i.e., if the adsorbate and the solvent do not interact with one another detectably on the surface, linear plots should be obtained and $\Delta_{H_2O}^{org} E_{\sigma=0}$ may be extrapolated reliably to $\vartheta = 1$. This model is apparently obeyed at the water–air interface.[114] If the model of two parallel capacitors holds, i.e., if the adsorbate and the solvent interact with one another detectably on the surface, then nonlinear plots are obtained and even an extrapolation of the initial portion of the curve may lead to misleading results.[115] Figure 6 illustrates the two cases on the basis of experimental results.

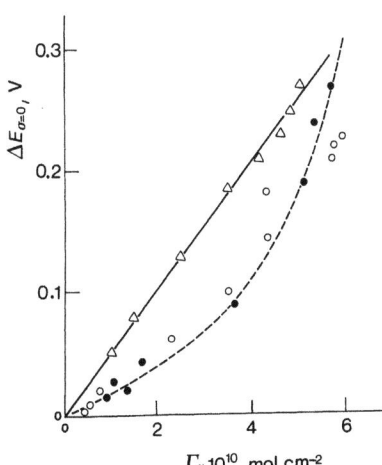

Figure 6. Potential shift[114] upon adsorption of propanol at the Hg–aqueous-solution interface (---) and at the free surface (——).

In the case of alcohols, linear adsorption potential shifts are obtained at the water–air interface, and nonlinear patterns are observed at the Hg–solution interface.[114,116] This suggests that the nonlinearity is probably not a consequence of direct solvent–adsorbate interaction but of the presence of the conducting surface. Any direct interaction would lead to nonlinear behavior at the solution–air interface as well. It is important to stress that if the model of two parallel capacitors is obeyed, $g^{H_2O}(dip)_0$ is not obtained by extrapolating the initial slope even for $g^{org}(dip)_0 = 0$. It is, in fact, necessary to measure directly the adsorption potential shift at $\vartheta = 1$, where the maximum $\Delta E_{\sigma=0}$ is independent of the model chosen.

Müller[112] suggested that $g^{H_2O}(dip)_0$ for Hg could be obtained from the adsorption potential shift with n-butanol. This alcohol almost certainly becomes adsorbed following a parallel-capacitor model[117] (or an intermediate one; either way a nonlinear pattern is observed). However, $g^{H_2O}(dip)_0$ may be derived from the initial slope of the $\Delta E_{\sigma=0}$ vs. ϑ plots, as suggested by Müller, only if the nonlinearity is related to some reorientation of n-butanol molecules with coverage from flat with $g^{org}(dip) = 0$ at $\vartheta \to 0$ to vertical in a more close-packed assemblage at $\vartheta \to 1$. Although such a reorientation mechanism is explicitly included in Frumkin and Damaskin's theory of organic molecule adsorption,[110,115,118] physically the possibility cannot be ruled out that also the more polarizable water molecules present on the surface undergo in fact some reorientation as ϑ increases. In such a case, the initial slope is no longer governed only by $g^{H_2O}(dip)_0$.

For n-butanol, $\Delta E_{\sigma=0} = 0.24$ V has been measured[119] at $\vartheta = 1$. The positive value has been taken[110] to indicate some contribution from the permanent dipole in the butanol molecule with the positive end (hydrocarbon tail) towards the metal. Energetically, this is in fact the more favorable position for adsorption as calculated by Bockris et al.[25] As $\vartheta \to 0$, $\Delta E_{\sigma=0} = 0.07$ V.[112] This value, *in the context of the above view*, may be a measure of $g^{H_2O}(dip)_0$. If this is the case, a positive value for $\Delta E_{\sigma=0}$ according to Eq. (15) implies a negative value of water dipole contribution to the potential of zero charge in Eq. (9). Thus, water molecules come out to be preferentially adsorbed at the pzc with the negative end of the permanent dipole towards the metal.

It is difficult to verify whether the derivation proposed by Bockris meets all the requirements for reliability. However, the present author[120] has recently shown that it is quantitatively possible to prove that n-butanol becomes adsorbed on Hg with a strong positive component of dipole perpendicular to the surface. In such an analysis, even as $\vartheta \to 0$, $\Delta E_{\sigma=0}$ values would represent a compromise between the positive shift associated with the adsorbed butanol and the replaced water, and the negative shift involved in some reorientation of water molecules surrounding the adsorbate molecule.

A determination of $g^{H_2O}(dip)_0$ based on the measure of the potential shift upon adsorption of ethylene glycol has been proposed by the present author.[3,72] The pattern of the $\Delta E_{\sigma=0}$ vs. ϑ plot up to $\vartheta = 0.6$ was found to conform to the series-capacitor model. It has been suggested[116,121] that linearity might be only apparent due to the very small curvature of the plot. However, the plot extrapolated to $\vartheta = 1$ yielded[7] $g^{H_2O}(dip)_0 = 0.07$ V. The same value was observed[7,122] with butan-1,4-diol, which, however, being less adsorbable, enabled higher coverages to be reached where some nonlinearity could be observed. In the light of recent work,[120] it is now possible to state that both butan-1,4-diol and ethylene glycol adsorb on Hg with a residual positive dipole toward the metal. This casts serious doubts on the reliability of the $g^{H_2O}(dip)_0$ suggested above. However, the coincidence of the various estimates with different substances is remarkable and may indicate that reorientation of water molecules may occur proportionally to the magnitude of the vertical component of the organic dipole so that a sort of compensation effect may take place at very low coverages, leading to apparent consistency between the values measured for the initial slope of $\Delta E_{\sigma=0}$ vs. ϑ plots with different adsorbates.

More recently, Conway and Dhar[5] have proposed pyrazine as a suitable molecule for evaluation of $g^{H_2O}(dip)_0$ by measuring adsorption potential shifts. They obtained linear $\Delta E_{\sigma=0}$ vs. ϑ plots and measured $g^{H_2O}(dip)_0 = 0.12$ V. However, pyrazine contains a benzene ring whose electrons, although less free to move around because of the presence of the electronegative nitrogen atom, could yet interact with the metal surface, therefore contributing a small additional potential drop. Further, the linearity in $\Delta E_{\sigma=0}$ vs. ϑ plots has not been proved experimentally up to ϑ

close to 1. Nonlinearity may become quite serious precisely at very high coverages.[116] It would be necessary to know the experimental value of the capacity at $\vartheta = 1$ to attempt some estimation of the vertical dipole component at the pzc.

In conclusion, no derivation of $g^{H_2O}(\text{dip})_0$ based on adsorption potential shifts now appears to be entirely reliable. It seems better to resort to some alternative procedures which rest on a less ambiguous basis.

(ii) Route through χ^{H_2O}

Despite what has been said above, adsorption potential shifts in some particular cases can nevertheless give quantitative information on the difference between χ^{H_2O} and $g^{H_2O}(\text{dip})_0$ on Hg.[13,61] It may in fact be written that

$$g^{H_2O}(\text{dip})_0 = \chi^{H_2O} + \delta\chi^{H_2O} \tag{16}$$

where $\delta\chi^{H_2O}$ is the possible change in χ^{H_2O} upon contact with a different phase. There is only one set of data enabling the estimation of $\delta\chi^{H_2O}$ with Hg electrodes to be made.

$\Delta E_{\sigma=0}$ data at $\vartheta = 1$ are available for a number of linear aliphatic alcohols. It has been found[119] that $\Delta E_{\sigma=0}$ at Hg is always some 60 mV less than that for the same alcohol at the free surface of water. If $\Delta E_{\sigma=0}$ is plotted against the number of carbon atoms in the alcohol, the points for the two interfaces run parallel, and linear plots are obtained (Fig. 7). If the orientation of a monolayer of alcohol on Hg is assumed to be the same as that at the free surface of aqueous solutions, then from Eq. (15) and from the following equation,

$$\Delta^{\text{org}}_{H_2O} E_{\text{air/sol}} = \chi^{H_2O} - g^{\text{org}}_{(\text{air})}(\text{dip}) \tag{17}$$

it follows* that

$$\Delta^{\text{org}}_{H_2O} E_{\text{air/sol}} - \Delta^{\text{org}}_{H_2O} E_{\sigma=0} = \chi^{H_2O} - g^{H_2O}(\text{dip})_0 = -\delta\chi^{H_2O} \tag{18}$$

It then results that

$$\delta\chi^{H_2O} = -0.06 \text{ V} \tag{19}$$

* It should be noted that in ΔE at the air–solution interface and in $\Delta E_{\sigma=0}$ the surface potential of the liquid phase appears with opposite signs in that the reference phase for the potential drop is the liquid and the metal, respectively, at the two interfaces.

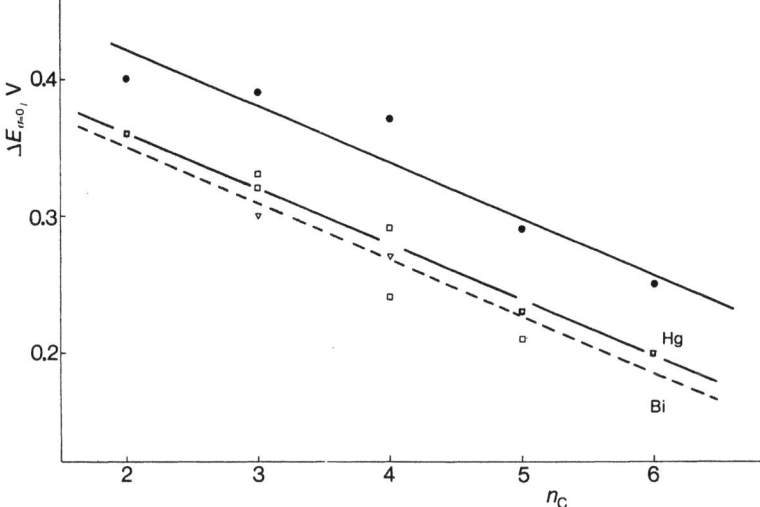

Figure 7. Potential shifts upon adsorption of alcohols[119] (●) at the free surface of water, (□) at Hg electrodes and (▽) at Bi electrodes[188] plotted against the number of carbon atoms in the alcohol chain.

$g^{H_2O}(dip)_0$ may now be derived from Eq. (16) if χ^{H_2O} is known. Although this is a surface potential as well as is $g^{H_2O}(dip)_0$, and thus not amenable to direct experimental measurement, it is nevertheless much more reliably derivable indirectly from experimental data in that complications arising from the presence of a conducting surface are excluded.

4. Surface Potential of Water at the Free Surface

The problem of the magnitude and sign of χ^{H_2O} has been discussed for a long time.[2,123–125] The earlier papers were reviewed by Frumkin et al.,[67] by Kamienski,[126] and by Conway.[17b] Two main routes to the derivation of χ^{H_2O} are available. The surface potential of a liquid phase can, in principle, be obtained from the difference between the real and chemical free energy of solvation of ions[127]:

$$(\alpha^S_{M^z} - \mu^S_{M^z})/ze = \chi^S \tag{20}$$

This route tends definitely to give negative values for χ^{H_2O}, although the latest estimates[128] do not rule out a probable, very small, positive value. A large uncertainty is, however, implied in Eq. (20) since $\alpha_{M^z}^S$ can be measured experimentally,[2,71,127] whereas $\mu_{M^z}^S$ cannot and can only be calculated on the basis of models.[128,129–132] It follows that χ^S is derived from a small difference between two large quantities. No consensus of opinion exists even for χ^{H_2O} values calculated by taking into account the electronic structure of the molecule of water and making assumptions on the modes of distribution of dipoles at the surface[133–135] and at the local water structure.[17b]

It appears that a reasonable route is probably that through adsorption potential shifts because ΔE vs. θ plots at the free surface of water are very often found to be strictly linear.[114] The most direct piece of evidence in favor of a positive value for χ^{H_2O}, as first suggested by Frumkin et al.,[67] is the negative value of its temperature coefficient. The most reliable value[136,137] is -0.40 mV K^{-1}. In terms of a simple model of oriented dipoles which are reorganized by entropy effects as the temperature is increased, χ^S and $d\chi^S/dT$ must have opposite signs.[127,138] This simple picture has been questioned[139] in terms of a model[55] assuming strong depolarization between dipoles. This model has been shown[139] to be able to predict the same sign for χ^S and $d\chi^S/dT$. Similarly, when dipole interaction effects are considered,[5] $d\chi/dT$ for an oriented layer has an apparently unexpected sign. However, in such an occurrence, nonlinear ΔE vs. ϑ plots should probably be expected upon adsorption of organic substances at the free liquid surface, which does not seem to be the case.

All of the evidence for a positive sign of χ^{H_2O} has been reviewed recently.[7,17b] χ^{H_2O} was suggested by Frumkin et al.[67] to be small and positive, probably 0.1–0.2 V. A value of 0.14 V may be derived from the potential shift upon adsorption of benzene, whose permanent dipole is certainly zero, at the free surface of the water.[140] 0.14 V was suggested as a maximum value on the assumption of constant $d\chi^{H_2O}/dT$ [136] up to the critical temperature. Randles[124] suggested a value of 0.08 ± 0.06 V on the basis of a number of considerations. 0.08 V is estimated from the potential shift upon adsorption of un-ionized surfactants.[141]

However, the adsorption was studied from 4 M NaCl and some lowering in χ^S may be expected by adsorption of anions at the free surface.[67,124,142,143] A semiquantitative estimation is that derivable from jet experiments by Kochurova et al.[144] The liquid surface is assumed to form within some finite time after the start of the jet. The value of $\Delta\chi^S$ over the jet lifetime may thus be taken as a measure of χ^S. It has been found that the evolution of χ^S is towards a more positive value, probably of the order of 0.1 V.

On the basis of all of the available experimental evidence, it has been suggested[7,13] that the most probable value of the surface potential at the free surface of water may be

$$\chi^{H_2O} = 0.13 \pm 0.02 \text{ V} \qquad (21)$$

Although there is agreement on the order of magnitude of this value, it has been suggested[145] that the estimated uncertainty is probably somewhat low. It may be added here that the value in Eq. (21) may receive some experimental support *a posteriori*, as will be shown later.

On the basis of Eqs. (16), (19), and (21), the conclusion is reached that

$$g^{H_2O}(\text{dip})_0 = 0.07 \pm 0.02 \text{ V} \qquad (22)$$

for Hg electrodes. This value is very close to those derived from adsorption potential shifts with ethylene glycol,[72] butan-1,4-diol,[7,122] and n-butanol.[112] Such a surprising closeness may be simply fortuitous. However, it might also be that the compensation effect often observed[74,146] between entropy and enthalpy of adsorption is paralleled by some compensation between the extent of reorientation of water upon adsorption of organic substances at $\vartheta \to 0$ and the magnitudes of the normal components of the dipoles of the organic adsorbates.

5. Surface Potential of Water at Other Metals

Values of $g^{H_2O}(\text{dip})_0$ for metals other than Hg can now be derived from the set of data in Table 1. Results are summarized

Table 2
Absolute Values of Surface Potential of Water at Metal/Solution Interfaces at the Potential of Zero Charge Based on 0.07 V for Hg Electrodes

Metal	$g^{H_2O}(dip)_0$ (V)	
	(a)[a]	(b)[b]
Au	0.00	0.07
Tl	0.02	(0.18)
Hg	0.07	0.07
Pb	0.11	0.08
Bi	0.12	0.10
Sn	0.16	0.17
In	0.18	0.19
Cd	0.22	0.28
Ga	0.32	0.39
d metals	0.40[c]	—

[a] From column (a) of Table 1.
[b] From column (b) of Table 1.
[c] From Fig. 5.

in Table 2.* It can be seen that the preferential orientation of water is always with oxygen towards the metal surface. The maximum value of 0.4 V for $g^{H_2O}(dip)_0$ should still be regarded as tentative. On the other hand, no evidence for a negative value of $g^{H_2O}(dip)_0$ can be found for any metals in the group investigated. Thus, the preferential orientation of water molecules at the free surface and at the interface with metals is qualitatively the same but quantitatively different.

III. RELATION OF SURFACE POTENTIAL TO STRENGTH OF WATER ADSORPTION

It is worthwhile stressing that the values of $g^{H_2O}(dip)_0$ in Table 2 have been derived without any detailed description of the

---

*The surface potential of water on Pb was overestimated in the previous papers[3,10,13–15] because the experimental charge–potential curve was not normalized[61] to that for Hg at strongly negative charges. The surface potential of water on Tl was calculated in the previous papers by using $\Delta E_{\sigma \ll 0}$ for Ga–Tl alloys[61] because the value for Tl[37] was still unavailable. This procedure apparently leads to different results, unlike the case[61] of In and In–Ga alloys.

structure of the water layer being made. The present approach is purely phenomenological and should be distinguished from those based on models[4,6,11] which require some *a priori* choice of parameters and detailed description of the molecular structure of the water layer as prerequisites. Without going into the detailed structure of the water layer it is, however, equally possible to formulate some hypotheses about the origin of the dependence of $g^{H_2O}(dip)_0$ on the nature of the metal.

1. Hydrophilicity of Metals

The value of χ^{H_2O} can be seen to be between that for the surface potential of water on Hg and that on Bi. If the surfaces of the liquid and solid phases are first assumed as far apart and then brought into contact to create the interface, the absolute value of $g^{H_2O}(dip)_0$ may be regarded as the result of combined metal–solvent and solvent–solvent interactions on the surface and towards the bulk of the liquid phase; also, the value of χ^{H_2O} is presumably the result of combined water–vacuum and water–water interactions integrated over all the surface region.

It has been suggested[11] (but cf. Ref. 17b) that a monolayer of water adsorbed on a metal should reproduce the properties of the first layer of liquid water in contact with the metal. Accordingly, the surface tension of Hg with a monolayer of water adsorbed on it[147] (44.00 μJ cm^{-2}) has been regarded as fairly close to the interfacial tension at the Hg–water boundary[148] (42.60 μJ cm^{-2}). However, if 44.00 − 42.60 = 1.4 μJ cm^{-2} may be negligible with respect to about 40 μJ cm^{-2}, it is certainly not so when compared with 5.8 μJ cm^{-2}, which is the surface tension drop for Hg in contact with pure water,[149] and to about 7.2 μJ cm^{-2}, which is the surface tension of pure water.[150] This means that the change in the interfacial structure is small when compared to that of Hg but it may be large if compared with that of water.

For metals for which $g^{H_2O}(dip)_0 > \chi^{H_2O}$, some extra attraction for the negative end of the water dipole is thought to be present at the interface compared to the free surface. Chemically, this factor may be identified as the affinity for the oxygen atom. In surface chemistry, surfaces with acid sites[151] (OH groups) are generally defined as hydrophilic[152] in that they bind strongly other water molecules by acting as proton donors.[153,154] In contrast, strongly dehydroxylated surfaces containing exposed oxygen groups[155,156] often exhibit hydrophobic behavior.[157,158] Consistently, a metal surface binding water molecules through the oxygen atoms may be defined as hydrophilic. If the χ^{H_2O} value is taken as the watershed separating apparent oxygen attraction from apparent oxygen repulsion, then metals with $g^{H_2O}(dip)_0 > \chi^{H_2O}$ may be classified as *hydrophilic* and metals with $g^{H_2O}(dip)_0$ as *hydrophobic*. In fact, only* Hg and Au (and probably Sb, provided excessive complications do not arise due to its semimetallic nature) can be classified as weakly hydrophobic among the metals investigated. Conway[17b,74] suggested that apart from specific interactions between lone-pair orbitals on the metal and acceptor orbitals at the metal surface, all metals should, in a sense, behave hydrophilically due to the continuation of effective hydrogen bonding through metal–H_2O interfaces on account of OH bond dipole imaging which should give a short-range attraction of H_2O dipoles to all metal interfaces. However, the limitation of image interaction relations at short distances from metal surfaces is well known.

Table 2 may be regarded as giving a scale of surface hydrophilicity of metals. Some support for this view may be obtained if a measure of the affinity for oxygen is identified in the enthalpy of formation of the metal oxide (ΔH°_{ox})[159] according to the equation

$$M(s) + \tfrac{1}{2}O_2(g) \rightarrow MO(s) \qquad (23)$$

It may be objected that ΔG°_{ox} values would be more relevant; however, bulk oxides are involved in Eq. (23). While similarity or parallelism in enthalpy terms may be reasonable, ΔS° terms are actually expected to differ in the two cases both for the different structure of the oxide phase and for possible entropy–enthalpy compensation effects at the surface.[146,160–163]

Figure 8 shows that a good linear correlation exists between ΔH°_{ox} and $g^{H_2O}(dip)_0$ for a group of metals. Some points are seen to fall a little farther from the linear relationship. They may constitute some tentative evidence for a parallel linear correlation for another group of metals. The actual value of the metal–water

---

* The exact location of Tl is still uncertain.

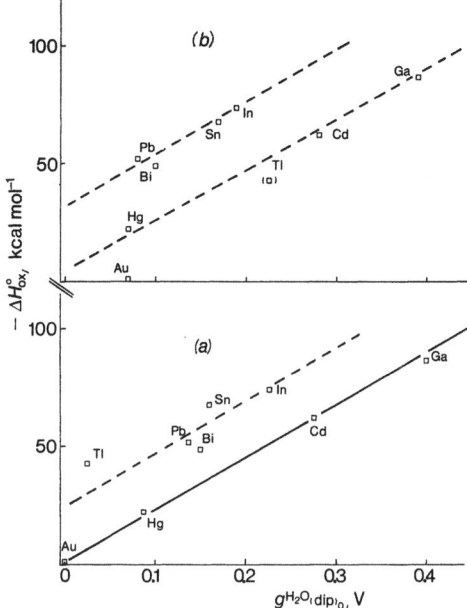

Figure 8. Surface potential of water on a number of metals at the pzc plotted against the enthalpy of formation of the oxide MO. (a) Surface potentials from column (a) of Table 2. (b) Data from column (b) of Table 2.

interaction strength is not known. However, the free energy of adsorption of water at Hg and Ga have been calculated[164] and also derived[149] from data of ion adsorption and of work of adhesion of the solvent. ΔG_{ad}° for water has been found to range from 1 to 3 kcal mol^{-1} for the above metals.

Figure 8 shows data only for *sp* metals. Although different ΔH_{ox}° are expected for different *d* metals[165] and hence some metal-dependent orientation of water should arise also in this group, the results of Fig. 5 imply however that the orientation is in fact metal invariant. This apparent inconsistency disappears if the constant value of $g^{H_2O}(dip)_0$ on *d* metals is taken to mean that the hydrophilicity is so strong on all surfaces that water molecules are always forced to a maximum orientation probably determined by structural constraints. The importance of the role of *d* electrons in adsorption phenomena can hardly be overemphasized.[166–168]

An alternative way of looking at the hydrophilicity scale is to relate the effect of metal surfaces on the water structure to the action of ions in hydration phenomena.[169–171] Structure-making ions interact strongly with water molecules which are thus strongly held and sharply oriented in the hydration shell. Structure-breaking ions interact weakly with water molecules. Opposite effects on the viscosity of water near ions have been observed for the above two classes.[172] Consistently, hydrophilic metals may be regarded as structure-making surfaces and hydrophobic metals as structure-breaking surfaces.

Strongly structure-making surfaces are expected[173–175] to extend their orienting effect on water molecules possibly further into the bulk liquid beyond the first monolayer. The Gouy–Chapman theory of the diffuse layer provides a tool to test the validity of the above argument. The original theory[2] involves the bulk dielectric constant of the liquid phase. This has been shown to be a very good approximation on a number of metals[66,90,91,100,176,177] except[178] Ga. It has been found[179] that with this metal the theory is obeyed only if the dielectric constant of bulk water is assumed to be about 120. This provides some definite evidence that a strong hydrophilic surface tends to structure the aqueous phase beyond the OHP.

A decrease in the apparent dielectric constant of a liquid phase is generally postulated for hydrophilic surfaces as a consequence of the decreased ability of molecules to follow an alternating field.[153,180–184] However, the dielectric constant of a polar liquid may become anisotropic[185,186] at an interface. As the charge on the metal is made less negative, a higher apparent dielectric constant may result for the water layer if the reorientation of molecules, smoothing down the relative change in potential, is increased by the combined action of the decrease in the electric field and of the favorable affinity of the surface for oxygen. This effect is, however, present only *perpendicular* to the metal surface, but it cannot be expected to occur also in planes parallel to the interface. In other words, ion–ion repulsions *along* the surface are expected to be less shielded if water molecules cannot reorient because they are held by strong perpendicularly directed chemical forces (Fig. 9). Thus, dielectric properties of water at the interface with metals are expected to become increasingly anisotropic with increasing hydrophilicity of the solid

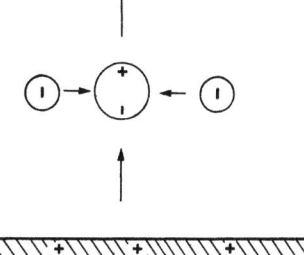

Figure 9. Sketch to illustrate the possible origin of anisotropy in the dielectric properties of water molecules in the inner layer.

surface. Unfortunately, no reliable tests of the validity of the Gouy–Chapman theory with d metals are available.

An alternative hypothesis[10] may be that there is no smooth transition from surface to bulk structure of water. Drost-Hansen[173] has suggested that on some surfaces water molecules are strongly oriented for some monomolecular layers (5–10) and then the structure evolves towards the bulk order via some completely disordered layers. Water molecules in such layers should possess a higher ability to reorient in any direction in that they are freed from any structural limitations. If such a view is applied to Ga electrodes with the limitation of a probably monomolecular layer of ordered molecules, it could as well account for the increased apparent dielectric constant in the diffuse layer, although on a somewhat different physical basis.

In terms of Eq. (15), $\delta\chi^{H_2O}$ may be regarded as a direct measure of hydrophilicity. It has been shown above that this parameter can be estimated experimentally in the case of Hg. It should be interesting to have some information on its magnitude for other metals. Unfortunately, the adsorption of alcohols on other metals does not offer the possibility of deriving $\delta\chi^{H_2O}$ as easily as with Hg electrodes. The problem is that especially the longer chain alcohols tend to adsorb flatter and flatter as the hydrophilicity of the surface increases as a result of increasing interaction of the metal with the OH groups.[187] While corroborating the view that the affinity for oxygen is a parameter of paramount importance, the adsorption behavior of aliphatic alcohols on sp metals other than Hg is such that the adsorption potential shifts are not directly applicable for the calculation of $g^{H_2O}(dip)_0$ because of the metal-dependent limiting orientation of the organic molecule. However, weakly hydrophilic metals like Bi still exhibit[188] a linear variation of $\Delta E_{\sigma=0}$ ($\vartheta = 1$) with the number of carbon atoms in the chain (Fig. 7), which tentatively gives a value of $\delta\chi^{H_2O}$ for this metal slightly higher than that for Hg. An interesting feature of alcohol adsorption is the value of C_1, the capacity at saturation coverage, which is observed to increase fairly regularly with increasing hydrophilicity (Table 3). It has been suggested[189] that this may be due to the presence of some water molecules even in the adsorbed monolayer of alcohols. However, since C_1 is observed to be metal dependent at strongly negative charges, this explanation is unsatisfactory. The remarkable parallelism between C_1 and $g^{H_2O}(dip)_0$ still requires some convincing explanation. It is to be noted that whereas with Hg electrodes C_1 may be assumed to be independent of charge, this is not the case with more hydrophilic metals like Cd in that C_1 increases sharply about the pzc as a result of the alcohol reorientation.[190] Thus, the Frumkin–Damaskin theory[110] of adsorption of organic substances requires the introduction of an additional parameter which is not considered in the original version.

Table 3
Values of Capacity at Saturation Coverage and Zero Charge Potential Shifts upon Alcohol Adsorption on *sp* Metal Electrodes

Alcohol	Parameter[a]	Hg[b]	Pb[c]	Bi[d]	Sn[e]	In–Ga[f]	Cd[g]	Zn[h,i]
Ethanol	C_1	6.40	—	6.6	7.40	7.5	6.40	—
	$\Delta E_{\sigma=0}$	0.36	—	0.36	—	—	0.13	—
n-Propanol	C_1	4.85	—	5.4	5.30	6.0	7.15	10.1
	$\Delta E_{\sigma=0}$	0.33	—	0.30	0.35	0.35	0.15	0.12
n-Butanol	C_1	4.80	—	4.8	5.1	5.4	7.20	9.4
	$\Delta E_{\sigma=0}$	0.24	—	0.27	0.25	0.25	0.15	0.10
n-Pentanol	C_1	4.20	4.20	4.2	4.6	4.8	7.20	9.1
	$\Delta E_{\sigma=0}$	0.21	0.22	0.23	0.22	0.23	0.09	0.12
Cyclohexanol	C_1	4.10	3.86	3.70	4.2	—	5.00	8.00
	$\Delta E_{\sigma=0}$	0.30	0.17	0.40	0.33	—	0.25	0.42

[a] C_1 in $\mu F\,cm^{-2}$; $\Delta E_{\sigma=0}$ in volts.
[b] From Refs. 119 and 481.
[c] From Refs. 270 and 482.
[d] From Refs. 188 and 483.
[e] From Refs. 89 and 486.
[f] From Ref. 479.
[g] From Refs. 190 and 480.
[h] (0001) face.
[i] From Refs. 484 and 485.

2. The Interaction Parameter of Metal Surfaces

No independent information is available about the interaction parameter $\delta\chi^M$. The Volta potential at the potential of zero charge may provide a clue to its estimation. For $\sigma = 0$, the condition[2]

$$\Delta\phi_{\sigma=0} = (\psi^M + \chi^M) - (\psi^S + \chi^S) \quad (24)$$

applies. From Eqs. (1) and (2), since $g(\text{ion}) = 0$ at the pzc,

$$g^M(\text{dip}) - g^S(\text{dip}) = (\chi^M - \chi^S) - \Delta\psi_{\sigma=0} \quad (25)$$

$\Delta\psi_{\sigma=0}$ is the Volta potential at the pzc and, as a rule, it differs from zero. In this respect, some opposite views have been held[57,62,191] but it is now generally understood[2,15,61] that at the pzc only the metal–solution interface is uncharged, whereas the metal–vacuum and the solution–vacuum interfaces need not be uncharged. $\Delta\psi_{\sigma=0}$ is an experimentally measurable quantity.[2] Measurements for aqueous solutions were first made by Klein and Lange.[192] More accurate values were obtained by Randles.[71] At the pzc its accepted[3,97] value is[68] -0.26 V.

The interaction terms may be introduced into Eq. (25), which now takes the form

$$\Delta\psi_{\sigma=0} = \delta\chi^M - \delta\chi^{H_2O} \quad (26)$$

Thus, $\Delta\psi_{\sigma=0} \neq 0$ whenever the contact between a metal and a solvent leads to some finite reorganization of the surface dipolar layers. Since for water on Hg, $\delta\chi^{H_2O} = -0.06$ V, Eq. (26) gives for $\delta\chi^{Hg}$ the estimated value of -0.32 V. This is taken to mean that the contact between Hg and water, irrespective of the orientation of dipoles which is accounted for in $\delta\chi^{H_2O}$, reduces the work function of the metal.

The derivation of $g^S(\text{dip})_0$ as made in this work is independent of any assumption about the nature and magnitude of $\delta\chi^M$. However, knowledge of a possible value for $\delta\chi^M$ is important for understanding the $E_{\sigma=0}$ vs. Φ relationships, especially for transition metals for which no independent determination of water orientation is available. If the linearity of the plot in Fig. 5 is accepted to a first approximation, then according to Eq. (9) this can only mean strictly that the term $(\delta\chi^M - \delta\chi^{H_2O})$ remains constant through the d metal group. It has been suggested[67] that $\delta\chi^M$

may be a result of image effects.[193] In such a case, $\delta\chi^M$ is expected to be more negative on more hydrophilic metals, but this is very unlikely because $\delta\chi^M$ and $\delta\chi^{H_2O}$ appear, on the contrary, to compensate for each other. However, the ultimate effect of imaging of dipoles on work function changes has been calculated[194,195] to be negligible.

Local change in electron density is to be expected in adsorption phenomena,[196,197] even without the formation of a real chemibond. Thus, an alternative suggestion[10] is that $\delta\chi^M$ may originate from some repulsion of surface electrons as a consequence of a change of the dielectric environment. The hypothesis of a metal-invariant $\delta\chi^M$ may be more reasonable in this case. Consistently, the idea that, for the same metal, $\delta\chi^M$ may be independent of the nature of adsorbates (H_2O or organic substances) is even more reasonable if the latter exhibit very similar high-frequency dielectric constants.

Some indirect information on the nature of $\delta\chi^M$ is thought to be derivable[15] from measurements on adsorption of rare gases on metals. The mechanism of such an adsorption process has been debated repeatedly.[198–204] An explanation in terms of surface field at the metal[205] has been proposed.[206] Roughness effects on an atomic scale have been suggested[198] as enhancing possible surface fields. According to the latter view, metals with the same melting point are expected to exhibit the same work function change upon adsorption of rare gases. This theory has been in turn questioned on the basis of two alternative models, based on surface-induced dipole formation[207,208] and charge-transfer–no-bond[209,210] interaction, respectively. It appears that although the physical bases of the two models are different, the practical outcome is the same. In fact, if a charge-transfer–no-bond mechanism is assumed, it has been calculated[209] that the observed $\Delta\Phi$ should correspond on the average to partial transfer of a charge of about $0.1e$. Operatively, this corresponds to very small displacements of electrons with respect to nuclei, i.e., to a small surface-induced dipole.

It is interesting to note that the adsorption of Xe on metals follows[209,210] a reactivity scale which parallels the hydrophilicity scale of electrodes. This is not surprising in view of the fact that the chemical nature of the surface is always the essential factor governing the reactivity. However, this does not clarify the role of $\delta\chi^M$ which is included, if at all, in the measured $\Delta\Phi$. The work function drop at Hg upon adsorption of rare gases has been found to be[198,211] -0.23 to -0.27 V and this value might be regarded as in satisfactory agreement with $\delta\chi^M$ found above, provided that $\Delta\Phi$ in such a case measures essentially $\delta\chi^M$, which requires the assumption of absence of molecular polarization. However, $\Delta\Phi = 0$ as found with Ca and Na can hardly be explained in the same context, whereas this result might be understandable in terms either of the charge-transfer–no-bond or the polarization model.

The question of the possibility of obtaining information on $\delta\chi^M$ from some experimental measurements thus remains unsettled, although the satisfactory agreement between calculated and measured heats of Xe adsorption on metals[209,210] provides some evidence that $\delta\chi^M$, explicitly neglected in calculations, may in fact be either constant or negligibly variable from metal to metal. A direct experimental method which may be suggested for the estimation of $\delta\chi^M$ on metals other than Hg is the measurement of the Volta potential between electrode and solution. Since $\delta\chi^{H_2O}$ has been shown to be derivable independently, $\delta\chi^M$ could easily be obtained from Eq. (26). Unfortunately, only liquid electrodes can be used with techniques suitable to ensure reliability with regard to surface contamination. Some direct measurements with Ga and In–Ga alloys, providing highly hydrophilic surfaces, could be a great help in deciding about the possible constancy of $\delta\chi^M$.

3. Additional Evidence for the Hydrophilicity Scale

Indirect additional evidence for the existence of a hydrophilicity scale of metals (as proposed in Table 2) rests on the basic idea that orientation of a given molecule on a surface and strength of its adsorption run parallel. Consistently, the wetting of Ga by water is certainly better than that of Hg as reflected by the surface tension drop upon metal–water contact.[212] Similarly, the heat of water adsorption on Pt has been found[213] to be greater than on Au,[214] and the latter value appears to be of the same order of magnitude as that on Hg.[215]

Work function drops upon water adsorption cannot be interpreted unambiguously. Qualitatively, the negative values[216-220] of $\Delta\Phi$ on Fe, Ni, Al, and Ga provide evidence that adsorption occurs preferentially through the oxygen atom. Moreover, the value is almost the same for the first three metals, in agreement with Eq. (13) and with the implicit idea[3,10,15] that the most electropositive sp metals tend to merge with the d metal group. A decrease in Φ upon water adsorption may also be inferred[221] for Ni from the shift of potential occurring when the electrode is immersed in aqueous solution, measured by a vacuum electrochemical technique. Also, $\Delta\Phi$ on Cu is lower[217] than on Fe and Ni,[216] in agreement with the much lower hydrophilicity predicted[222] from the pzc of Cu. However, quantitatively, the observed $\Delta\Phi$ are higher (about -1.1 eV for the transition metals) than expected from $0.40 + 0.32 = 0.72$ eV, viz., $e(g^{H_2O}(dip)_0 + \delta\chi^M)$. One reason may be sought in the absence of a bulk effect on a monolayer of water adsorbed on a metal surface from the gas phase. No quantitative account may, however, be given for this. Another reason may be the lack of any electrical control of the state of the surface during water adsorption from the gas phase. Such a process thus occurs at $q = 0$ (q is the total charge density[26,223]) rather than at $\sigma = 0$. If some partial charge transfer takes place, this certainly will contribute to the final value of $\Delta\Phi$. Proof of this may be found[224-227] in the *increase* in $\Delta\Phi$ upon water adsorption on such d metals as Pt, W, Ta, Cr which is understandable in terms of oxidation of the surface with H_2 evolution.[228] The same phenomenon occurs[217,229] on Fe at higher temperatures. Decrease in Φ for adsorption of H_2O on Pt has been calculated[230] theoretically but the magnitude of the $\Delta\Phi$ has been found to be about half that suggested in this work.

There exists contradictory evidence for the hydrophobicity of Au. Electrochemical data provide[3,149,214,231] evidence for weak Au–H_2O interaction. Constrasting results are, however, obtained in studies on water adsorption from the gas phase. Some of them[232] give $\Delta\Phi \simeq -1$ eV. These and other results[233,234] would place Au together with d metals. However, the surface of Au certainly shows[235] very little affinity for oxygen although this may be regarded[236,237] as a consequence of high activation energy for molecular dissociation. Recent results[238] suggest that Au adsorbs water quite weakly. The inconsistency of the above results may be due to the critical features of the surface of Au. It is known[239-241] that (100) surfaces can evolve by reconstruction to (111) surfaces under certain conditions. On the other hand, the reverse may occur in electrochemical experiments.[242] It has been suggested[15,231] that this may be the main reason for the disagreement between "electrochemical" and directly measured work functions for Au. Au is certainly an sp metal when negatively charged since the Fermi level is inside the sole sp band.[243] Inorganic chemistry suggests,[244] however, that Au should be regarded as a transition metal and this is certainly true since Au possesses empty d levels when in its ionic forms. Consistently, on very rough surfaces, it may be[245-247] that at some sites Au atoms exhibit transition metal characteristics. Similarly, when positively charged, because of the closeness of the top of the d band to the Fermi level, d levels may become emptied at the surface.

More substantial indirect evidence for the hydrophilicity scale may be obtained from studies on adsorption of ions and organic substances. The relationship between ion adsorption and hydrophilicity has been discussed recently.[149] If adsorption is envisaged as a solvent replacement reaction,[25] then metals binding water molecules more strongly are expected to adsorb ions more weakly. Inspection of ΔG°_{ad} and adsorption potential shifts shows that the adsorbability scale for a given ion on a series of metals is reversed with respect to the hydrophilicity scale. The case of I^- adsorption from aqueous solutions is shown in Fig. 10. Au can be seen to be located at the lower end of the hydrophilicity scale, in agreement with the expectation based on the $E_{\sigma=0}$ value.

Adsorption of organic substances is often more relevant[54,111] than ionic adsorption for gaining insight into the energetics of water adsorption because ΔG° values in the latter case are usually comparable so that possible differences from metal to metal are hardly obscured, as in fact may happen[149] with adsorption of ions. Experimental data from the Russian school show[54] that adsorption of organic substances is consistently weaker on more hydrophilic metals. Some data are collected in Table 4.

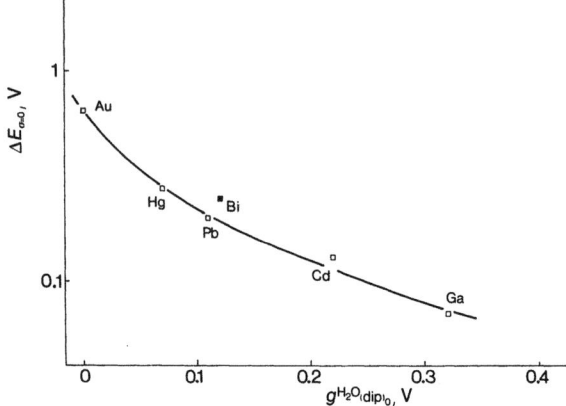

Figure 10. Semilog plot of potential shifts upon I^- adsorption on electrodes from $0.1\,N\,I^-$ aqueous solutions against the surface potentials of water at the same metals. For the source of data of potential shift, see Ref. 149.

Data for adsorption of organic substances on Au are quite scanty. It has, however, been shown[53,54] that desorption of a given substance on various metals occurs approximately at the same negative charge. As Fig. 4 shows, the same negative charge is reached[14] on different metals at different rational potentials

Table 4
Standard Free Energy of Adsorption (kcal mol^{-1}) of Some Organic Substances at sp-Metal/Aqueous-Solution Interfaces at the Potential of Zero Charge

Substance	Hg^a	Pb^b	Bi^c	Sn^d	In–Ga^e	Cd^f	Zn^{g,h}
Ethanol	1.92	—	1.87	1.76	1.48	1.80	—
n-Propanol	3.05	—	2.65	2.84	2.73	2.30	2.4
n-Butanol	3.83	—	3.47	3.68	3.49	3.07	3.02
n-Pentanol	4.58	4.3^i	4.26	4.27	4.11	3.86	3.72
Cyclohexanol	4.62^j	4.48	4.16^k	4.48^l	—	4.0^m	3.8^n
Phenol	6.78^o	5.4	5.18^k	—	—	—	—
Aniline	5.31^p	5.30	—	—	—	—	—

^{a} From Ref. 119. ^{g} (0001) face. ^{l} From Ref. 486.
^{b} From Ref. 482. ^{h} From Ref. 484. ^{m} From Ref. 480.
^{c} From Ref. 188. ^{i} From Ref. 270. ^{n} From Ref. 485.
^{d} From Ref. 89. ^{j} From Ref. 481. ^{o} From Ref. 487.
^{e} From Ref. 479. ^{k} From Ref. 483. ^{p} From Ref. 488.
^{f} From Ref. 190.

depending on the extent of the curvature of the initial portion of the charge–potential relationships. In particular, the more hydrophilic the metal, the more negative the rational potential for desorption of organic substances. This has been clearly rationalized[14,15,248] through correlations. The concept may be better illustrated by Fig. 11 where theoretical coverage–potential curves are drawn for two metals with the same absorbate. If the maximum coverage is the same, the rate at which the coverage decreases with potential will be higher on the more hydrophilic metal because[111,249] the difference in polarizability between adsorbate and solvent on the surface is higher. Since the desorption peak develops at about half the maximum coverage value,[110] it is easy to understand why this will occur at different potentials on the two metals, the value observed with the less hydrophilic solid being the more negative. Now, in the case of Au, some data on adsorption of pyridine, whose polarizability[250] is of the same order of magnitude as that of long-chain alcohols[119] (similar C_1),[111] show[251,252] that desorption takes place at rational potentials close to that of Hg as expected on the proposed hydrophilicity scale. It has been suggested[53] that desorption peaks are governed not only by the difference in hydrophilicity but also by particle–particle lateral interactions on the surface, so that quantitative comparisons should be made cautiously. However, lateral

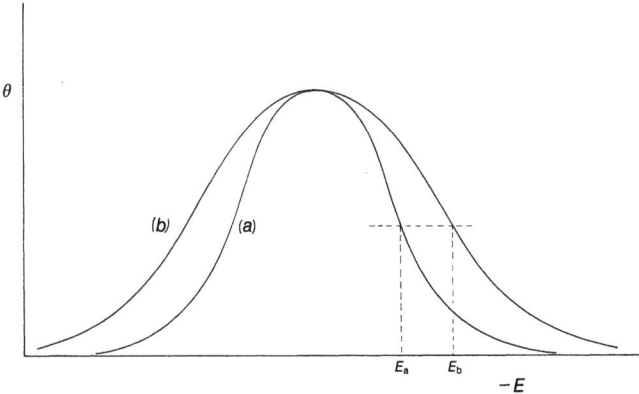

Figure 11. Sketch to show the relation between hydrophilicity of metals and potential where cathodic peaks for desorption of organic substances develop. (a) More hydrophilic metal; (b) less hydrophilic metal. Coverage–potential curves have been normalized at the point of maximum adsorption.

interactions are expected to depend on the state of water at the interface, so that they should be a function of the metal. Thus, at least qualitatively, the general picture emerging from adsorption studies on different metals seems to be rather clear-cut. Metal–water interactions must be accounted for in any aspect of the electrochemical behavior of metals. Thus the adsorption of any species is expected to be affected by the metal–water bond strength. Effects on kinetics are thus expected as well. Analysis of adsorption data of H and O on metals confirms[248] that d metals bind water more strongly than sp metals. This aspect has been discussed extensively elsewhere[15,248] and will not be dwelt on further here. Attention will be restricted to double-layer aspects only.

4. Dielectric Aspects

One of the fundamental aspects that a modern view of the double layer must be able to explain is the increase in capacity as the charge on the metal approaches zero from the negative side.[53,54] With reference to Eqs. (1) and (2) and the relevant discussion, the following relation may be written:

$$\partial \Delta\phi/\partial\sigma = \partial g(\text{ion})/\partial\sigma + \partial g(\text{dip})/\partial\sigma \tag{27}$$

Since $g^M(\text{dip})$ is assumed to be charge invariant, then

$$\partial \Delta\phi/\partial\sigma = \partial g(\text{ion})/\partial\sigma - \partial g^S(\text{dip})/\partial\sigma \tag{28}$$

It follows that

$$1/C_i = 1/K_{\text{ion}} - 1/C_{\text{dip}} \tag{29}$$

where C_i is the experimental inner layer capacity and the differential capacity C_{ion} has been replaced by the integral capacity K_{ion} [cf. Eq. (5)].

According to Eq. (29), any change in C_i with charge must necessarily be attributed to some change in C_{dip}, regarding K_{ion} as a constant. Otherwise, it is necessary to start from a model including some additional components of the potential drop. For any model to be reliable, it must be able to predict the observed dependence of C_i at constant charge on the nature of the metal, as Table 5 shows for the potential of zero charge. Figure 12 shows the comparison of the negative branches of the capacity curves for Hg,[64] Ga,[36] and In–Ga.[99] A potential shift associated with specific adsorption of ions has been considered[25] as a possible additional component of the interfacial potential drop. However, although even F^- ions are known[253] to become adsorbed on Hg, there is overwhelming evidence[254] that this occurs only at charges more positive than about $5\,\mu C\,cm^{-2}$. This rules out that specific adsorption may have any effect on capacity curves at charges negative to the potential of zero charge.[64] It has been shown above that ionic adsorption decreases on more hydrophilic metals so that the possibility of interference is even lower in that case. Actually, F^- is known[53,100,255] to adsorb on Ga, In–Ga, and Sn presumably because of specific chemical reasons, but the data for Ga in Fig. 12 refer to electrolytes for which any specific adsorption can be ruled out at $\sigma \leq 0$.

If the model implied in Eq. (29) is retained, any increase in C_i along a series of metals can be understood only in terms of an

Table 5
Values of Inner-Layer Capacity of sp-Metal/ Aqueous-Solution Interfaces at the Potential of Zero Charge

Metal	C_i ($\mu F\,cm^{-2}$)
Au	30^a
Tl	24^b
Hg	28^c
Pb	33^d
Bi	26^e (36 corr.)^f
Sn	39^g
Cd	53^h
In–Ga	60^i
In	80^j
Ga	$135^{k,l}$

^{a} From Ref. 403.
^{b} From Ref. 37.
^{c} From Ref. 64.
^{d} From Ref. 91.
^{e} From Ref. 404.
^{f} From Ref. 8.
^{g} From Ref. 490.
^{h} From Ref. 489.
^{i} From Ref. 101.
^{j} From Ref. 92.
^{k} Experimental capacity in $1\,N\,Na_2SO_4$.
^{l} From Ref. 35.

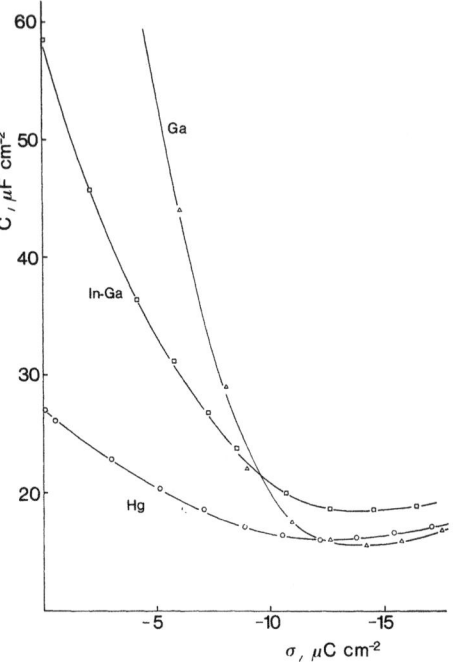

Figure 12. Experimental capacity–charge curves in 0.5 M Na_2SO_4 for some metals. Original data, by the courtesy of Dr. I. A. Bagotskaya.

increasing contribution from water- reorientation. As Table 5 shows, the hydrophilicity scale in terms of C_i parallels the hydrophilicity scale in terms of $g^{H_2O}(dip)_0$ so that the whole picture is entirely consistent, apart from some quantitative details. Although the apparent dielectric constant can be quantitatively introduced[4] only after the differential capacity has been turned into the integral capacity, it is, however, possible to state qualitatively that, parallel to Eq. (28), the following equation holds:

$$\varepsilon_i = \varepsilon_m + \varepsilon_o \tag{30}$$

where ε_m is the apparent dielectric constant related to molecular polarization effects and ε_o is related to orientation polarization effects. Now, since ε_m has been assumed to be metal independent, it follows that ε_o increases with the metal hydrophilicity. This is in keeping with the idea that water can become reoriented as the charge is changed, the more hydrophilic is the metal surface. Experimental data for *sp* metals seem thus to conform well to a chemical rather than to an electrostatic view of the double layer.

It has been suggested above that ε_i may be anisotropic. Since ε_m has been assumed to be isotropic, anisotropy may arise for ε_0 as a result of the preferential reorientation of water molecules in a plane perpendicular to the electrode. Lateral repulsion between adsorbed ions should thus be expected to decrease as ε_o increases, rather than the opposite. The very few available data,[256-258] however, suggest tentatively that the second virial coefficient in the isotherm for I$^-$ adsorption may follow the opposite pattern. This particular aspect is thought to deserve further careful experimental analysis.

In the above context, it is surprising that the interaction constants for the adsorption of a given alcohol on different metals do not depend[190] on the nature of the metal. However, it is more expedient to compare the slope of the linear change[259] of the interaction term with potential. A higher slope is, in fact, associated[190] with more hydrophilicity of the metal.

As discussed elsewhere,[122] the interaction constant a may be better regarded as the resultant of three combined interactions: particle–particle, particle–solvent, and solvent–solvent. The interaction term may thus be envisaged as a measure of the ability of the organic substance to *dissolve* in the surface layer of water. An increase in a as the potential is made less negative corresponds to an increased tendency for dissolution of the adsorbate to occur in the surface water layer. The slope of a as a function of E is higher for Cd than for Hg and this provides some evidence for a more marked transition of water from a disordered to an ordered state on Cd than on Hg. Other data for adsorption of alcohols on metals unfortunately do not include such parameter.

An organic substance which lends itself especially well to be used as a probe for the study of the state of water at electrodes is thiourea.[260] The molecule is rigid and becomes strongly adsorbed through the S atom with a resulting linear adsorption potential shift and almost linear dependence of ΔG_{ad}° on charge. If the dipole is assumed to adsorb perpendicularly to the surface, calculations can be made leading to an estimation of ε_i as a function of charge.[261-263] Although ΔG_{ad}° is not observed to change very much from metal to metal because the sulfur–metal interaction increases in parallel with the oxygen–metal interaction, the picture emerging from all the data[260-269] conforms well to the expected hydrophilicity scale. In particular, the second virial coefficient of thiourea adsorbed on sp metals increases in the series[261,262,265] $Hg < Bi < Sn$. Since the hydrophilicity increases in the same direction, this result may mean that particle–particle repulsions are less screened because the apparent dielectric constant of the solvent increases in the above series perpendicularly to the electrode. Alternatively, the increase in the virial coefficient with hydrophilicity may provide some evidence for increasing attractive interaction between thiourea and water due to the increasing opposite orientations of the two species.

It is emphasized here that the procedure adopted in the present work to derive the hydrophilicity scale gave correct predictions[3] before the publication of relevant experimental data for some of the metals.[53] A particular case is that of Pb, for which it was suggested[91,270,271] for some time that it behaves quite similarly to Hg, whereas recent measurements[272,273] by optical techniques indicate beyond any doubt that it is more hydrophilic. Experimental evidence on the basis of electrochemical measurements has become clearer only recently.[14,15,54]

IV. DESCRIPTION OF THE STRUCTURE OF INTERFACIAL WATER

Hitherto, no detailed description of the surface water layer has been introduced. It is useful to give at this point some discussion of models for electrode/water interfaces. Such models must be able to reproduce a number of things: the shape of capacity curves in the absence of specific adsorption, the sign (if not magnitude) of $g^{H_2O}(dip)_0$ and of its temperature coefficient, the temperature coefficient of the capacity, the rise of capacity at strongly positive and negative charges, and the effect of the nature of the metal. The reliability of a given model increases with the increasing number of facts it is able to account for. For this reason, models able to reproduce only a limited number of features will not be discussed here. Levine et al.[55] proposed a sophisticated model able to predict the hump at positive charges coexisting with a positive value of $g^{H_2O}(dip)$. However, the model now appears[274] to be obsolete in the light of the new developments. The most peculiar feature of capacity curves is, in fact, the rise at $\sigma > 0$ increasing as the hydrophilicity of the metal surface increases. For a detailed discussion of the earlier models readers are referred to Reeves' chapter[17] and the review by Conway.[17b]

1. Theoretical Structural Models

A conceptually new model was proposed first by Damaskin and Frumkin[4] (DF). The inner layer is treated as a monolayer of water. Two possible species of water are considered, aggregates reorienting freely under the field and molecules chemisorbed through the oxygen atom in a fixed orientation. Reorientation of aggregates is treated with the equation first proposed by Watts-Tobin.[49] The energy of adsorption of chemisorbed molecules is assumed to be a linear function of charge, analogous to the case of thiourea.[260] The equation proposed for the dipole potential drop has the form

$$g^{H_2O}(dip)_\sigma = (\gamma_1/K_1)\tanh(\sigma/\gamma_1) + (\gamma_2/K_2)\exp(\sigma/\gamma_2) \quad (31)$$

where γ_1 and γ_2, K_1 and K_2 have the dimensions of charge and capacity, respectively, and depend on both the population of the given species and the normal component of its dipole. The first term on the right-hand side of Eq. (31) accounts for the reorientation of the aggregates, and the second term for the contribution given by chemisorbed molecules.

The capacity is given by

$$1/C_i = 1/K_{ion} - (1/K_1)\text{sech}^2(\sigma/\gamma_1) - (1/K_2)\exp(\sigma/\gamma_2) \quad (32)$$

It follows from Eq. (32) that in the absence of water chemisorption, the total capacity is represented by a bell-shaped curve. In the absence of aggregates, the capacity increases monotonically as the charge becomes more positive, from its minimum value at some negative values of charge (Fig. 13). Combination of the two effects leads to a capacity curve with a hump at small positive charges even though the preferential orientation of dipoles at the

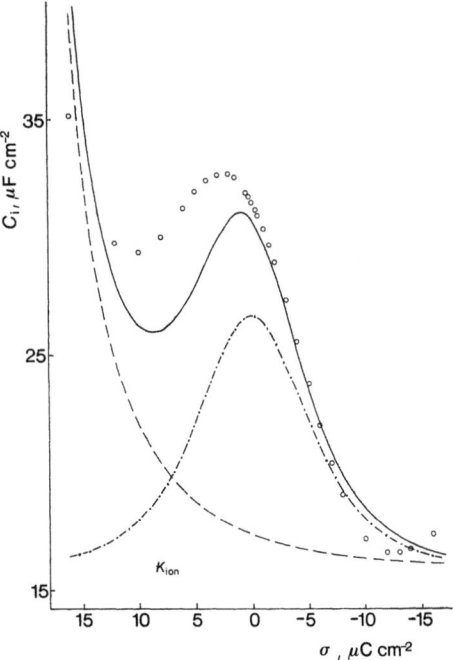

Figure 13. Experimental and calculated inner-layer capacity–charge curve for Hg in contact with aqueous solutions of not specifically adsorbed electrolytes at 0°C. Capacity associated with clusters (–·–), with chemisorbed molecules (– – –), and total capacity (——) according to the DF model.[4] (○), experimental points. K_{ion} is the capacity associated with clusters at fixed dipole orientation.

pzc is with oxygen towards the metal. This model has been shown to predict correctly the increase in capacity for $\sigma \geq 0$ when passing from Hg to Cd.

It is very interesting that Eq. (32) is formally similar to the equation derived by Watts-Tobin[49] from his own model. The difference is in the physical significance attached to the exponential term. Watts-Tobin suggested that the capacity rise at positive charges could be associated with adatom formation (incipient dissolution). The DF model rests on the idea that the capacity rise is associated with chemisorption of water molecules. It has been shown in Fig. 8 that water chemisorption can be regarded as a precursor to surface oxidation. It is thought that the similarity between the results of the two above models is due to the fact that they treat the same physical process but from different points of view.

Water chemisorption presumably occurs on a surface where the ion cores are very little screened by surface itinerant electrons.[205] This means that water chemisorption and adatom formation run parallel, the latter being in some way a prerequisite for the former. In this context, Watts-Tobin's model can even predict that the capacity rise will be steeper on more hydrophilic metals. A measure of the trend to form adatoms may be given by the standard potentials of metals. E^0 depends[13,93,94,275] in fact on the evaporation and ionization free energies of metals and on the hydration free energy of the metal ions formed. Ultimately, the formation of adatoms involves qualitatively the same factors.

Table 6 shows that if metals are arranged in the order of increasing hydrophilicity, E° [276] values run parallel, apart from obvious quantitative differences due mainly to the choice of the ionic form that the electrode equilibrium should be referred to. It is thought that, in the present context, the early idea of Watts-Tobin on the origin of the capacity rise turns out to be revalidated.

The DF model has been improved by Parsons[6] (P). This author has raised the question that the water population in the above model is left to increase freely as the charge is made more positive. Further, he pointed out that the DF model is unable to

Table 6
Standard Potentials[a] of Some sp Metals in Aqueous Solutions

Metal	$E^°$ [V(nhe)]
Au	1.50
Hg	0.79
Pb	−0.13
Sn	−0.14
In	−0.34
Cd	−0.40
Ga	−0.53

[a] From Ref. 276.

predict the capacity rise at negative charges. In order to allow for the above shortcomings, Parsons proposed a four-state model for water at the interface. Components are clusters and free molecules, both freely reorienting under the field but with different interaction energies with the metal and the bulk liquid. The equilibrium between the different species is properly accounted for by the Boltzmann distribution law. The surface potential associated with the solvent is given by

$$g^S(\text{dip})_\sigma = -\rho\mu N^c_+/\varepsilon + \rho\mu N^c_-/\varepsilon - \mu N_+/\varepsilon + \mu N_-/\varepsilon \quad (33)$$

where N^c and N, with subscripts, indicate the number of water molecules in the two states and in the two allowed opposite orientations, respectively. N^c and N depend on both charge and metal–particle interaction energy. ρ is the ratio of the dipole moment per molecule in clusters to that of a free molecule.

This model predicts capacity curves having shapes quite similar to those of the experimental ones and with satisfactory proportions between minimum and maximum values. The position of the characteristic points is, however, somewhat shifted, as Fig. 14 shows. However, the model predicts correctly the capacity rise at negative charges. This is due to reorientation of free molecules combined with cluster disruption. The model has not been tested by comparison with data for other metals.

The original DF model has been revised first by Damaskin, Palm, and Salve[8] (DPS) and again by Damaskin[9] (D). Both the DF and P models do not account for lateral interaction between water molecules in the monolayer. This was accounted for in pairs in an earlier model by Bockris et al.[25] (BDM). DPS have accounted for this by pointing out that in the expression for the electric field, the effect exerted on a dipole by all other dipoles in the layer must also appear. This is introduced by means of the equality between field in terms of σ and field in terms of potential drop:

$$X = \Delta\phi/d_i = 4\pi\sigma/\varepsilon_i - g^S(\text{dip})/d_i \quad (34)$$

Equation (34) is a direct consequence of the splitting of the potential drop in Eq. (1) into two components; d_i is the thickness of the inner layer.

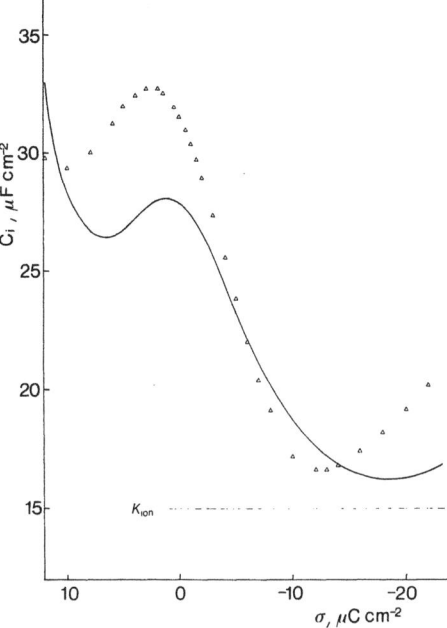

Figure 14. Experimental (△) and calculated (——) inner-layer capacity–charge curve for Hg in contact with aqueous solutions of not specifically adsorbed electrolytes at 0°C. Theoretical curve according to the P model.[6]

Equations derived for this model are similar to those obtained in the DF model although parameters are a little different. Figure 15 shows that the agreement between calculated and experimental capacity values is more satisfactory. In particular, the agreement is remarkably good for Hg and Cd. Calculated capacities are higher in the case of Bi. Improvement is obtained by introducing a capacity term for the metal surface associated with some space-charge effect when $\sigma \geq 0$. Possible interferences by the semiconducting properties for Bi were actually envisaged in previous papers.[53,149] However, the model again is unable to predict the rise at negative charges.

Damaskin[9] (D) has gone over the model again recently. Following Parsons' suggestion,[6] he has introduced equilibrium exchanges between the different species on the surface. A number of other corrections have also been made: (i) the total

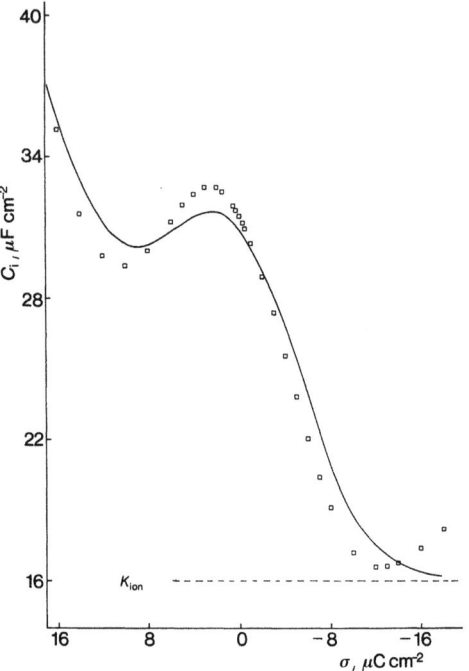

Figure 15. Experimental (□) and calculated (——) inner-layer capacity–charge curve for Hg in contact with aqueous solutions of not specifically adsorbed electrolytes at 0°C. Theoretical curve according to the DPS model.[8]

number of molecules has been calculated on the basis of the sum of the areas of associated and of chemisorbed molecules; (ii) a discreteness coefficient has been introduced to optimize the field generated by dipoles; (iii) the possibility for a temperature-dependent size of aggregates has also been taken into account.

Results[9] do not attest definite improvements with respect to the DPS model. At a given T the agreement with experimental data is apparently less satisfactory. The distortion observed in the curves calculated by the P model is again found here, although the general shape is correct. Again, the model neglects the rise at negative charges. No comparison is made with other metals.

A quite different model has been proposed by Bockris and Habib[11] (BH). It is substantially a modification of the Bockris–Devanathan–Müller (BDM) model,[25] which in turn was the result of the development of some ideas put forth first by Bockris and Potter[277] about 25 years ago. The monolayer of water is described in terms of a three-state model: monomers freely reorienting under the field with different metal-particle interactions in the two opposite orientations, and dimers originating from the juxtaposition of two "up" and "down" monomers. The net dipole of dimers is assumed to be zero. Lateral interaction between dipoles is accounted for in pairs. Dimers are assumed not to be disrupted by the field in the usually investigated range of charges.

The surface potential associated with the solvent at the electrode is given by

$$g^S(\text{dip})_\sigma = \frac{4\pi N_m \mu}{\varepsilon} \frac{\exp(2x+b)-1}{\exp(2x+b)+1} \tag{35}$$

where N_m is the monomer concentration on the surface, μ the normal component of its dipole moment, x contains the energy of both monomer-monomer and field-monomer interaction, and $b = -\Delta \Delta G_c/kT$ accounts for the difference in metal-monomer interaction energy in the two limiting orientations. The capacity comes out to be a very complicated expression but the relevant point is that, in this case, the choice of parameters is such that C_{dip} comes out high enough (series relationship) not to have any effect on C_i whose behavior must thus be associated not with dipole reorientation but rather with some other factors. The interest in this model is in the fact that it provides an alternative explanation for the hump. This aspect will be discussed in more detail later.

2. Comparison with Experiments

As Table 7 shows, all models predict a positive value for $g^{H_2O}(\text{dip})_0$ corresponding to some preferential orientation of water with oxygen towards the metal. Calculated values are lower than that estimated in this work. The lowest values are, surprisingly, obtained with models[6,9] allowing for aggregate-monomer exchange equilibrium. Where different metals have been investigated, models[4,8] correctly predict a value of C_i increasing in the sequence Hg < Bi < Cd. The P model is as well expected to be

Table 7
Surface Potential of Water at Metal/ Aqueous Solution Interfaces as Calculated from Models at 0°C

Model	$g^{H_2O}(dip)_0$ (V)		
	Hg	Bi	Cd
DF[a]	0.04	—	0.16
P[b]	0.005	—	—
DPS[c]	0.03	—	—
D[d]	0.003	—	—
BH[e] (25°C)	0.03	—	0.06
This work	0.10	0.15[f]	0.25[f]

[a] From Ref. 4.
[b] From Ref. 6.
[c] From Ref. 8.
[d] From Ref. 9.
[e] From Ref. 11.
[f] Calculated from the value at 25°C by assuming $\Delta_{Hg}^{M} E_{\sigma=0}$ to be independent of T.

able to reproduce the correct order of hydrophilicity because the metal–water bond strength is properly accounted for in Eq. (33).

According to the BH model[11] $g^{H_2O}(dip)_0$ is a linear function of $\Delta\Delta G_c$ in agreement with the findings in Fig. 8. $\Delta\Delta G_c$ is essentially computed in terms of difference in image and dispersion energies between metal and "up" and "down" monomers. The model predicts[11,278] some increase in $g^{H_2O}(dip)_0$ in the sequence Hg < Cd < Zn which, however, is much lower than found in this work or predicted by other models. It is possible that image and dispersion energies do not account entirely[279] for the complex chemical nature of metal–water interactions.

The expression for the dispersion energy contains[11] the polarizability of the solid surface. Different interaction energies for different metals may thus be obtained in principle. Similar calculations have been performed by Gardner[280] to support the metal-dependent water orientation theory.[3] He has shown that it is in fact possible to calculate that a decrease in electronegativity along a transition row is paralleled by some increase in water–metal interaction energy. The role of electronegativity[281] in water adsorption has been discussed at length elsewhere.[3,15] However, Gardner's calculations turn out in fact to be hardly useful quantitatively in support of the above theory since the latter actually assumes constancy in water orientation on d metals. Calculations would give much more useful indications in the case of sp metals. Now, Gardner's calculations predict a higher interaction energy for Cu than for Zn, which is contrary to the experimental evidence.[3,15] This means that a model based on a classical view of the metal–water interaction energy is probably unable to give a complete quantitative account of it.

(i) Inner-Layer Capacity at the Minimum

K_{ion} turns out to be a determining factor in the procedure suggested in this work. It has been derived from the limiting slope of the charge–potential curve at negative charges (Fig. 3). For NaF solutions, K_{ion} is thus about 17 $\mu\text{F cm}^{-2}$. It is negligibly dependent on cation size.[25] This has been explained[25,282] in terms of cations embedded in a medium with higher apparent dielectric constant than that of the first water monolayer. This view has led BDM[25] to formulate the hypothesis that two water layers with different ε lie in fact between the metal surface and the OHP. However, such a model leads to consequences for which no convincing experimental evidence exists. First of all, if ε is not constant within the inner layer, $\Delta\phi$ cannot be taken as a linear function of thickness. One objective difficulty with the models of Bockris is that separately they are apparently able to account for single aspects of double-layer behavior, but they are, in fact, not entirely consistent when treated together. For example, ε_i is assumed[283] to vary[284] within the inner layer across more than one monomolecular layer, but $g^S(\text{dip})$ and C_{dip} are then calculated[11] by taking into account only the first monlayer of solvent. It seems possible that if the constant value of the minimum of C_i at negative charges is explained in terms of a bilayer of solvent molecules, then the second layer must also contribute to both $g^S(\text{dip})$ and C_{dip}.

In the other models, K_{ion} is taken almost equal to the capacity minimum and precisely 16 $\mu\text{F cm}^{-2}$ in the DF[4] and D[9] models and 15 $\mu\text{F cm}^{-2}$ in the P model.[6] These values are in reasonable agreement with the value that can be calculated with the plane capacitor formula $K_{ion} = \varepsilon_i/4\pi d_i$ if d_i is taken as the thickness of one water molecule and ε_i is given the high-frequency value. Additional evidence in favor of an inner layer one water molecule thick has been presented and discussed elsewhere.[10] The bilayer model is expected to predict lower capacity drops upon adsorption of organic substances in that only the monolayer closer to the metal should be replaced by the adsorbate. The idea of a water bilayer has been put forth also in connection with adsorption of some organic substances.[285,286] The value of C_1 is in this case expected to be charge dependent. Results apparently supporting this may be criticized because C_1 values were derived from a not entirely justified[287] linear extrapolation of C_i vs. ϑ plots.

The presence of some water molecules in the adsorbate monolayer has been postulated[189] to explain the change in C_1 for alcohol adsorption on different metals. However, this view does not imply a water bilayer at $\vartheta = 0$. Presently, the experimental evidence[10] is in favor of a monomolecular inner layer but it is thought that this point should deserve further careful attention. In fact, if the OHP is only one water molecule distant from the metal surface, at full coverage with an organic adsorbate ions in the OHP would touch the organic layer with an additional decrease in energy approximately equal to half the free energy of transfer from water to the given organic medium.[288]

(ii) Thickness and Apparent Dielectric Constant of the Inner Layer

The inner-layer thickness is taken as 0.31 nm in the P model[6] and 0.33 nm in the DF model.[4] Results based on thiourea adsorption suggest[260] a somewhat higher value consistent with the sum of a water molecule diameter and an ionic radius, but the derivation of these data may be vitiated by some approximations about the structure of the adsorbed layer. Further, if cations in the OHP are envisaged as being located in voids between water molecules in the first layer, it is very probable that $d_i < d_{H_2O} + r_{ion} \simeq d_{H_2O}$. The constancy of K_{ion} with cation size may be understood in terms of discreteness of charge[289] at the OHP. The complex path followed[50] by the electric field may be such that the lines of force from the metal surface arrive at the ions in the OHP crossing layers of water beside or even behind the charged particles, which certainly are characterized by higher apparent dielectric constant (Fig. 16).

If the above view is correct, ε_i must be somewhat higher than the value pertaining to high-frequency measurements,[84,85,290] ε_∞. Actually, it has been pointed out[25] that water molecules are largely oriented at the minimum in capacity, hence dielectric saturation should be achieved there. Different values have been suggested[25,48,63] for ε_∞ depending on the available experimental evidence. Thus, the value of 6 has been accepted[11,25,63] for long, and this value is used also in the DF model.[4] The P model suggests[6] $\varepsilon_i = 5.3$. A value of 5.2 has been suggested by Müller[112] as the dielectric constant related to the sole molecular polarizability. More recent measurements[84,85,290-292] give for ε_∞ a value close to 4.5. This if introduced in the formula for K_{ion}, would give for d_i an unreasonably low value.

The value of 4.5 does not correspond, in fact, to the minimum value of the dielectric constant of water. The latter equals the square of the refractive index involving the relaxation of electrons only[84,85] and amounts to about 1.9. It has been suggested[49] that ε_i in the double layer should be equated to 3.1, the limiting value of dielectric constant for ice. However, an additional relaxation phenomenon is in fact related to molecular reorientation.[291]

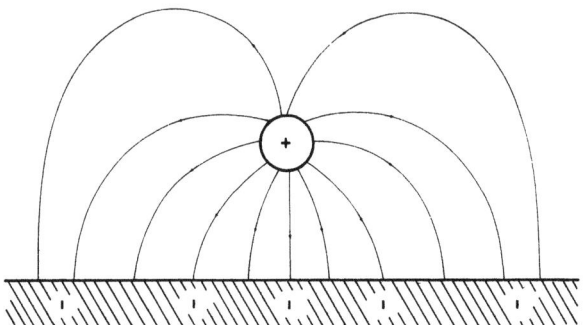

Figure 16. Sketch to illustrate how the electric field between an electrode and an ion can cross solvent layers beside and even behind the charged particle.

Actually, two-bonded molecules can still rotate around the axis of the hydrogen bonds.[290] However, such a relaxation phenomenon can hardly be effective at such small distances of ions from the electrode surface as in the double layer.

The dielectric behavior of water close to electrodes may resemble[293] that of water around charged particles in solution.[294] It is an undisputed fact that ε decreases as the ionic concentration increases[295] as a consequence[296] of the increasing number of water molecules oriented around ions. It seems, however, reasonable to think that the charge concentrated in an ion is always higher than that smeared out on an electrode even under extreme conditions. Further, some ions are able to reorient water molecules detectably, others are unable to do so.[296,297] Consistently, a negatively charged metal may be better compared to a weakly hydrated anion. If discreteness of charge[289] is considered, a strong field will be located around cations in the OHP, but a considerably lower field should be operative elsewhere.[63] In conclusion, on the whole, water molecules at the capacity minimum need not be considered largely oriented[49] with consequent breakage of hydrogen bonds, although they may equally well be considered in a limiting position of orientation and then with reduced possibility of rotation.

If a value within 0.3–0.4 nm is assumed for d_i, then a value ranging from 5.8 to 7.7 results for ε_i from the plane capacity formula. Values close to the upper limit have been suggested on the basis of various indirect evidence,[3,260,298,299] but a value closer to the lower limit may be structurally more probable at the capacity minimum. Here, water molecules should be regarded as still retaining some relaxation possibility; hence dielectric saturation should not be so strong that electrostriction[48,63] would be an unavoidable occurrence.

(iii) *Effective Dipole Moment of Water in the Inner Layer*

The effective dipole moment of water within the double layer is another critical quantity. μ may be taken[6,9] equal to the value of isolated molecules in the gas phase. Cooperative and correlation effects of hydrogen bonds[300] generally lead to an increased μ in the condensed phase.[49,301] Such effects are mainly responsible for the high value of ε in polar liquids.[84,290] They cease to be operative, however, at boundary regions where μ may thus be reasonably expected to be closer to the value for isolated molecules. It is in fact possible to give an explanation for the low ε in the double layer without necessarily invoking saturation due to the electric field simply in terms of lack of correlation effects. Under such circumstances the dielectric constant may be calculated as ranging from 10 to 15, a value close to that usually obtained for water on Hg at the pzc.[260,302,303]

The value of μ at the interface may be expected to depend on the strength of interaction of the molecule with the solid surface. An increase in μ may, in principle, be postulated for chemisorbed molecules consistently with observations[153,304] on water adsorption through hydrogen bonds[305,306] on oxide surfaces. As the strength of interaction increases, the increase in μ should be reflected by some enhancement of hydrogen bond strength towards the next water layer. The nonapplicability of the Gouy–Chapman theory with Ga[179] when using the normal bulk value of ε for water may be related to the above point.

When hydrogen bonding is enhanced, a closer packing of molecules may be expected with a consequent increase in local density of the liquid phase. This point has been experimentally corroborated[173,307,308] for some especially hydrophilic surfaces, but it is doubtful if it occurs with metals to such an extent.[309] Nevertheless, optical measurements[272,273] can distinguish different hydrophilic metals precisely in terms of different increases in density. The water population at the interface[29,73] is seen[272,273] to increase with positive charge more on Pb than on Hg. Although only 5% increase is apparent[240] on Hg from one extreme to the other of the charge range, this aspect cannot perhaps be neglected on more hydrophilic metals, and it probably turns out to be a determining factor for the capacity rise at positive charges.

Chemisorbed molecules are likely to be pulled closer to the adsorbing surface with a parallel increase in μ.[9] Field-induced chemisorption is known[310–312] to occur in adsorption from the gas phase. Transition between electrostatic adsorption and chemisorption of some ions has been suggested[313] in the case of Hg electrodes. A potential-dependent d_i may thus be a not unreasonable occurrence,[70] although specific adsorption of ions may affect such a determination experimentally. It has been suggested[70] in this respect that water at electrodes may reach a minimum in density at small negative charges where its structure is more relaxed.

(iv) Intrinsic Dielectric Constant of Water Molecules

A further difficult point in theoretical calculations is the value of ε to introduce in the Helmholtz formula for surface potential drops due to dipole layers[2,314]:

$$g^S(\text{dip}) = 4\pi N\mu_\perp/\varepsilon \qquad (36)$$

where N is the surface concentration of dipoles. Physicists usually[209,216] adopt a value for ε in Eq. (36) pertaining to vacuum. Actually, ε in Eq. (36) represents the dielectric constant of the space *between* the two ends of the dipoles, and as such it is expected to differ from that characterizing the entire thickness of the dipole layer. In practice, nevertheless, no different ε is used[4,6,8,9] for the two cases in calculations. If, between the ends of dipoles, only electrons are envisaged as being present, the square of the refractive index may, at the most, be appropriate in Eq. (36). It is, however, possible that some modification in the whole structure of the molecule results from some variation in the charge at the ends of the dipole. Thus, the polarizability of the *whole* molecule would in fact be involved and the use of ε_∞ also in Eq. (36) would be more appropriate. However, the value of ε to be introduced into Eq. (36) may still differ from that resulting from the capacity minimum.

One of the reasons why ε_i is introduced into Eq. (36) is that otherwise the equation would predict unreasonably high values of potential drops. This is, however, true only if perpendicular dipoles are assumed for the limiting orientation of solvent molecules. The problem of the correct choice of ε to be introduced into the plane capacitor formula and Eq. (36) should not be underestimated and is thought to deserve further careful study.

(v) Limiting Orientations of Water Molecules in the Inner Layer

In the P model[6] use of $\mu = 1.84$ D is made and perpendicular dipoles are assumed in the limiting orientation. In the DF[4] and DPS[8] models the value for $\mu_\perp$ is not postulated but it is derived from some partial fitting to experimental data. μ is again taken as 1.84 D in the D model[9] but chemisorbed molecules are allowed to have higher dipole moments. Damaskin finds that μ for the latter may be as high as twice that for isolated molecules. This result is reasonable in principle but it should be consistent with the chemical nature of the chemisorption bond for water molecules which is in fact not discussed.

In the BH model[11] $\mu_\perp$ is taken equal to $\mu/3^{1/2}$. This implies that water molecules in the limiting position are not perpendicular to the surface. The limiting orientation is taken as such from the model proposed by Watts-Tobin,[49] who in turn based his formulation on the magnitude of the experimental heat for water adsorption on Hg.[215] However, the high adsorption heat on Hg is surprising when compared to that measured[315–317] for strongly hydrophilic surfaces such as those of some oxides. Further, the idea[49] that the magnitude of ΔH_{ads} may be related to the formation of three bonds per molecule with the surface is questionable. It may correspond equally well to the formation of one bond with the metal and two hydrogen bonds with other molecules on the water surface layer. If so, the limiting orientation could be very much less.

(vi) Water Population in the Inner Layer

The water population N in Eq. (36) has not yet been uniquely established in calculations. Its value may be derived from considerations involving the bulk density.[318] As a rule, a hexagonal close-packed surface layer is assumed. However, the surface excess entropy of pure water is positive,[150,319,320] which implies a somewhat more relaxed structure. In spite of this, the bulk coordination number per molecule is almost entirely retained on the surface,[321] which implies that hydrogen bonds are not broken there. The value of the surface tension does not justify[321] assumptions of compression in the surface of more than 1%.

In bulk water, dead space is present[28] and should be accounted for when calculating the density at the interface. Between the surface layer and the metal there is much more space left than between two layers in the bulk where some copenetration is possible in the valleys between neighboring molecules. If a square surface lattice is assumed,[10,15] N may be as low as 0.5×10^{15} molecules cm^{-2} for almost flat molecules, and $A \simeq 0.2$ nm^2 without invoking aggregate formation. Actually, such an apparent projected area has been found in water adsorption on Fe[216] and Pd.[322] It is possible that on these metals the low water density is a result of some constrained adsorption on fixed sites. However, it does not seem unreasonable to envisage that N may actually be substantially less than about 10^{15} molecules cm^{-2} and then it increases as chemisorption starts. In the BH model[28] the total water population is again calculated on the basis of a close-packed arrangement (1.31×10^{15} molecules cm^{-2}) but the number of monomers contributing effectively to the dipole potential is calculated as only 0.42×10^{15} molecules cm^{-2}.

3. The Idea of Water Clusters

The most interesting innovation in the latest models of the double layer in comparison with earlier simpler models based on "up" and "down" dipoles, is the introduction of the idea of water clusters. This implies some recognition that the complex features of the capacity curve require a more detailed description of water at interfaces to be made consistently with the complex nature of bulk water. On the whole, models employing the idea of clusters apparently work. This does not necessarily imply, however, that the models should physically correspond to the real structure. They merely suggest what water at interfaces *may* be like but not what in fact it *must* be. This step from possibility to reality still has not been taken even for bulk models.[323] It is thought that the step is harder for models at interfaces where a number of additional complications may arise.

A prerequisite for models for water in a boundary layer is that they should be consistent with those models which are regarded as more acceptable for bulk water. In addition, they must introduce reasonable modifications to account for the presence of the extra factor generated by the metal surface. In this respect, reasons given to justify the existence of clusters do not seem to be entirely convincing as discussed below.

Models proposed for bulk water[87,324-338] are usually classified into three groups[323]: continuum (uniformist), mixture, and interstitial models, respectively. Models in each of these classes are able to explain something but not all of the behavior of bulk water. Continuum models envisage the structure of water as a continuous three-dimensional network of hydrogen-bonded molecules. Recent results,[339,340] however, suggest that unbonded molecules are almost certainly present together with bonded ones. This is in favor of mixture as well as interstitial models.

The various models proposed for bulk water have recently been critically reviewed by Frank.[323] The conclusive suggestion is that X-ray diffraction data[339,340] actually seem to be in favor of interstitial rather than mixture models. It now seems obvious that models for interfacial water claiming the existence of water clusters drive from a simplified view of mixture models. The most immediate effect of introducing clusters is the reduction of the limiting orientation of dipoles because it is generally assumed that μ for clusters is always less than μ for free molecules. Limitingly, $\mu = 0$ for clusters is assumed in the BH model.[11] It is possible, however, that a quite similar effect may be described in terms of interstitial models. These envisage[323,336,338] water as a continuous network of hydrogen-bonded molecules, some of which have left the lattice sites to reside in interstices. These models imply the presence on the surface layer of more hydrogen bonds than do models assuming independent clusters.

It has been suggested[61,341] that S-shaped isotherms, as observed for water adsorption on Hg,[147] indicate the formation of clusters. However, other data[342] do not plot as S-shaped isotherms. Moreover, such isotherms do give evidence[110] that lateral interactions are operative in the surface layer but they are not necessarily restricted to pairs or to small groups of molecules.

Another reason adduced[61,341] to support the idea of water clusters is the value close to one generally observed for the size ratio in organic adsorption (ratio of A_{org} to A_{H_2O}) which implies an area per solvent molecule apparently two to three times larger than that usually accepted for a monomeric solvent species. The problem of the size ratio is important.[343-347] Some change in this parameter necessarily involves some modification in other parameters, which makes its choice critical to quantify isotherm parameters.[348] However, it is stressed here that in interstitial models a piece of the network must necessarily become detached having an area approximately equal to that of the adsorbate, and it must then be transferred to the bulk and assimilated into the lattice. The whole effect is thought to be quite similar to that associated with cluster replacement. The noticeable difference is that such aggregates are not physically independent.

DF[4] and DPS[8] have calculated areas for the aggregates of 0.2 and 0.4 nm², respectively. No detailed structural description is given for them. Damaskin[9] has suggested that the best fit is found with $A = 0.3$ nm² and postulates trimers. Such species were proposed earlier by Parsons,[6] who supported his findings with the apparent consistency between the dipole moment derived from calculations and that predictable on the basis of the structural model proposed for trimers by Del Bene and Pople.[335] However, calculations by the latter authors are entirely theoretical in nature and apply to what water *might* be like and not for what water *must* be. Moreover, the same authors have calculated for the cyclic trimer higher stability than for the open one. Others[349] claim that the latter should be more stable than the former but all calculations consistently suggest[300] that trimers are the least stable among open and cyclic polymers possibly existing in bulk water.

The model suggesting surface dimers is even more difficult to justify on the basis of data for bulk water. The existence of dimers is postulated[11,215] on the basis of the high heat[215] and the negative entropy[11,350] for water adsorption on Hg. However, the magnitude of the entropy drop is compatible with the condensation of water at room temperature.[159] Further, the entropy change for water adsorption on Hg from the bulk liquid is positive around the pzc,[28,29,75-78] which does not seem to support the idea[350] of immobile molecules on the metal surface.[351] The behavior of a monolayer of water on Hg may possibly differ from that of the first layer of bulk water at Hg, since the latter may be closely bound to the bulk. Dimers are postulated[11,28] to have $\mu = 0$ as a result of juxtaposition of one "up" and one "down" molecule whose permanent dipoles form the same angle to the surface. Hydrogen bonds are expected to be strongly bent under such circumstances[352] and this involves some destabilization of the system,[300,334,335] making such a configuration highly unlikely.[353-361] Moreover, in Watts-Tobin's[49] "up" and "down" orientation adopted in the BH model,[11] molecules are envisaged as possibly hydrogen bonded towards the bulk liquid but not on the surface, precisely because of the unfavorable position of the free hydrogen bond left.

Some further difficulty may be encountered when considering the possibility of cluster reorientation under an electric field. Although the effect of frequency on capacity measurements on Hg may always be vitiated by spurious factors,[362-364] the possible relaxation time for water dipole reorientation at the interface, it has been suggested,[365] may range from 10^{-5} to 10^{-7} sec. The rotation of a cluster as large as three water molecules can hardly be envisaged as involving such low relaxation times,[153,182] although dipole rotation may be cooperative.[366] Fluctuations of solvent molecules in different orientations may in fact certainly resemble cluster rotation. However, the existence of clusters is not essential in such a case in that they may be envisaged as being a part of a more branched structure.

V. ELECTRIC FIELD EFFECTS

The charge variation of $g^S(\text{dip})$ may be derived still without recourse to detailed models for the water layer by applying the procedure suggested at the beginning of this chapter, i.e., based on the conceptual acceptance of Eq. (1).

1. Charge Dependence of Water Dipole Orientation

The procedure[12] for derivation of $g^S(\text{dip})$ as a function of charge is implied in Fig. 2 and illustrated in detail in Fig. 17. In the case of Hg electrodes, a straight line parallel to the line of limiting slope of the charge–potential curve, and starting from a potential, at $\sigma = 0$, 0.07 V more positive than the experimental one, is drawn on the experimental curve. The straight line simulates the hypothetical charging curve of a perfect water capacitor across which $\Delta\phi = 0$ when $\sigma = 0$ (no preferential orientation of dipoles) with dielectric properties equal to those of water at a fixed molecular orientation. The difference in potential at constant charge between the experimental curve and the straight line measures the dipole contribution at the given charge. The procedure may be automatically repeated with any metal for which it is conceptually applicable provided the appropriate value of $g^{H_2O}(dip)_0$ is employed.

Figure 18 shows the water dipole contribution to the electrode potential as a function of charge for some metals. Data have been restricted to $\sigma \leq 0$, where the absence of specific adsorption is certain.[253] For more hydrophilic metals it is difficult to reach positive charges[36] because the metal starts to dissolve or oxidize.[40] Results of the present derivation show that as the

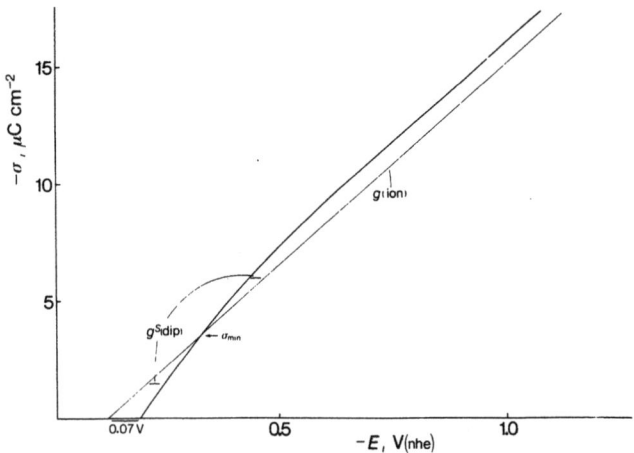

Figure 17. Procedure to derive from experimental charge–potential curves the solvent dipole contribution to the electrode potential. The curve is that for Hg in NaF aqueous solutions at 25°C. The arrow indicates the point where the net dipole orientation is zero.

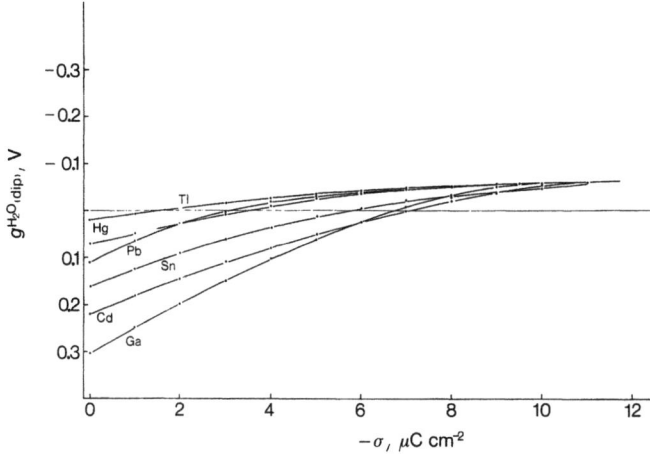

Figure 18. Surface potential of water on a number of metals as a function of charge.

charge is made increasingly negative $g^{H_2O}(\text{dip})$ becomes progressively less positive, ultimately reversing its sign. The latter situation corresponds to an average orientation of water molecules preferentially with the positive end towards the metal.

Figure 18 shows that the rates of decrease in $g^{H_2O}(\text{dip})$ with charge increases with increasing hydrophilicity. The limiting orientation is, however, the same for all metals at strongly negative charges. This is implicit in the procedure adopted for the derivation of $g^{H_2O}(\text{dip})_0$ in that it is a postulate of the model. Such a limiting orientation cannot be derived independently for each of the metals. Almost the same value observed[53,54] for the capacity at the minimum may be a good support to the above postulate, although the same molecular polarizability may in fact be associated with different limiting orientations. An effect of the nature of the metal should thus be possible in principle also at strongly negative charges. Some metals exhibit, in fact, a higher value of capacity at the minimum (in particular, In,[92] Pb,[91] In-Ga[99]). Reasons for this are at present obscure,[102] so that capacity values are in practice normalized to that of Hg. This aspect may become more understandable when more data become available on the effect of free charge on the surface electron distribution of metals.

Plots in Fig. 18 end at about $-15\,\mu\mathrm{C\,cm^{-2}}$. It is obvious that if more negative charges were included without introducing conceptual modifications in the approach, $g^{H_2O}(\mathrm{dip})$ would necessarily be observed to increase again. This effect was described[48,52,63] in earlier models in terms of electrostriction effects. Parsons[6] has been able to reproduce the same effect by introducing some residual affinity of free molecules to the negatively charged surface. The large increase in capacity results from a sharp increase in free-molecule population coming from cluster disruption. It follows that the value of the capacity at the minimum in the P model may not be associated simply with a layer of fully oriented clusters but may result from a decreasing capacity due to the increasing orientation of clusters combined with an increasing capacity due to the increasing orientation of free molecules. The minimum in capacity may thus not necessarily coincide with K_{ion}, as shown in Fig. 14, unlike the result of the present derivation and the other models.[4,8,9,11] However, the difference in value has in fact been found[6] to be small.

In view of the small limiting potential drop associated with dipole orientation at the capacity minimum (about 60 mV as found here), electrostriction effects are thought to be insignificant. Molecules are likely to be largely unoriented. In Fig. 19

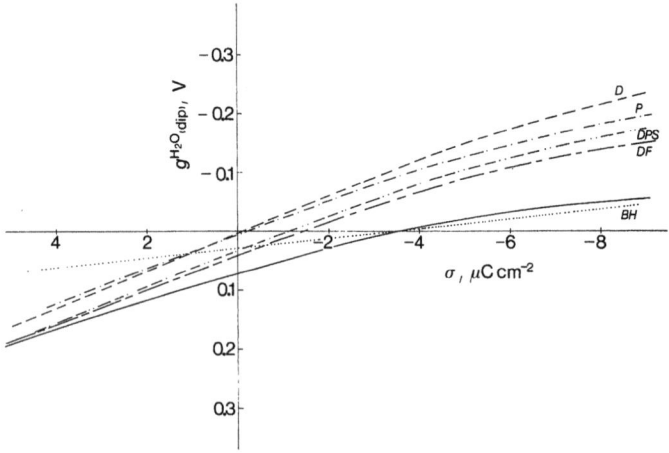

Figure 19. Surface potential of water on Hg electrodes according to various models. (———), according to the present work.

g^{H_2O}(dip) values obtained in this work are compared with those calculated from models. The results of the DF[4] and DPS[8] model are seen to approach best those of the present derivation, especially at small positive charges. The g^{H_2O}(dip) value, as obtained in this work, has been called[12] "experimental." This definition is not to be taken as rigorous in that[9] some model assumptions are involved anyway. However, the model adopted here is merely that implied in Eq. (1), so that the slope of g^{H_2O}(dip) vs. σ plots, once $1/K_{ion}$ is taken into account, coincide necessarily with the slope of the experimental capacity curve. The only true postulate is that the capacity minimum measures K_{ion}, but this is implicit in all other models, except that of Parsons.

A distinct feature of the behavior of all the theoretical models is the larger dipole potential drop at negative charges. It is thought that this is a simple consequence of the large orientation allowed for free molecules or clusters. The BH model[11] does not exhibit any appreciable approach to saturation of g^{H_2O}(dip) at high charges. Since C_{dip} is always much higher than K_{ion}, the capacity curve derived from such a model is apparently featureless. Since no specific adsorption can be invoked[253] at $\sigma \leq 0$, no reasons are apparently given for the increase in capacity from the minimum up to the pzc in fluoride solutions.

2. The Capacity Hump

The hump arises in all models except BH's,[11] as a consequence of a dielectric maximum. This was first suggested[260,261] on the basis of data for thiourea adsorption on Hg. However, the capacity hump should not be regarded as necessarily the place where the orientation of water is at a minimum. As illustrated in Fig. 13, the hump comes out as a result of a decreasing ε due to cluster reorientation (maximum ε at $\sigma = 0$) combined with an increasing ε due to chemisorption of free molecules. On the negative side of the capacity curve the two effects add to one another and since both exhibit a monotonic decrease there, it ensues that the point where g^{H_2O}(dip) = 0 is dielectrically nonspecific.

A different explanation for the hump is offered in the BH model.[11,25] Dielectric effects are ruled out. The hump is suggested as resulting from an inflection on the increase of specifically adsorbed ions with charge. Such an inflection is attributed to ion–ion repulsion,[164] operative at higher coverages. Such a theory was first proposed[25] in 1963 and has been resumed and enlarged, together with a theory on isotherms for ionic adsorption in a number of recent papers.[283,367,368] According to the BH theory, the factor determining first the rise in capacity and then appearance of the hump, is[25] $d\sigma_i/d\sigma$, where σ_i is the specifically adsorbed charge.

The model is, in principle, successful also in respect to the effect of temperature[283] (see Section IV) on the hump. However, an entirely satisfactory model must be able to reproduce a number of things together and not just single features. Experimentally, apart from the hump, the capacity is seen[53,54] to increase more steeply close to $\sigma = 0$ in the sequence Hg < Pb < Cd < Ga. In terms of the BH model, this should reflect an increasing value of $d\sigma_i/d\sigma$ along the series. Figure 20 shows some results for ionic adsorption. The slope of the σ_i vs. σ curves in fact *decreases*[257] in the series Hg > Pb > Cd. Results[369] also show that such a slope is lower for Ga than for Hg. This is[149] probably an effect of water chemisorption increasing with positive charge[310] in the sequence Ga > Cd > Pb > Hg.

Another difficulty encountered with the BH model is that the hump does not disappear but is simply reduced as the experimental inner layer capacity is extrapolated[253,370,371] to $\sigma_i \to 0$, although some uncertainty may arise in the calculation of C_i and in

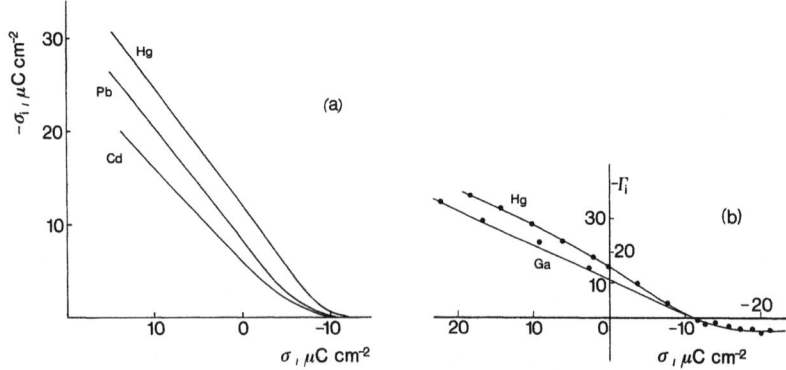

Figure 20. Amount of specifically adsorbed I⁻ as a function of charge for a number of metals. (a) 0.1 N I⁻ solutions[257]; (b) 1 N I⁻ solutions.[369]

the extrapolation. Actually, both effects may in fact be operative.[372] A third factor may be particle–solvent interactions on the surface.[373,374] Some reduction in preferential orientation of water at $\sigma = 0$, due to ionic adsorption, has been suggested.[72] This, in terms of the P model,[6] would correspond[372] to increased stability of clusters. Thus, the capacity rise is less and the hump may be more pronounced. Coupling of all these effects may well explain the large variety of humps observed with different electrolytes.

3. The Charge for Zero Net Dipole Orientation

(*i*) *Derivation*

The charge where the net orientation of dipoles vanishes is not evidenced by capacity curves (unless the early theories[48–50,63] on the hump are accepted), but nevertheless it is a very important structural parameter for the description of the state of water at electrodes. This charge (σ_{min}) is readily derivable[12] from the plots in Fig. 18 in that it is given by the intercept of the given curve on the charge axis. σ_{min}, according to Fig. 2, also coincides with the charge at which the straight line describing K_{ion} intersects the experimental charge–potential curve. Thus, σ_{min} may be obtained without any additional assumptions besides those made at the beginning.

σ_{min} can be seen in Fig. 18 to become increasingly negative as the hydrophilicity of metals increases. This is because a higher negative field is needed to overcome the spontaneous orientation of dipoles at the pzc. A close relationship is expected between $g^{H_2O}(dip)_0$ and σ_{min}. Table 8 summarizes σ_{min} for a number of metals.

σ_{min} does not necessarily define the situation where the potential drop across the liquid portion of the double layer is zero and $g(ion)$ is not necessarily zero when $g^S(dip) = 0$. The charge for zero net potential drop across the solution side of the double layer is that at which $g(ion) = g^S(dip)$. According to Fig. 17, such a charge can be derived by drawing the perpendicular to the starting point of the straight line associated with $g(ion)$, then seeking the charge at which the distance of the perpendicular from the straight line equals the distance of the latter from the experimental curve.

Table 8
Charge Where the Net Orientation of Water Dipoles at Metal/ Aqueous-Solution Interfaces is Zero at 25°C

			$-\sigma_{min}$ ($\mu C\, cm^{-2}$)			
Metal	This work	DF model[a,b]	P model[a,c]	DPS model[a,d]	D model[a,e]	BH model[f]
Tl	1.6	—	—	—	—	—
Hg	3.4	1.3	0.1	1.0	0.1	3.5
Pb	3.1	—	—	—	—	—
Sn	5.9	—	—	—	—	—
In	6.0	—	—	—	—	—
Cd	7.0	4.1	—	5.9	—	—
Ga	6.6 (7.2)	—	—	—	—	—

[a] 0°C. [c] From Ref. 6. [e] From Ref. 9.
[b] From Ref. 4. [d] From Ref. 8. [f] From Ref. 11.

(ii) Physical Meaning with Reference to Models

A physical description of the situation at σ_{min} requires the introduction of a more detailed model for water at a molecular level, since from the choice of the model ensues the kind of experimental evidence that σ_{min} is expected to give rise to. The BH model[11] gives the simplest explanation in this case. σ_{min} corresponds to the point where the numbers of "up" and "down" molecules are equal (provided the normal components of the dipoles are equal and opposite for the two positions, which is in fact the case). The model additionally predicts[25] that water molecules are least bound to the metal at σ_{min}. Some evidence for this may be offered by the maximum value found[70] for the thickness of the inner layer around the same charge. With the BH model a charge of $-3.5\ \mu C\, cm^{-2}$ is calculated for σ_{min}.

The other models give much lower values for σ_{min}. The DF[4] model gives about $-2\ \mu C\, cm^{-2}$ for Hg and correctly predicts an increase to about $-4\ \mu C\, cm^{-2}$ for Cd. Models[6,9] allowing for an exchange equilibrium between the water species give much lower σ_{min} values, practically close to zero, which is consistent with the very small values predicted for $g^{H_2O}(dip)_0$. No sharp separation is achieved in these models between σ_{min} and σ_{hump}, although the two charges are in fact quite distinct conceptually, unlike earlier models where $\sigma_{min} = \sigma_{hump}$ was a characteristic feature.

In mixture models for water at electrodes, the physical interpretation of σ_{min} is much more involved. For example, in the DF[4] model the situation at σ_{min} is such that the net value of the dipole potential drop is zero, but this by no means corresponds to a symmetrical distribution of differently oriented molecules. Further, σ_{min} does not at all correspond to the position where the number of chemisorbed molecules vanishes. Rather, some of the chemisorbed molecules are balanced by some excess clusters oriented the other way round. Since[4] $\mu_{chem} > \mu_{clust}$, the number of dipoles oppositely orientated can hardly be envisaged as equal. The situation is better described by Fig. 21, where contributions to $g^S(dip)$ by chemisorbed molecules and clusters are separated.

Four water species are involved in the P model.[6] If however, monomeric forms with the positive end towards the metal are very unstable at moderately negative charges, the situation practically reduces to that of the DF model. It is interesting, however, that the increase in water population as measured experimentally[272,273] starts on Pb at more negative charges than on Hg. If such a charge is identified with the point of incipient chemisorption of water, then it should be expected to be slightly more negative than σ_{min}. Actually, such a charge is[273] about $-4\ \mu C\ cm^{-2}$ on Hg and $-6\ \mu C\ cm^{-2}$ on Pb. It would be interesting to do similar optical measurements on such a hydrophilic metal as Ga.

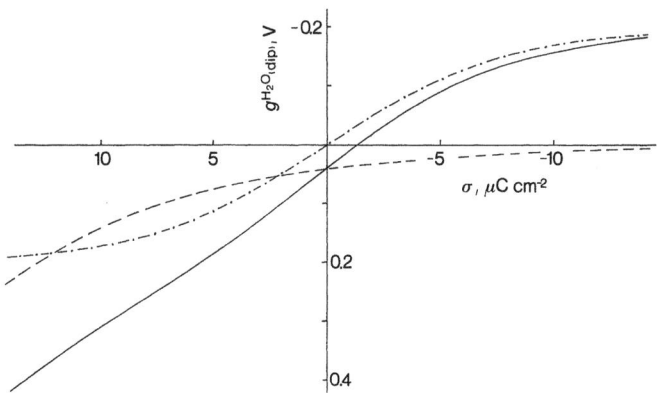

Figure 21. Theoretical surface potential of water at Hg (———) calculated by means of the DF model.[4] (---), potential associated with chemisorbed molecules; (-·-), potential associated with water clusters.

(iii) Relation to the Charge of Maximum Adsorption of Organic Substances

BDM[25] first suggested that, provided the normal component of the organic dipole is zero, the charge for maximum adsorption of organic substances (σ_{max}^{org}) should be a measure of σ_{min}. At the time the BDM theory was proposed, n-butanol was regarded[112] as an organic substance meeting the prerequisites for a reliable derivation of σ_{min}. Thus, the σ_{max}^{org} at about $-2\,\mu\mathrm{C\,cm}^{-2}$ was suggested as the possible σ_{min}. This is the value that Conway and Dhar[5] found recently in studies of the adsorption of pyrazine regarded as a rigid, nondipolar probe molecule for determination of water orientation.

It has been pointed out by Damaskin and Frumkin[115] that the value of σ_{max}^{org} does not depend[111] only on the polarity of the adsorbate, but also on its polarizability. These authors have been able to show that there exists a wide range of σ_{max}^{org} predictable on the basis of the model[110] of two parallel capacitors. The present author[120] has gone over this point recently to show that polarizability effects can be disentangled from polarity effects, irrespective of the model chosen for the adsorption layer. σ_{max}^{org} has been shown to vary from 0 to σ_{min} depending on C_1 (a measure of the organic polarizability) being equal to 0 or to C_{H_2O}, respectively. This means that the value of σ_{max}^{org} may help to gain insight into the structure of the water layer only if $\mu_\perp$ for the organic molecule is independently known (cf. the case of thiourea adsorption for which $\mu_\perp$ is reasonably taken[260] equal to the molecular dipole moment). Otherwise, nothing quantitative about σ_{min} can be known from σ_{max}^{org}. However, σ_{max}^{org} and σ_{min} are expected to have the same sign if the effect of μ_{org} is negligible or allowed for.

In spite of the lack of correspondence between σ_{max}^{org} and σ_{min}, comparison between σ_{max}^{org} for adsorption of the same organic substance on different metals is expected to provide evidence for possible differences in σ_{min}. Published quantitative data in a form useful for derivation of σ_{max}^{org} are unavailable because all of the data for adsorption of organic substances on metals other than Hg come from the Russian school where representation of the results at constant potential is preferred over that at constant charge. However, n-propanol can be seen[187] to absorb on Hg with $\sigma_{\max}^{\mathrm{org}} \simeq -2\,\mu\mathrm{C\,cm^{-2}}$ and on Cd with $\sigma_{\max}^{\mathrm{org}} \simeq -3\,\mu\mathrm{C\,cm^{-2}}$. This is a mere qualitative comparison because C_1 on hydrophilic metals is higher than on Hg and, moreover, some reorientation is possible around the pzc. The trend is, however, that expected.

$\sigma_{\min}$ is a structural parameter and as such it bears some definite relationship to the excess entropy of the double layer. This point will be postponed to the discussion of the effect of temperature on $g^{H_2O}(\mathrm{dip})$.

VI. TEMPERATURE EFFECTS

1. Temperature Dependence of Water Dipole Orientation

(i) Mercury

The effect of temperature on double-layer structure has been studied in detail only for Hg electrodes. The now well-known capacity curves at different temperatures measured by Grahame[65] in NaF solutions are shown in Fig. 22. The more striking effect of

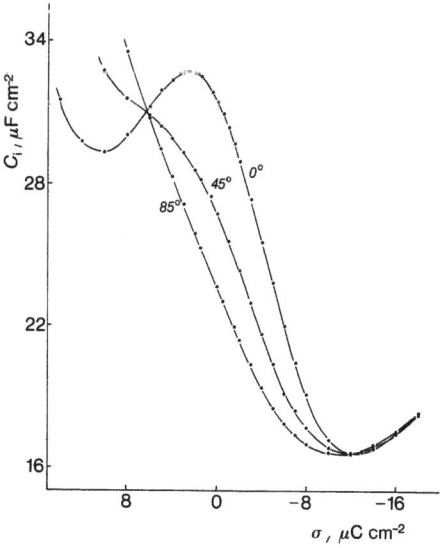

Figure 22. Experimental inner-layer capacity[65] of Hg in contact with NaF aqueous solutions at different T.

temperature is the depression of capacity in the region of the hump. A marked increase in capacity is, on the contrary, seen at more positive charges.

The procedure proposed in this work enables $g^{H_2O}(\text{dip})$ to be derived as a function of temperature. The temperature coefficient of $E_{\sigma=0}$ has been measured by Randles and Whiteley[375] and found to be $0.57\,\text{mV}\,\text{K}^{-1}$. The positive value has been taken as evidence for dipole disorganization as the temperature is raised. This is consistent with the usual view of entropic effects on an oriented dipole layer. The fact that the temperature coefficient of the potential drop across the Hg–water interface is higher than that at the free surface of water has been taken[136] to mean that possibly $g^{H_2O}(\text{dip})_0 > \chi^{H_2O}$.

It is to be noted[3,7,13] that $\partial E_{\sigma=0}/\partial T$ is not a function of $g^{H_2O}(\text{dip})$ only, but also of the work function. From Eq. (9) is follows, in fact, that

$$\partial E_{\sigma=0}/\partial T = \partial(\Phi'/e)/\partial T - \partial g^{H_2O}(\text{dip})_0/\partial T \quad (37)$$

where $\Phi' = \Phi + e\delta\chi^M$. In principle, even the sign of the temperature coefficient of $E_{\sigma=0}$ may be not directly related to structural aspects of the dipole layer.

The positive sign of the temperature coefficient of $E_{\sigma=0}$ for Hg has been confirmed by a number of other authors.[376–380] Quantitatively, $\partial E_{\sigma=0}/\partial T$ has been claimed not to be a property of the sole ion free layer but to depend on the local ion–solvent interactions. In this respect, the temperature coefficient of $E_{\sigma=0}$ has been shown[379] to increase as the hydration of ions decreases. This might involve some aspects of entropies of ionic hydration. In some respects, this may also be related to the problem of the possible dependence of both d_i and ε_i on the nature of the cation.[381] However, it should be noted that some measurements have been made with the reference electrode in a lateral compartment kept always at constant room temperature. This introduces an unknown thermal-diffusion potential which may well depend on single-ion entropies. However, Randles and Whiteley[375] used a reference electrode at the same temperature as the test electrode, the temperature coefficient of which was allowed for in the calculations. Thus, Eq. (37) applies correctly to Randles' data.

A way to resolve $\partial E_{\sigma=0}/\partial T$ into the two components according to Eq. (37) has been proposed by the present author.[382] At negative rational potentials it may be written, from Eq. (8), that

$$\partial E/\partial T = \partial g(\text{ion})/\partial T + \partial(\Phi'/e)/\partial T - \partial g^S(\text{dip})/\partial T \quad (38)$$

Figure 23 shows some E vs. σ curves at different temperatures. The temperature independence of the limiting slope is taken as indicating that $\partial g(\text{ion})/\partial T = 0$, which is tantamount to assuming a temperature-independent molecular polarizability for water. In principle, this may not be justified in that ε_∞ has in fact been observed[85,290] to be slightly temperature dependent. However, the orientation polarizability of water obviously decreases as the temperature is increased so that the constancy in K_{ion} may be the result of some decrease in the residual ε_o combined with some increase in ε_m, according to Eq. (30). Consistently, a temperature-independent limiting orientation may be assumed for water. Thus, Eq. (38) reduces to

$$\partial E/\partial T = \partial(\Phi'/e)/\partial T \quad (39)$$

in the range of charge where the charge–potential curves are parallel.

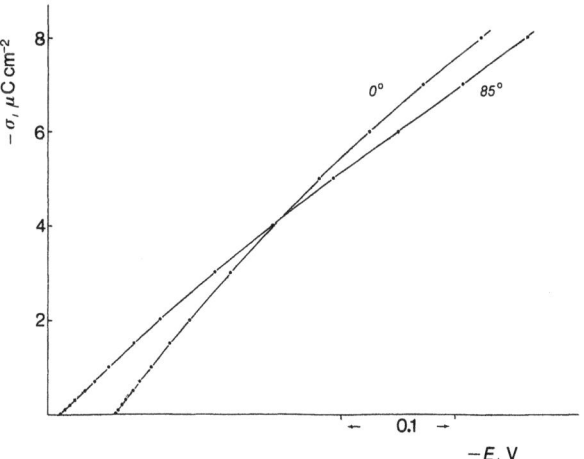

Figure 23. Charge–potential curves for Hg by integration of curves in Fig. 22. Potentials of zero charge have been separated as though the experimental T coefficient[375] were constant all over the T range.

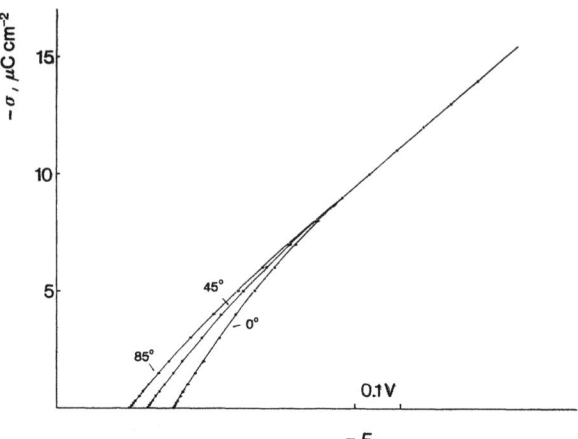

Figure 24. Charge–potential curves at constant work function, obtained from those of Fig. 23 by superimposing the linear portions.

If charge–potential curves[65] at different temperatures are superimposed on the linear portion by sliding them along the potential axis, changes in work function are allowed for according to Eq. (39). Figure 24 shows a resulting family of such curves at constant work function. At a given charge, the difference in potential between two curves now represents the true relative variation of $g^{H_2O}(dip)_0$. Since $g^{H_2O}(dip)_0$ at 298 K has been estimated, values at other temperatures can be readily obtained. They are summarized in Table 9, where some values predicted by models are also listed.

Figure 25 shows the dependence of $g^{H_2O}(dip)$ on charge at different temperatures. The procedure employed to derive the data in the figure is quite similar to that used at 298 K. At each charge, $g^{H_2O}(dip)$ can be seen to decrease as the temperature is increased. This is undoubtedly a consequence of the postulated constancy in $g^{H_2O}(dip)$ at strongly negative charges. Some reversal in the sign of the temperature coefficient of $g^{H_2O}(dip)$ may be obtained with the present derivation only if the limiting orientation is allowed to increase with increasing temperature. This may not be unreasonable, as will be discussed later, but the sign of the temperature coefficient of $g^{H_2O}(dip)_0$ would certainly not change. On the contrary, $g^{H_2O}(dip)_0$ would decrease even more rapidly

Table 9
Surface Potential of Water at Hg/Aqueous-Solution Interfaces at the Potential of Zero Charge at Different Temperatures

	$g^{H_2O}(dip)_0$ (V)		
T (°C)	This work	P model[a]	D model[b]
0	0.100	0.005	0.003
25	0.070	—	—
45	0.042	—	—
65	0.019	—	—
85	0.004	0.023	0.015

[a] From Ref. 6.   [b] From Ref. 9.

with temperature. At present, no way is envisaged for distinguishing which is the case in fact.

Figure 26 shows that the temperature coefficient of $g^{H_2O}(dip)_0$ is negative and greater than that of $E_{\sigma=0}$, and thus also that[136] of χ^{H_2O}. Nevertheless, $g^{H_2O}(dip)_0$ on Hg has been found[375] to be greater than χ^{H_2O}. This may be taken[382] to mean that the contact with Hg has some disordering rather than ordering effect on the free surface of water, and this may be evidence for the weak hydrophobic character of such a metal surface.

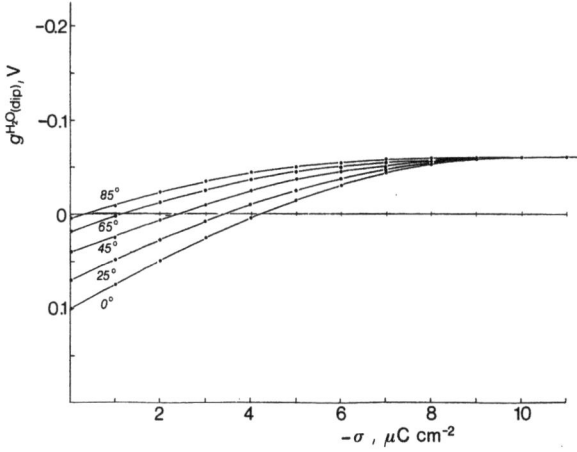

Figure 25. Surface potential of water on Hg as a function of charge at different T.

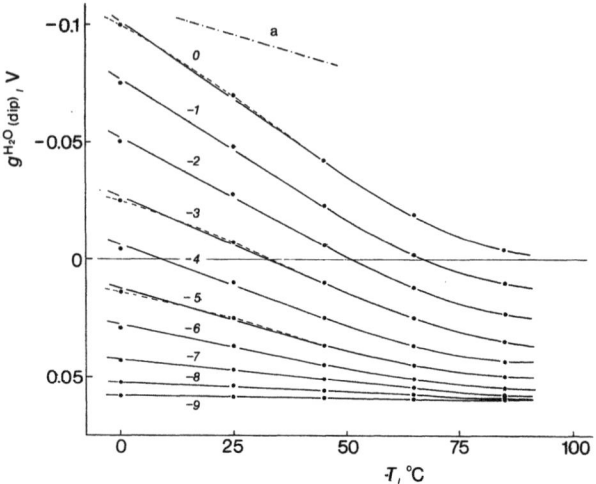

Figure 26. Surface potential of water on Hg as a function of T at different charges. (a) Experimental[375] T coefficient of the potential of zero charge.

According to the present model, σ_{min} is a temperature-dependent quantity. Table 10 summarizes values at different temperatures. The decrease in σ_{min} is a direct consequence of the decrease in $g^{H_2O}(dip)_0$ with temperature. Both observations depend closely on what really happens at strongly negative charges. This behavior conforms to the simple picture expected[5,318] for an oriented dipole layer at charges close to zero but not at more

Table 10
Charge Where the Net Orientation of Water Dipoles at Hg/Aqueous-Solution Interfaces is Zero at Different Temperatures

T (°C)	$-\sigma_{min}$ ($\mu C\ cm^{-2}$)		
	This work	P model[a]	D model[b]
0	4.2	0.1	0.1
25	3.4	—	—
45	2.9	—	—
65	1.1	—	—
85	0.2	0.9[c]	0.6[c]

[a] From Ref. 6.   [b] From Ref. 9.   [c] Estimated.

negative charges where reversal in orientation occurs. Actually, at constant charge, some disorientation of dipoles as the temperature is raised should always take place. Thus, the limiting orientation should be achieved at higher negative charges the higher the temperature. However, if the orienting field operates against interaction forces which are weakened by an increase in temperature, then a more rapid orientation with temperature may be expected at every charge as shown in the treatment by Conway *et al.*[5,318] Among the cases they discussed, the results of Fig. 26 apparently conform surprisingly to the case of absence of dipole–dipole interaction.

(ii) Other Metals

What has been done for Hg cannot be applied to other metals straightforwardly. Capacity curves at different temperatures have been determined for Pb,[383] Cd,[351,384] Ga,[100] and In–Ga.[100] For the last two metals, capacity values depend on temperature close to $\sigma = 0$, but at more negative charges they appear to be temperature invariant. At strongly negative charges, where the capacity rises again, opposite temperature dependences have been observed. Further study is thought to be necessary in this region. Around $\sigma = 0$, the temperature coefficient of the capacity is higher on In–Ga than on Ga, and this may be further evidence in favor of stronger chemisorption of water on the latter metal.

The effect of temperature on capacity is rather complex with Pb and Cd. Around the pzc, capacity values decrease with temperature in the series Cd < Pb < Hg, in agreement with the hydrophilicity scale. However, the temperature coefficient of capacity changes sign at strongly negative charges and this commences at a more negative charge with Cd than that with Pb. It has been suggested[383] that, on Pb, changes in C_i as a result of solvent density changes are quite negligible and the sign of the temperature coefficient of C_i should coincide with the sign of the temperature coefficient of ε_i, a reversal of which should be operative at some negative charges. A change in the sign of $\partial \varepsilon_i / \partial T$ would provide evidence for transition in the water layer from a nonordered to an ordered state. However, this transition is almost not apparent with Hg (-13 μC cm^{-2} after some authors[383]); it occurs at -6 μC cm^{-2} on Pb and -11 μC cm^{-2} on Cd. No regularity is thus observed, nor also is there any detectable evidence for such a transition with Ga and In–Ga.

Some temperature dependence of C_i at high negative charges could be related to some reported[85] increase in ε_∞. Alternatively, some change in the small residual orientation polarization of the solvent may be invoked. However, theoretically[50,385–387] some inversion in the temperature coefficient is obtainable in terms of C_i but not in terms of K_i, the integral capacity. One difficulty is the absence of any detectable effect on Hg. Moreover, at strongly negative charges, no specific effect of the metal surface is expected. A possible explanation may be in terms of the minimum in capacity being the resultant of one descending and one ascending capacity component. In such a case, an effect of temperature may be expected. However, an effect of the metal surface at such negative charges must again be invoked. It may be that the role of the solid surface at negative charges has so far been underestimated. Undoubtedly, if the capacity at the minimum were intrinsically temperature dependent, the procedure adopted here would not be unambiguous.

(iii) *Comparison with Predictions of Models*

Models[6,9] allowing for exchange equilibrium between different solvent species at the metal surface reproduce capacity curves at different temperatures in remarkable agreement with the shape of experimental curves. Some depression of the hump is also predicted by the BH model,[283] although no quantitative comparison with experimental curves was attempted over all the charge range. It should be noted that a correct temperature dependence of C_i was also obtained by Watts-Tobin[49] in terms of free dipole reorientation and adatom formation.

The correct shape of curves derived from the P[6] and D[9] models is a result of allowing for some disruption of clusters at higher temperatures. An inversion in the temperature coefficient of capacity is predicted at negative charges but this is much more than observed on Hg. Nevertheless, the agreement between calculated and experimental charge dependence of the coefficient of C_i is remarkably good all over the charge range investigated. However, the sign of the temperature coefficient predicted for $g^{H_2O}(dip)_0$ is opposite to that obtained in the present work. $g^{H_2O}(dip)_0$ has been calculated to be $-3\,mV$ at 273 K and $-14.6\,mV$ at 358 K with the D model and $-8\,mV$ at 298 K and $-23\,mV$ at 358 K with the P model. Undoubtedly this is a consequence of the formation of chemisorbed molecules from cluster dissociation outweighing the disordering effect of temperature on the oriented layer itself.

It has been suggested[9] that any model based on chemisorption of molecules as the cause for the capacity rise at positive charges must necessarily lead to a positive rather than negative temperature coefficient of $g^{H_2O}(dip)_0$. This is, however, true only if the model also rests on cluster dissociation as a coeffect of increasing temperature. Although the experimental temperature coefficient of $E_{\sigma=0}$ for Hg is[375] positive [which corresponds to a negative temperature coefficient for $g^{H_2O}(dip)_0$], it has been argued[9] that even its sign may be doubtful because the experimental quantity contains the unknown temperature coefficient of the potential of the reference electrode. However, this is not the case with Randles and Whiteley's measurements because they allowed for this effect thermodynamically on the basis of an acceptable value[137] for the entropy of H^+_{aq}. A change in the sign of the temperature coefficient has been shown[375] to imply an unreasonable value for $S^0_{H^+}$. Moreover, allowance for the temperature coefficient of Φ has been shown above to lead to an increase of the temperature coefficient of $g^{H_2O}(dip)_0$, a trend opposite to that shown by the P and D models. It is stressed again here that, in principle, no relation is expected between the temperature coefficient of C_i and that of $g^{H_2O}(dip)$. A close inspection of Fig. 23 shows that if the curves are shifted vertically, the temperature coefficient of $g^{H_2O}(dip)$ can be reversed without necessarily affecting the temperature coefficient of C_i detectably.

The temperature coefficient of $E_{\sigma=0}$ for Cd has been found[351] to be $0.15\,mV\,K^{-1}$. This is less than for Hg. A direct quantitative comparison is impossible because of lack of data of $d\Phi/dT$ for Cd. Values of $\partial E_{\sigma=0}/\partial T$ for Pb, Ga, and In–Ga are not reported in the papers. Such quantities would be of interest, especially for Ga and In–Ga, because the temperature independence of the capacity at the minimum could enable the procedure adopted for Hg to be extended to those metals.

2. Surface Excess Entropy in the Inner Layer

The effect of temperature on the behavior predicted from models is thought be be particularly relevant to decide about their validity. A number of parameters have been shown above to be experimentally available, which can throw light on the structure of the double layer and help to decide about models. These parameters are $g^{H_2O}(\text{dip})$, σ_{max}^{org}, and σ_{min}. Here, one more parameter is added: the entropy of the double layer. A thermodynamic analysis of this quantity has been given by Hurwitz and D'Alkaine.[388] Depending on the model, a different interconnection among the above parameters is expected.

The model ensuing from the present work predicts a decreasing positive $g^{H_2O}(\text{dip})_0$, hence a decreasing negative σ_{min}, as the temperature is increased. σ_{min} has been shown[115,120] to be not directly related to σ_{max}^{org}, but the two parameters are, however, expected to change with temperature following the same pattern. Data supporting this view are quite scarce. Studies of the effect of temperature or adsorption appear to be underdeveloped for Hg and quite absent for other metals. However, the adsorption potential shift for butanol on Hg at the pzc has been shown[119] to decrease by about 80 mV in the range 298–348 K, in good agreement with the data in Fig. 26. With pentanol, $\Delta E_{\sigma=0}$ is[119] only 40 mV but the change is in the expected direction. Quantitatively, such a comparison implies that alcohols retain the same orientation at different temperatures. If a close-packed perpendicular orientation is assumed[389,390] at 298 K, as suggested[391] also by photoemission measurements, little effect of temperature on the monolayer orientation is expected. This is apparent from the constant projected area found[119] experimentally for the adsorbate molecule. Some decrease in $\Delta E_{\sigma=0}$, together with a negligible change in C_1, necessarily involves a decreasing σ_{max}^{org} with increasing temperature and hence a decreasing σ_{min} within the framework of the present model.

The P[6] and D[9] models give values for σ_{min} very close to zero and almost temperature independent, which is not consistent with the temperature-dependent σ_{max}^{org} found experimentally. In the context of the above models, a conceptual assessment is required to reconcile σ_{min} and σ_{max}^{org}. Results supporting the possible decrease in σ_{max}^{org} with temperature may be found in the paper by Conway and Dhar.[5] $\Delta E_{\sigma=0}$ for pyrazine adsorption is seen to decrease with T and σ_{max}^{org} consistently shifts from about $-2\,\mu\text{C cm}^{-2}$ at 298 K to about -1 to $-2\,\mu\text{C cm}^{-2}$ at 318 K. Thus, if the view of BDM[25] on the conceptual relationship between σ_{min} and σ_{max}^{org} is qualitatively accepted, then the available experimental evidence would be somewhat against cluster models.

Cluster models, however, predict[9,78] a maximum in the temperature coefficient of $1/C_i$ at small negative charges, in agreement with the experimental results. This parameter has been associated[29,73] with the entropy of formation of the double layer. It is positive for such models to predict an entropy maximum where it is in fact observed. Entropy maxima have been found[29,75-77] experimentally in the range -3.5 to $-4\,\mu\text{C cm}^{-2}$, after allowance[17,75] for ionic effects.

The nature of the entropy change with charge was first discussed by Conway and Gordon.[74] They found a maximum in the negative entropy of adsorption of pyridine on Hg and explained this in terms of a charge-dependent structure of the water layer being replaced. However, the maximum (rather than a minimum in ΔS_{ad}°) suggested that at that charge interfacial water should possibly be more structured than at other charges. It is thought that the shape of the ΔS_{ad}° curve may be affected by possible pyridine–water interactions.[392] In fact, a reasonable *minimum* in ΔS_{ad}° was later found by Conway et al.[146] at about $-2\,\mu\text{C cm}^{-2}$ (i.e., in correspondence of σ_{min}) for pyrazine adsorption on Hg and interpreted in terms of a *minimum* in the ordering of the solvent layer.

Conway and Gordon[74] were the first to attempt a theoretical treatment of the entropy change of the water layer with charge in terms of (a) configurational entropy and (b) field-dependent librational contributions. A similar treatment due to Cooper and Harrison,[393] apparently in ignorance of Conway's work, predicts a maximum for the entropy without quantifying the charge where it occurs (henceforth denoted with σ_{max}). A more developed treatment was given subsequently by Bockris and Habib.[28,79] These authors, first[79] in terms of a two-state water model, later[28] of a three-state water model, calculated the total entropy of formation of the double layer as the sum of configurational, vibrational, rotational, librational, and structure-breaking terms. A maximum was found at $-3.5\,\mu\text{C cm}^{-2}$ for Hg. Within the framework of the model, this implies a preferential orientation of water at the pzc with the oxygen towards the metal. The presence of uncharged dimers is ineffective in respect of the position of the maximum but it may affect the absolute value of the entropy.

According to the BH model the entropy maximum necessarily occurs at σ_{min}. The present derivation does not require *a priori* that σ_{min} coincide with σ_{max} but, as Fig. 18 shows, this in fact happens to be the case. It has been remarked[393] that the BH model predicts the correct behavior of the double-layer entropy but cannot predict the hump and the steep rise at positive charges. This limitation does not hold for the present work, although the coincidence of σ_{min} and σ_{max} is in agreement with the BH model. This is possible in that the present derivation is independent of the choice of a detailed molecular model for water.

The possible coincidence of σ_{min} with σ_{max} depends closely on the physical meaning given to the maximum of the double-layer entropy. One definite piece of experimental evidence is that the same σ_{max}^{org} is derived[146] from the charge dependence of both ΔH_{ad}^0 and ΔS_{ad}^0. Since various σ_{max}^{org} but only one σ_{min} may exist, data on entropy of adsorption can hardly be referred straightforwardly to structural aspects of the properties of water. Also, in view of compensation effects in adsorption processes,[163] it must be concluded that ΔH_{ad}^0 and ΔS_{ad}^0 run parallel as a function of charge, so that ΔS_{ad}^0 cannot be naively used to derive structural information about the water layer.

An interesting feature of the P[6] and D[9] models is that the maximum of the temperature coefficient of $1/C_i$ does occur at the charge corresponding to σ_{max} but σ_{max}, in turn, does not coincide with σ_{min}. It is very important to realize whether this is an inconsistency of the models or rather it possesses an apparently concealed physical meaning. If the above inconsistency is simply apparent, then the entropy maximum requires a more complicated model to be employed than BH's.[28] Alternatively, if the observed inconsistency is real, then this means that the above models may give correct results on an unreasonable physical basis. Thus, with reference to Fig. 25, for $g^{H_2O}(dip)_0$ to exhibit a negative temperature coefficient, as predicted by the above models, the limiting orientation at 358 K must be less positive than that at 298 K, perhaps even opposite in sign, which is physically unlikely.

A possible explanation for the noncoincidence between σ_{min} and σ_{max} in results from models may be that σ_{min} actually does not identify the point of maximum disorder. This is possible if, at σ_{min}, the difference between positively and negatively oriented clusters is balanced by a very small number of chemisorbed molecules with higher $\mu_\perp$. In such a case, σ_{max} would define the point where the configurational entropy of a system of *three* rather than *two* components is at a maximum. This suggestion requires some quantitative confirmation. However, if the above situation is true, the problem of reassessing some aspects of double-layer theories hitherto considered as conceptually settled, must be faced. Consistently, adsorption potential shifts should also be regarded in the light of the quite inhomogeneous nature of water at the interface.

Entropy measurements with other metals are quite scanty. Some data[351] ave available for Cd. The characteristic maximum is not apparent although a small hump may be recognized in the range -5 to $-10\ \mu C\ cm^{-2}$. At more negative charges the entropy increases again steadily. This is certainly closely related to the unusual temperature dependence of the capacity at the minimum. Such a pattern for the entropy change with charge has been explained in terms of stronger localization of water molecules on Cd than on Hg surfaces. A difficult point is the apparent effect of the metal at such negative charges in the case of Cd, whereas this is absent in the case of Ga,[100] a more hydrophilic metal. The two cases require a consistent explanation.

The temperature dependence of the charge where the entropy of the double layer is at a maximum is a somewhat unresolved point. Some data[29] seem to indicate that σ_{max} is possibly temperature invariant. Others[77] suggest a possible small variation with temperature. This matter has actually never been investigated in detail. If the $1/C_i$ data of Grahame[65] are plotted as a function of temperature and differentiated at different temperatures, then the entropy maximum can be seen,[382] in fact, to be located at lower negative charges as the temperature is increased. In the context of the present work, a decrease in σ_{max} is consistent with a decrease in σ_{min}. Since the two quantities are ultimately derived from processing the same original data, this may be taken to mean that the two quantities go, in fact, parallel. In the case of cluster models, if the agreement of the temperature coefficient of $1/C_i$ with the experimental one is assumed to hold at all temperatures, a different trend should be obtained for σ_{max} and σ_{min}, the latter becoming slightly more negative as the temperature is increased. Such a divergence calls for some support on a physical basis.

The BH[28] model gives the following expression for σ_{max}:

$$\sigma_{max} = \Delta \Delta G_c \varepsilon_i / 8 \pi \mu_\perp \tag{40}$$

Apparently, $\Delta \Delta G_c$ is a temperature-independent quantity according to the way it is calculated. σ_{max} is thus predicted to be temperature independent. This point has been discussed recently.[382,394] In reality, $\Delta \Delta G_c$ cannot be temperature invariant because this quantity already includes the configurational entropy actually contributing to the stability of the given structure. As the temperature is raised, $\Delta \Delta G_c$ is expected to decrease, which in fact involves a decrease in the negative value of σ_{max}.

VII. SUGGESTIONS FOR A POSSIBLE ALTERNATIVE MODEL

Comparison between results from the present derivation and predictions of models may give suggestions which future models may take into consideration. It seems definite that the structure of water is more complicated than the simple model of "up" and "down" molecules at interfaces implies. Also, it must be admitted that all the components of the water layer contribute to the capacity, otherwise capacity cannot be reproduced even at charges negative to the pzc.

One of the most striking aspects observable in models is the reduction of $\mu_\perp$ for water molecules in the limiting orientations. Values of $\mu_\perp$ equal[318] to that pertaining to perpendicular molecules now seem unrealistic. Reduction in $\mu_\perp$ is achieved either by dipole–dipole interaction,[5,25] or by introduction of clusters grouping water molecules with lower orientation,[4,6,8,9] or by apolar dimers[11] whose effect is essentially that of subtracting orientable monomeric species. It follows that what seems difficult to accept currently in existing models is apparently that water molecules could be only partially oriented at large negative charges, even if behaving as monomers.

The bulk structure of water seems now to be better described by an interstitial model.[323] Thus, the liquid may be envisaged[338] as being constituted of molecules bound to each other on a continuous network plus some interstitial free molecules. However, interstices may conceptually be envisaged in the bulk but hardly on the free surface of the phase. There, conditions are unfavorable for the existence of free molecules and all surface molecules may be imagined to be in bound sites, while some distortion of bonds gives rise to the surface potential χ^{H_2O}.

At the metal–solution interface, the presence of interstices is again conceptually possible, although they require quite different energetic conditions. Thus, the surface layer may now be envisaged as a continuous network of molecules plus some free molecules whose number and orientation are governed by the strength of metal–water interactions. The resulting model is thus qualitatively similar to those proposed by the other authors with the difference that clusters are independent of each other in the latter models, which conforms to the pattern of mixture models.

The effect of charge on an interstitial model for water at the interface may be envisaged as follows. If the charge is made increasingly positive, the network will be distorted without any necessary disruption of hydrogen bonds while an increasing number of molecules jump into the position pertaining to chemisorption. Both free and bound molecules are expected to contribute to the capacity of the interface so that the ultimate effect corresponds to cluster models.

A similar description holds at negative charges. The network will be distorted the other way around and an apparent saturation may be reached as hydrogen bonds are so bent as to impede further rotation. For more negative charges, some hydrogen bonds could actually be broken and some molecules jump into positions corresponding to a sharper orientation of the dipole with the positive end towards the metal. Ultimately, this model coincides with the four-state model of Parsons[6] although the description of aggregates is physically quite different. The continuous distortion of hydrogen bonds[17b,334,395] could explain the absence of a true steady value of C_i at $\sigma \ll 0$ replaced by a shallow minimum.

The statistical description of such a model requires the identification of possible limiting orientations. These may be defined not in terms of the relevant $\mu_\perp$, but directly in terms of the value of $g^{H_2O}(\text{dip})$ corresponding to full coverage with similarly oriented molecules. Thus, at $\sigma \ll 0$, the maximum potential drop may be taken to be 60 mV (Fig. 18), and the same opposite value may be postulated for the network contribution at strongly positive charges, which implies reasonably some feasibility in the symmetric distortion of hydrogen bonds. Ultimately, the network may be described in terms of an ensemble of "up" and "down" molecules with the above limiting orientations. Quantitatively, the same effect results from the presence of clusters but the physical background is different.

It is more difficult to give a molecular picture of chemisorbed molecules. It may be assumed that the limiting value of 0.4 V for $g^{H_2O}(\text{dip})_0$ on strongly hydrophilic metals (cf. Table 2) corresponds to a monolayer of chemisorbed water molecules. According to the present experimental evidence, this may be taken to be a limiting orientation. At the far negative end of the charge range, water molecules may be assumed to be in the opposite orientation but with the same limiting value of $g^{H_2O}(\text{dip})$ in that symmetrically distributed bonds are normally expected for water molecules even at the interface. If the Helmholtz formula, defined by Eq. (36), is applied, a molecularly significant picture may be derived only with some assumptions about ε and N. To a first approximation, ε may be taken equal to that relevant at the capacity minimum, viz., in the range 5.8–7.7, as shown above. N is usually taken[6,11,318] to be of the order of 10^{15} molecules cm^{-2}.

However, if voids and a possible square close-packed distribution are taken into account, it is possible[10,15] that N can be in the range 0.5 to 1×10^{15}. With $\mu = 1.84$ D, the angle made by chemisorbed molecules to the surface to give 0.4 V at full coverage is found to be in the range 20–62°. If μ is possibly higher for chemisorbed molecules, then the resulting limiting orientation may be even smaller. The same happens if the actual limiting orientation potential on d metals is higher than 0.4 V.

The picture which emerges from the above model is that water in the maximum allowed orientation may, however, be little oriented, which rules out[396] the possibility[61] of significant dipole–dipole repulsion. Molecules are on the contrary expected to be able to give hydrogen bonds on the surface and towards the bulk. The resulting molecular position seems reasonable. Some interaction with the metal through one of the lone pairs, with a possible angle of the dipole of about 30–50° to the surface, may be consistent with calculations from Eq. (36). Such a position is usually that calculated for water around ions[129,353,396–399] and for proton acceptor molecules with respect to proton donor molecules in hydrogen bonds.[335,353–355,357] In the case of water adsorption on ionic solids[400,401] some concrete evidence is available for oxides[402] that the plane of the molecules makes an angle of less than 45° to the surface.

The model outlined above assigns some structuring role to both the bulk liquid behind the monolayer and the metal surface facing it. The structure of the boundary region is thus expected to be a result of the two combined effects. The effect of temperature at negative charges consists in some easier distortion of the network under the action of the field. Thus, the limiting orientation is achieved with increasing charge more rapidly at higher than at lower temperature, as Fig. 25 shows. For a monolayer of freely orienting dipoles, an opposite effect is obtained, as the D^9 model in fact gives.

At charges close to zero, the disorienting effect of temperature on chemisorbed molecules is expected to prevail. But at positive charges, bonded molecules achieve the saturation orientation more rapidly, whereas free molecules jump more easily into the position of chemisorption because the network becomes softer. In other words, this model may predict an increase in the population of chemisorbed molecules at strongly positive charges but some decrease in orientation at charges close to zero as the temperature is increased. This is possible if the temperature has no effect on the number of chemisorbed molecules at low field, i.e., when the network is weakly distorted. It is interesting that theoretical calculations applied to X-ray diffraction measurements do show[339] that the population of free molecules is almost temperature invariant in the range 273–373 K.

The above view is apparently not consistent with a Boltzmann distribution of particles as assumed in the P and D models.[6,9] These unavoidably predict disruption of clusters as a coeffect of temperature, together with oriented dipole disorganization. If the former outweighs the latter, the population of chemisorbed molecules may increase with temperature. A Boltzmann distribution implies that all hydrogen bonds, or alternatively all like species, are energetically equivalent. This is the very point which may be questionable. The number and energy of hydrogen bonds are very likely to be interrelated[300,335] so that if the former is changed, so is the latter. This suggests that free molecules present at the interface may possibly come from the bulk, being adsorbed as foreign species. This implies that bulk-surface exchange equilibria should be accounted for rather than exchange equilibria between different species on the surface. This view could be consistent with Bewick's findings[272,273] that the density in the water layer increases as the charge is made more positive, a phenomenon apparently proportional to the hydrophilicity of the metal. It has been actually found[307] that, at especially hydrophilic surfaces, the density of water is much higher than in the bulk. In the above context, some preference could possibly be given to the DPS[8] model which leaves chemisorbed molecules to increase freely without allowing for a parallel decrease in aggregates.

The model outlined above is necessarily approximate because again it tends to simplify the structure of water probably beyond some allowable limits. It should be taken only as a first step towards an interstitial model for water at the interface. However, the following additional parameters may be useful to better focus the picture. It has been shown[10,14,15,53,54] that the inner-layer capacity is some function of the hydrophilicity at the

Double-Layer Potential Drop at Electrodes

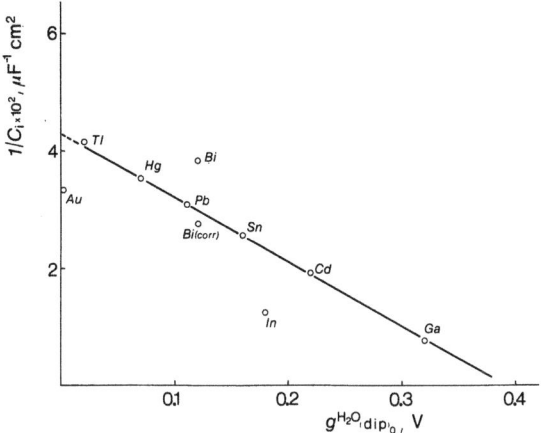

Figure 27. Plot of the reciprocal of the inner-layer capacity at the pzc as a function of the surface potential of water on the various metals. Data of C_i from Table 5.

pzc. The value of C_i in the hypothetical absence of chemisorption may be derived by extrapolating a plot of $1/C_i$ vs. $g^{H_2O}(\text{dip})_0$ to $g^{H_2O}(\text{dip})_0 = 0$. As Fig. 27 shows, a value of C_i of about 23 μF cm^{-2} is found. Au experimentally exhibits[403] a somewhat higher capacity than expected for absence of preferential orienta-

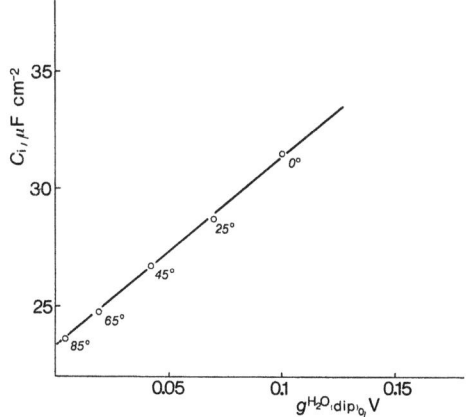

Figure 28. Plot of inner-layer capacity of Hg at the pzc in NaF aqueous solutions[65] as a function of T.

tion. The experimental value of C_i for Bi^{404} is too low, but a correction according to DPS^8 for the series capacity due to some space-charge in the metal, also places this metal correctly on the line. C_i at $g^{H_2O}(dip)_0 = 0$ can also be derived by plotting C_i at $\sigma = 0$ for Hg^{65} as a function of $g^{H_2O}(dip)_0$ at different temperatures and extrapolating to zero surface potential. This is shown in Fig. 28. A value of about 23 $\mu F\,cm^{-2}$ is again found for C_i, consistent with the result from Fig. 27.

VIII. NONAQUEOUS SOLVENTS

For solvents other than water, knowledge in the field of boundary layer potentials is more limited because the great number of possible solvents has led to a large spread of studies with little or insufficient focusing onto single systems. Available essential parameters like the temperature coefficient of surface potentials, charge of maximum adsorption of neutral substances, adsorption potential shifts, charge of the entropy maximum, etc. have, in general, not been evaluated for the same solvent. Thus, only limited quantitative conclusions can be drawn from the available experimental data. For the above reasons, attention will be focused only on those solvents for which the amount and quality of available data are such as to enable some definite conclusions to be reached.

1. Qualitative Aspects

Capacity curves for Hg in nonaqueous solvents have been reviewed by Payne,[405] who has also given some discussion of the degree of adhesion of a given solvent to the Hg surface. Capacity curves exhibit grossly the same essential features for all solvents, in that a shallow minimum can be observed at negative charges while the capacity rises much more steeply at more positive than at more negative charges. In detail, curves in different solvents differ noticeably, however, especially with regard to the presence or otherwise of some characteristic humps. Qualitatively, humps have in general been related to solvent reorientation[48] as implied in a naive two-state model for the solvent. A much more complex and sophisticated interpretation is needed[406] on a quantitative basis since the nature of humps may differ substantially in this respect.

(i) General Behavior

Parsons[78] has extended his four-state model to a variety of other solvents in an attempt to reproduce capacity curves on Hg electrode. He has classified capacity curves into three main groups, according to the qualitative shape as outlined in Fig. 29. (i) Water-like capacity curves are included in the first group. Simple and monosubstituted amides are assigned to this class. These curves are characterized by a pronounced maximum between two minima. (ii) Purely humped capacity curves have been classified into the second group. Various solvents with the common feature of being cyclic give similar curves. Only a maximum is detected in this case while minima, if any, are presumably outside the explorable range. (iii) Finally, water-unlike capacity curves form the third group. They have the form of an inverted parabola with the minimum usually at negative rational potentials.

The fit of calculated data to experimental points, as carried

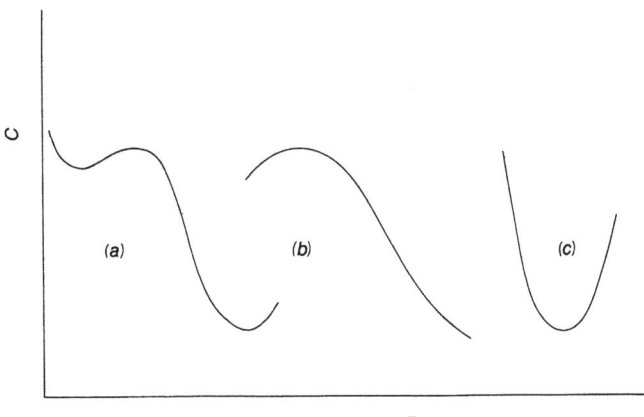

Figure 29. Sketch[78] of some typical capacity curves in nonaqueous solvents. (a) Water-like, (b) purely humped, and (c) water-unlike curves.

out by Parsons, is apparently satisfactory. Physically, as admitted by the author himself, it is much less significant. Curves of the second class are usually fitted with sometimes unreasonable parameters, which suggests some defects in the primitive model or alternatively some inadequate physical meaning attached to the hump. Group 3 curves, once fitted, also are found to be much more featured than appears at first sight. Further, the above classification has been suggested on the basis of curves with Hg electrodes. Other metals could help to throw light on possible concealed features. For instance curves in ethyl alcohol with Bi[407] apparently belong more to the class of purely humped curves, although a minimum is observable at negative charges. This may mean that the three classes are more interrelated than they appear at first sight in that some peculiar aspects may be obscured rather than missing.

The best fit is attainable[78] with the water-like curves, possibly as a consequence of the fact that the model works satisfactorily already with water.[6] Whereas the bulk structure of water has been studied extensively, this is not the case for most other solvents. Hence, some physical justification for the application of the original model, as such, to nonaqueous solvents is required. Some solvents have, however, been fairly well studied so that some more definite discussion is possible to illustrate the above view in these cases.

(ii) Formamide

Capacity curves for Hg in formamide exhibit[408–410] a well-developed, almost symmetrical hump at about $-8\,\mu\mathrm{C\,cm^{-2}}$, which has been attributed[410,411] to solvent reorientation. In terms of a cluster model,[6] since the hump is between two minima of approximately the same value, reorientation of aggregates should be involved. A definite preferential orientation of clusters at the pzc is required,[78] however, for the hump to appear at a negative charge. A model based on four limiting orientations for the solvent, regarded as constituted by a continuous network of single molecules, may be more appropriate in this case. The two minima would thus correspond to the two possible lower orientations.

The above description implies that, at the pzc, formamide is oriented with the negative end of the dipole facing the electrode. This is supported by a number of other indirect data. Parsons[412] has shown that the temperature coefficient of $E_{\sigma=0}$ for Hg in formamide is positive and higher than for water. He has suggested that the preferential orientation of the solvent should be more marked than that of water, but it has been shown above that this is not a generally valid conclusion. More simply, full disorganization of dipoles may be achieved more rapidly with increasing temperature in the case of formamide. The comparable preferential orientations for the two solvents is indicated[413] by the undetectable potential shift at $\sigma = 0$ for formamide adsorption on Hg from aqueous solutions. Further, the value of C_i at the hump[408] is much lower than that in the case of water,[64] while the values at the minimum are almost the same. This is taken to mean that the extent of dipole rotation with changing charge is lower for formamide than for water. It is thought that the absolute value of $g^S(dip)_0$ cannot directly be related to the magnitude of both its temperature coefficient and the height of the hump.

Measurements of excess entropy of the double layer confirm the reliability of the reorientation theory for formamide. The excess entropy exhibits[414] a maximum precisely at the charge of the hump. Formamide seems to have a simpler behavior than water, for which the maximum disorder does not occur at the hump. Data for other metals are quite scanty. The hump has been found[415] also with Pb electrodes but it appears to be shifted to more negative potentials. The adsorption of thiourea on Hg from formamide solutions has been studied by Dutkiewicz and Parsons,[411] who were unable to obtain an unambiguous dielectric hump presumably due to the small change in dielectric characteristics. The adsorption of diethylether, however, shows[411] clearly that at low coverage the maximum depression of capacity occurs precisely at the hump. In this case all available parameters give a consistent picture for the behavior of formamide. It is possible that the method devised here to derive $g^S(dip)$ from charge-potential curves is most suitable with this solvent in that σ_{max} and σ_{min} coincide both conceptually and physically. Enough data are unavailable to build up a solvophilicity scale for metals.

It seems[415] tentatively for formamide that changes in the nature of the metal produce less striking effects than in the case of water. Capacity curves in N-methyl formamide[416,417] exhibit the same features as those in formamide.

(iii) Methanol

A number of experimental measurements are also available for methanol. Grahame[418] suggested that, in methanol, Hg exhibits a capacity curve resembling that in water at higher temperatures. The positive shift of the pzc in the pure solvent upon adsorption of decalin on Hg[5] and the positive values of adsorption potential shift from water[419–421] suggest that methanol molecules are oriented with the negative ends of the dipoles facing the electrode, the extent of orientation being less than that for water molecules. This view is supported[422] by the positive temperature coefficient of $E_{\sigma=0}$ (with correction for the Φ effect). Also, thiourea adsorption suggests[422] that the minimum orientation probably occurs in the range -1 to $-2\,\mu C\,cm^{-2}$ where a minimum value of 5.6 has been derived for ε_i. As further support, the adsorption of dibutyl ether has been found[422] to be at a maximum in the same range of charge.

The behavior of methanol has been studied also on Bi electrodes.[423] Adsorption of thiourea suggests[424] that the apparent dielectric constant may be at a minimum at a small negative charge, consistent with the findings[422] for Hg. Some data are available also for Pb.[425] They show that the shape of the capacity curve is similar for Hg, Bi, and Pb.

Grahame[418] suggested that the idea of a negative dipole orientation at $\sigma = 0$ for methanol was misleading because, at the minimum in orientation, a maximum rather than a minimum of capacity should be observed. This is not true, in principle, since with water the minimum in orientation is not evidenced by any capacity maximum. It is thought that the behavior of methanol in this respect is similar to that of water, although the capacity curves with the former have been included[78] in the group of water-unlike curves. In other words, the values of C_i (and of the apparent ε_i) are determined by the rate of change of $g^S(\text{dip})$ with charge and not by the more or less pronounced orientation of the dipole layer. This is taken to mean that methanol is rapidly reoriented at both positive and negative charges as the charge is changed and this is responsible for the steep rise of capacity on both sides of the minimum.[426,427] Some Hg–methanol chemical interaction through the oxygen atom is probably to be taken into account.

The above view may be corroborated by the fact that methanol is the sole solvent for which an opposite orientation is likely to exist at Hg compared with the free liquid surface. Measurements[127] of $\Delta\chi$ between H_2O and CH_3OH, and of the temperature coefficient[428] of χ^{CH_3OH}, suggests that the orientation at the free surface is probably with the oxygen towards the bulk. Reasons for this might be sought in some relative freedom of methanol molecules in the liquid phase so that the interaction with Hg through the oxygen atom[429] may easily overcome the natural orientation with the methyl group towards the exterior of the liquid phase. This would imply that chemisorption of methanol at positive charges is relatively stronger than that occurring with H_2O molecules. The steep rise of capacity at negative charges, unlike the behavior in aqueous solutions, may be taken to be a consequence of the easy reacquisition of the natural orientation as the charge on the metal is no longer favorable.

(iv) Other Aliphatic Alcohols

Observations made with other aliphatic alcohols, especially ethanol, support the view taken for methanol. Hg with ethanol exhibits[426,427] a qualitatively similar curve but the negative end rises less. Data[127] suggest that the orientation of ethanol at the free surface is with the aliphatic chain towards the exterior of the phase and that such an orientation is retained[13,119] at Hg. Moreover, the ability of the molecule to reorientate under the field is expected to be lower with ethanol than with methanol due to size and steric effects. Linear aliphatic alcohols are known,[187] however, to reorient on some metals as the charge comes from the negative side close to zero as a result of increasing metal–OH interaction. No hump is observed with Hg [426,427] and Pb,[415,430] but something similar is apparent with Bi[407] at about 5 $\mu C\,cm^{-2}$, whereas the rise at negative charges seems to decrease in the same sequence. The capacity rise at the extreme negative end is apparently less steep the longer the aliphatic chain of the alcohol,[431] which indicates some increasingly rigid orientation. The rise at positive charges is always observable even though it decreases abruptly from butanol to pentanol. A U-shaped curve is observed[432–434] also with ethylene glycol but less markedly than with methanol.

(v) Dimethylformamide

Payne[435] has investigated a number of substituted amides. Capacity curves in dimethylformamide with Hg,[435,436] Pb,[415] and Bi[437] do not exhibit the characteristic hump. It has been suggested[438,439] that such a solvent should exhibit a negative χ^S. The same orientation is apparently retained at Hg electrodes[413] in view of the positive potential shift for adsorption from aqueous solutions. Payne[413] has suggested that the positive shift may simply be a result of oriented molecules being replaced, while DMF itself would adsorb flat. However, the magnitude of the shift at full coverage exceeds that expected for a monolayer of water molecules. A definite preferential orientation of DMF on Hg is suggested also by Salem,[440] and could also be inferred[441] from the negative shift of potential upon the adsorption of naphthalene and diphenyl, although effects due to π electrons cannot be excluded.

Adsorption of thiourea on Hg from DMF solutions has been studied by Ganzhina and Damaskin.[442] They have found that application to this system of Parsons' analysis gives for ε of the inner layer a minimum at about -2 to $-3\ \mu C\,cm^{-2}$ and a value of about 5 at the pzc. This is much less than the value of about 8 derived[443] from measurement of adsorption of thiourea on Bi. It is possible that the approximation of using the molecular value of $\mu_\perp$ for thiourea is not equally valid on all metals. It has been suggested that the orientation of DMF on Bi should change from essentially vertical when $\sigma \ll 0$ to highly planar when $\sigma > 0$.

(vi) Acetonitrile

Studies from more than one laboratory are available for acetonitrile.[101,444,445] The capacity curve with Hg resembles that in alcohols but the minimum is not actually reached and only a steady decrease of C_i is observable. A steeper increase of capacity has been found at positive charges. The nature of the metal surface is an important factor in this range and the capacity at constant charge can be seen[54,101,446] to increase in the series Hg < In–Ga < Ga, as shown in Fig. 30. The study of adsorption of thiourea,[447] which is complicated[448] by the breakdown of the classical Gouy–Chapman theory, shows that such a substance becomes less adsorbed on In–Ga than on Hg.

χ at the free surface is probably small and negative.[449] The same orientation is presumably retained at Hg when the molecule is adsorbed from aqueous solutions.[450] It is not known whether this is conditioned by the presence of bulk water. The same orientation is, however, very likely to exist also at the Hg–pure-acetonitrile interface. In such a case, the low constant capacity all over the negative range of charge may be taken as indicating

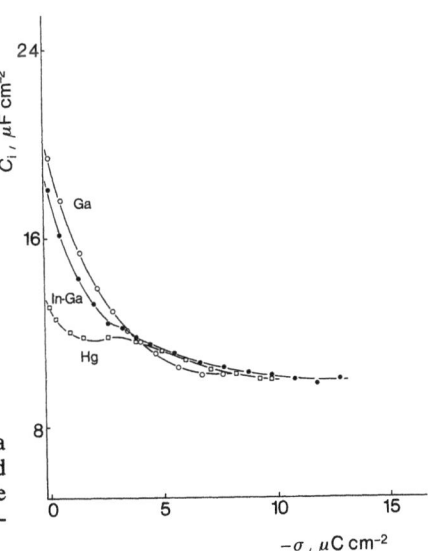

Figure 30. Inner-layer capacity as a function of charge for Hg, Ga, and In–Ga in 0.1 M LiClO$_4$ acetonitrile solutions. Original data, by the courtesy of Dr. I. A. Bagotskaya.

some rigid orientation. The absence of capacity rise at the extreme negative end is consistent with the idea that this is due to reorientation rather than to electrostriction.

The rise of capacity at positive charges may presumably be related to some very small rotation of the molecule (probably, along its longer axis parallel to the surface). The increase in capacity from Hg to Ga may be understood in terms of some increased interaction of the metal with the CN group which facilitates the reorientation under the positive field. It would be interesting to investigate the entropy change for this solvent and the charge of maximum adsorption of neutral substances.

An interesting feature is that capacity values at strongly negative charges for Hg, In–Ga, and Ga coincide in acetonitrile,[101] whereas In–Ga consistently exhibits[99] somewhat higher values of C_i at the minimum in aqueous solutions. It has been suggested[102] that the latter behavior could be due to an effect of the surface electronic distribution of metals. However, for this to be possible, other metals should contribute a lower capacity from the surface region.

A capacity contributed by the metal surface was actually taken into account by Watts-Tobin[49] but later this aspect was abandoned. The case of In–Ga and In may in some respects support Watts-Tobin's idea, although it is really difficult to envisage why the above metals should contribute a higher metallic capacity than Hg. This aspect is considered to be still unsettled and obscure. It is important to note that if the metal surface were to contribute to the capacity, then the correction applied[101] to charge to obtain parallel linear sections in charge–potential curves would no longer be justified. Moreover, even C_i at $\sigma = 0$ would no longer be directly relatable to the hydrophilicity scale.

(vii) Dimethylsulfoxide

Dimethylsulfoxide (DMSO) is another solvent for which data for different metals exist. Payne[451] found a hump at about 8 μC cm^{-2} in the capacity curves at Hg. Thus, this solvent may be classified[78] into the second group with regard to the shape of the capacity curves. The effect of temperature is small on the latter. The hump has been explained in terms of reorientation. This

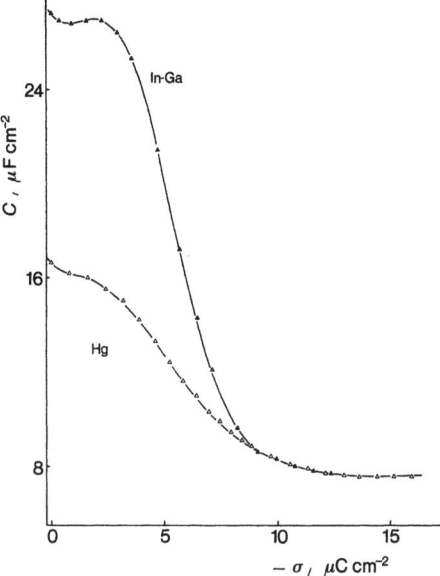

Figure 31. Inner-layer capacity as a function of charge for Hg and In–Ga in 0.1 M LiClO$_4$ dimethylsulfoxide solutions. Original data, by the courtesy of Dr. I. A. Bagotskaya.

would imply some preferential orientation of molecules with the positive ends of the dipoles towards the metal at the pzc. A positive shift of $E_{\sigma=0}$ has been actually found[421] when DMSO is adsorbed on Hg from aqueous solutions. The capacity at constant charge is observed (Fig. 31) to increase in the series[446] Hg < In–Ga < Ga, whereas the adsorption potential shift decreases in the same sequence. Chemical interaction between the metal surface and the functional group is probably responsible for this, which gives some further support to chemical theories of double-layer solvent behavior.

(viii) Sulfolane

Sulfolane has been studied by Lawrence and Parsons.[452] The positive shift of $E_{\sigma=0}$ upon adsorption from aqueous solutions[421] apparently indicates some preferential orientation with the positive ends of the dipoles towards the metal. The capacity has been found to be affected very little by an increase in temperature, although some decrease is apparent.

(ix) Formic Acid

The opposite orientation with respect to sulfolane has been postulated[453] for formic acid molecules at Hg. This is supported by the maximum adsorption of ethylether found at about $-4\ \mu C\ cm^{-2}$. That the adsorption of the solvent is rather strong in this case may be inferred from the linear charge dependence of ΔG_{ad}^0 observed for the above adsorbate at positive charge, presumably as a result of desorption of strongly oriented solvent dipoles. The shape of the capacity curve on Hg is the same as that for alcohols and the interpretation may be similar. The orientation of formic acid at the free surface is, however, possibly[127] with the positive ends of the dipoles towards the exterior of the phase. This behavior resembles that of methanol.

(x) Ethylene Carbonate

The structure of the Hg–ethylene-carbonate interface has been investigated by Fawcett and Mackey.[454] The hump observed at positive charges has been attributed to reorientation of solvent molecules in the inner layer placed with the positive ends of the dipoles towards the metal at more negative charges. This work is interesting because an attempt was made to discuss the hump in terms of the model by Levine et al.[55] Actually, only qualitative agreement has been obtained at the hump and the shape of the calculated curve does not seem to be entirely satisfactory. Adsorption of thiourea shows[455] that the dielectric constant of the inner layer decreases monotonically from about 10 at $+4\ \mu C\ cm^{-2}$ to about 6.5 at $-6\ \mu C\ cm^{-2}$.

Capacity curves in propylene carbonate[456,457] exhibit the same features as those in ethylene carbonate.

2. Quantitative Aspects

Only a very limited number of quantitative data are available for nonaqueous solvents. Further, insufficient data are available to derive absolute quantities independently so that only parameters relative to water can be obtained. Thus, the accuracy of the parameters for nonaqueous solvents depends on the accuracy with which the same parameters are known for water.

Jakuszewski et al.[458-460] attempted to derive quantitative relationships between $E_{\sigma=0}$ and Φ for a number of metals in several solvents. Linear parallel relationships with unit slope were shown[458] to be obeyed by all metals. The authors claimed that this was clear experimental support for the thermodynamic derivation[57,62] of constancy in the quantity $E_{\sigma=0} - \Phi/e$). As has been discussed above such a proposal has no sound thermodynamic basis because it rests on the idea that the Volta potential difference between a metal and a solvent at the pzc is zero. As a matter of fact, if values[461] of $E_{\sigma=0}$ for Hg in ethanol and methanol are introduced into Jakuszewski's graphical relationships,[458] while expected to fit best on the lines, they are found to fall far from them. Much more experimental data for different metals are necessary for reliable $E_{\sigma=0}$ vs. Φ plots in nonaqueous solvents to be made.

Values of χ may be derived for a few solvents. Values relative to water have been measured by a number of authors.[127,438,439,449,462-473] The absolute values summarized in Table 11 have been obtained from the experimental $\Delta\chi$ by adding 0.13 V as the χ value for water according to Eq. (21). The scatter of data for ethanol is less than for methanol. For the latter, the value derived from a $\Delta\chi$ of about -0.30 V is preferred for the reasons given below.

Reliable data for $E_{\sigma=0}$ are now available[101,446,474,475] for some metals in a number of solvents. Values of the Volta potentials may be obtained[475] if real free energies of hydration of H^+ are known in the respective solvents. Equation (9) may in fact be written as

$$E_{\sigma=0} = \Phi/e + \delta\chi^M - \delta\chi^S - E_k(H^+/H_2) \tag{41}$$

The last term on the right hand side of Eq. (41) has been shown[145] to be entirely calculable. It consists of the sum of dissociation, ionization, and real hydration free energies for hydrogen. It has been found[476] to be 4.44 V in water. Volta potentials can be obtained from Eq. (41) by recalling Eq. (26), viz.

$$\Delta\psi_{\sigma=0} = E_{\sigma=0} - \Phi/e + E_k(H^+/H_2) \tag{42}$$

Table 11
Surface Potential of Nonaqueous Solvents at the Free Surface

Methanol	Ethanol	Isopropanol	Acetone
$-0.17 \pm 0.06^{a,b}$	$-0.19 \pm 0.05^{a,c}$	-0.26 ± 0.03^{d}	$0.18 \pm 0.06^{a,e}$
-0.18^{f}	-0.25^{f}		-0.33 to $-0.41^{a,g}$
-0.21^{h}	-0.26^{h}		
-0.22^{i}	-0.26^{j}		
-0.29^{k}	-0.286 ± 0.03^{d}		
	-0.29^{k}		

Acetonitrile	Formamide	N-Methylformamide	Dimethylformamide
-0.10 ± 0.06^{l}	$0.05 \pm 0.05^{a,m}$	$-0.11 \pm 0.05^{a,n}$	-0.51^{o}
			-0.52^{p}
			-0.72^{q}

^a From the difference between real and chemical free energy of ion solvation.
^b From Ref. 466.   ^h From Ref. 67.   ^m From Ref. 464.
^c From Ref. 468.   ⁱ From Ref. 462.   ⁿ From Ref. 467.
^d From Ref. 470.   ^j From Ref. 491.   ^o From Ref. 438.
^e From Ref. 463.   ^k From Ref. 472.   ^p From Ref. 472.
^f From Ref. 127.   ^l From Ref. 449.   ^q From Ref. 439.
^g From Ref. 465.

More simply, Volta potentials may be obtained from the potentials of zero charge measured with respect to the same reference electrode in aqueous solution. In this case Eq. (41) gives

$$\Delta_{M_2}^{M_1} E_{\sigma=0} = \Delta_{M_2}^{M_1}\Phi/e + \Delta_{M_2}^{M_1}\Delta\psi_{\sigma=0} \tag{43}$$

With reference to Eq. (26), $\delta\chi^S$ cannot be experimentally disentangled from $\delta\chi^M$. If, however, $\delta\chi^M$ is assumed to be solvent invariant, $\delta\chi^S$ can be derived from the Volta potentials obtained by using -0.32 V for $\delta\chi^M$ for Hg in water. Values of $\delta\chi^S$ are shown in Table 12, column 3. Some estimate of $g^S(dip)_0$ can now be obtained by using values of χ^S reported in Table 11. Table 12, column 4, summarizes the results.

An interesting feature is that DMF and ethanol retain at Hg the preferential orientation exhibited at the free surface but the extent is lowered by the presence of the metal. The potential shift due to ethanol adsorption on the surface of Hg from aqueous solutions has been found[119] to be 0.36 V. However, Table 12 shows that $g^{H_2O}(dip)_0 - g^{EtOH}(dip)_0 = 0.24$ V. This is taken to mean that ethanol adsorbed on Hg from aqueous solutions retains the same orientation as that at the free surface, but the extent of orientation is less if the molecule is adsorbed on Hg from the bulk liquid. Presumably, the hydrocarbon chain is more repelled from aqueous solutions and tends to orientate perpendicularly to the surface. It is interesting that $\Delta_{H_2O}^{EtOH}g(dip)_0$ has been estimated by Jakuszewski et al.[470] by means of an extrapolation procedure to be in the range 0.26–0.276 V, in good agreement with the value independently derived above. On the contrary, methanol exhibits opposite orientations at the free surface and at electrodes. However, in order to obtain a positive value for $g^{CH_3OH}(dip)_0$, a value of $\Delta\chi$ in Table 12 close to -0.30 V should be preferred. Reasons for the scatter of data with methanol are not known. Also, some other considerations[462] seem, however, to suggest that a more probable value of χ for methanol is around -0.20 V.

Once values of $g^S(dip)_0$ are known for Hg, corresponding values for Bi may be easily derived with the aid of the many data available for this metal.[474] It is only necessary to assume that Eq. (11) holds irrespective of the solvent, a situation which will be later shown to be reasonably the case. Values of $g^S(dip)_0$ for Bi are shown in Table 12, column 5. More positive surface potentials are always observed on Bi than on Hg. This is further support for the idea that the lyophilicity scale should be qualitatively independent of the solvent, provided that adsorption forces are qualitatively similar. The electronegativity of the surface is thus the primary factor[3,15] governing such a scale. It is interesting

Table 12
Absolute Values of Surface Potential of Nonaqueous Solvents at Metal/Solution Interfaces at the Potential of Zero Charge[a]

Solvent	Hg		$g^S(dip)_0$ (V)	
	$\Delta\psi_{\sigma=0}$ (V)	$\delta\chi^S$ (V)	Hg	Bi
Methanol	−0.51	0.19	0.01	0.05
Ethanol	−0.44	0.12	−0.17	−0.13
Dimethylformamide	−0.67	0.35	−0.17	−0.13

[a] From the data of Refs. 474 and 475.

that $\Delta_{Hg}^{Bi}g^S(dip)_0$ is almost constant for the various cases (Table 13). This may, however, be only an approximation in view of the small difference in lyophilicity between the two metals.

Quantitative comparisons for nonaqueous solvents on the basis of charge–potential curves have been made only recently for Ga and In–Ga by Frumkin, Bagotskaya, and co-workers.[54,101,446] Some of the results are shown in Fig. 32. The differences in $E_{\sigma=0}$ between In–Ga and Hg have been found[101] to be higher in H_2O than in CH_3CN. At strongly negative charges, ΔE is, on the contrary, almost solvent independent. C_i at the pzc is 16 $\mu F\,cm^{-2}$ on Hg and as high as 28.5 $\mu F\,cm^{-2}$ on In–Ga, which points to a rapid reorientation with charge on In–Ga when passing from negative to positive charges.

Quantitative data are also available[101] for Ga in acetonitrile (ACN). $\Delta E_{\sigma=0}$ in relation to Hg is 0.29 V in ACN and 0.50 V in water, whereas $\Delta E_{\sigma \ll 0} = 0.24$ V, in agreement with the value found in water according to Fig. 4. The same value is practically observed[446] also in DMSO. No absolute value can be derived for $g^S(dip)_0$ since no quantitative data are independently available for Hg. Possibly,[450] for ACN, $g^S(dip)_0$ is small and positive so that it may be close to zero on In–Ga and Ga. Solvent adsorption potentials referred to Hg are summarized in Table 13.

Table 13
Relative Values of Surface Potential of Nonaqueous Solvents at Metal/Solution Interfaces at the Potential of Zero Charge

Solvent	$\Delta g^S(dip)_0$ (V)		
	Bi/Hg[a]	In–Ga/Hg[b,c]	Ga/Hg[c]
Water	0.05	0.14	0.25
Methanol	0.04	—	—
Ethanol	0.04	—	—
Dimethylformamide	0.04	—	—
Dimethylsulfoxide	0.10	0.20	0.47
Acetonitrile	0.03	0.05	0.06

[a] From Refs. 474 and 475.
[b] From Ref. 101.
[c] From Ref. 446.

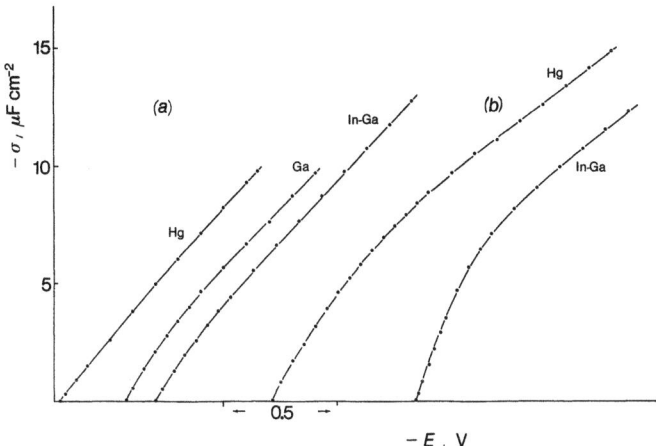

Figure 32. Charge–potential curves for Hg, Ga, and In–Ga in nonaqueous solvents by integration of curves in Figs. 30 and 31. (a) Acetonitrile; (b) dimethylsulfoxide. Original data, by the courtesy of Dr. I. A. Bagotskaya.

In DMSO,[446] $\Delta E_{\sigma=0} = 0.63$ V between Hg and In–Ga and 0.71 V between Hg and Ga. It is interesting that for both metals the value of $\Delta E_{\sigma=0}$ with respect to Hg is higher than in water (0.48 and 0.50 V, respectively[71]). Presumably, this is a consequence of the particularly activated oxygen atom in DMSO. The above trend is confirmed by the fact that C_i at the pzc is 24 μF cm^{-2} for Hg and 54 μF cm^{-2} for In–Ga as compared to 28 μF cm^{-2} for Hg and 60.5 μF cm^{-2} for In–Ga in water. Thus, a greater increase in C_i in DMSO than in water is observed, which is consistent with the expected stronger adsorption of DMSO on In–Ga than on Hg.

In conclusion, quantitative data for nonaqueous solvents are still relatively scarce but they may corroborate the belief that the chemical approach to the properties of the double layer is a correct one because it enables a general rational picture of the situation to be given. It seems that broad aspects of the structure of the electrode–solution interface can now be approached with more confidence, although a number of details call for further explanation. The shape of the capacity curves and the solvent dipole contribution to the electrode potential may actually depend on the combined action of three kinds of interaction: solvent towards the bulk liquid, solvent towards the metal, and solvent in the field. Depending on which factor prevails, capacity curves may present one or even more than one hump, or none at all.

Although some rationalization beyond that attempted by Parsons[78] is impossible at present, it may be tentatively suggested that water-like curves should be typical of solvents with strong interaction between molecules on the surface and in the bulk which may restrict the freedom of rotation in the field. If the metal–solvent interaction is strong enough, the capacity rise at positive charges starts at less positive values and may produce an apparent shift of the hump as a combined effect. If weak metal–solvent interactions are operative, the hump remains totally uncovered as in the case of formamide[408,410] and N-methyl formamide. Alcohols may be envisaged, on the basis of capacity curves, as solvents with relatively free molecules. Under such circumstances, interactions with both the field and the metal may be responsible for the much more pronounced reorientation phenomena with a steep capacity rise at both ends of the curve. This appears[431] to be reduced for long-chain alcohols, especially at negative charges and with hydrophobic metals.

Professor Frumkin ended one of his last papers[54] before his death with a critical remark about a statement by Grahame[66] reported by Payne[477] in a review, that "nearly everything one desires to know about the electrical double-layer is ascertainable with Hg surfaces, if it is ascertainable at all." We may now state with confidence that if the problem of the structure of the double layer at electrodes is perhaps going towards a more definite comprehension, an undoubtedly decisive impulse has certainly been given by the experimental work carried out on metals other than Hg; this is because that made at Hg gives insufficiently general information, yet it is more than necessary to disentangle the different factors which have been described above and to go into quantitative details.

ACKNOWLEDGMENTS

Thanks are due to Dr. A. De Battisti for stimulating discussions during the writing of this chapter. The author is indebted to Dr. I.

A. Bagotskaya for making her original capacity data in different solvents available. Financial support by the National Research Council (C.N.R., Rome) is gratefully acknowledged.

NOTATION

$\Delta E_{\sigma=0}$	Adsorption potential shift at zero charge.
$E_{\sigma=0}$	Potential of zero charge.
C_1	Capacity at full coverage with organic adsorbate.
ϑ	Coverage with organic adsorbate.
$g^S(\text{dip})_0$	Surface potential of solvent S at electrodes at the pzc.
$g^S(\text{dip})_\sigma$	Surface potential of solvent S at electrodes at a charge σ.
χ^M	Surface potential of metal M at the free surface.
χ^S	Surface potential of solvent S at the free surface.
$\Delta\psi_{\sigma=0}$	Volta potential between metal and solution at the pzc.
$g^M(\text{dip})$	Surface potential of metal M at interfaces.
$\delta\chi^M$	Interaction term due to contact of metal M with another phase.
$\delta\chi^S$	Interaction term due to contact of solvent S with another phase.
σ	Free-charge density on a phase.
q	Total-charge density on a phase.
$g(\text{ion})$	Component of the metal–solution potential drop due to free charges.
$g(\text{dip})$	Component of the metal–solution potential drop due to dipolar layers on the phases.
E	Electrode potential measured with respect to a given reference electrode.
$\alpha^S_{M^z}$	Real potential of ion M^z in phase S.
$\Delta\phi$	Electric potential drop between two phases.
$\mu^S_{M^z}$	Chemical potential of ion M^z in phase S.
A	Area projected by an adsorbate molecule.
ΔH^0_{ox}	Standard enthalpy of formation of the oxide MO.
Φ	Work function of metals.
ε_i	Apparent dielectric constant of a solvent in the inner layer.
ε_o	Orientational part of the apparent dielectric constant of a solvent.

ε_m	Molecular part of the apparent dielectric constant of a solvent.
d_i	Thickness of the inner layer.
μ	Dipole moment of a molecule.
$\mu_\perp$	Normal component of the dipole moment of a molecule of adsorbate.
K_{ion}	Integral capacity associated with a layer of solvent at fixed dipole orientation.
C_{dip}	Component of the inner-layer capacity associated with reorientation of solvent dipoles.
C_{ion}	Differential capacity associated with a layer of solvent at fixed dipole orientation.
C_i	Inner-layer capacity.
a	Particle–particle interaction term in the Frumkin isotherm.
σ_{min}	Charge where the net dipole orientation of solvent molecules at the metal is zero.
σ_{max}	Charge where the entropy of formation of the inner layer is a maximum.
σ_{max}^{org}	Charge where the adsorption of organic substances is a maximum.
$g^{org}(dip)$	Surface potential of a monolayer of adsorbate at a metal–solvent interface.
$\Delta E_{air/sol}$	Potential shift upon adsorption of substances at the free surface of a solvent.
X	Electric field.
N	Population of solvent molecules at metal–solution interfaces.

ABBREVIATIONS

DF	Damaskin and Frumkin's model for water at electrodes.
P	Parsons' model.
DPS	Damaskin, Palm, and Salve's model.
D	Damaskin's model.
BH	Bockris and Habib's model.
BDM	Bockris, Devanathan, and Müller's model of double layer.

ACN   Acetonitrile
DMSO Dimethylsulfoxide
DMF   Dimethylformamide
OHP   Outer Helmholtz Plane

REFERENCES

[1] A. Frumkin, *Z. Phys.* **35** (1926) 792.
[2] R. Parsons, in *Modern Aspects of Electrochemistry*, Ed. by J. O'M. Bockris, Butterworths, London, 1954, Vol. 1.
[3] S. Trasatti, *J. Electroanal. Chem.* **33** (1971) 351.
[4] B. B. Damaskin and A. N. Frumkin, *Electrochim. Acta* **19** (1974) 173.
[5] B. E. Conway and H. P. Dhar, *Croat. Chem. Acta* **45** (1973) 109.
[6] R. Parsons, *J. Electroanal. Chem.* **59** (1975) 229.
[7] A. De Battisti and S. Trasatti, *Croat. Chem. Acta* **48** (1976) 607.
[8] B. B. Damaskin, U. V. Palm, and M. A. Salve, *Elektrokhimiya* **12** (1976) 232.
[9] B. B. Damaskin, *J. Electroanal. Chem.* **75** (1977) 359.
[10] S. Trasatti, *J. Chim. Phys.* **72** (1975) 561.
[11] J. O'M. Bockris and M. A. Habib, *Electrochim. Acta* **22** (1977) 41.
[12] S. Trasatti, *J. Electroanal. Chem.* **64** (1975) 128.
[13] S. Trasatti, *J. Chem. Soc. Faraday Trans. 1* **70** (1974) 1752.
[14] S. Trasatti, *Gazz. Chim. Ital.* **106** (1976) 219.
[15] S. Trasatti, in *Advances in Electrochemistry and Electrochemical Engineering*, Ed. by H. Gerischer and C. W. Tobias, Interscience, New York, 1977, Vol. 10.
[16] R. Payne, in *Progress in Surface and Membrane Science*, Ed. by J. F. Danielli, M. D. Rosemberg, and D. A. Cadenhead, Academic Press, New York, 1973, Vol. 6; previous reviews are also quoted in this paper.
[17] (a) R. M. Reeves, in *Modern Aspects of Electrochemistry*, Ed. by B. E. Conway and J. O'M. Bockris, Plenum Press, New York, 1974, Vol. 9; (b) B. E. Conway, *Adv. Colloid Interface Sci.* **8** (1977) 91.
[18] W. Lorenz and G. Salié, *Z. Phys. Chem. (Leipzig)* **218** (1961) 259.
[19] R. Parsons, in *Advances in Electrochemistry and Electrochemical Engineering*, Ed. by P. Delahay and C. W. Tobias, Interscience, New York, 1970, Vol. 7.
[20] J. W. Schultze and K. J. Vetter, *J. Electroanal. Chem.* **44** (1973) 63.
[21] A. Frumkin, B. Damaskin, and O. Petrii, *J. Electroanal. Chem.* **53** (1974) 63.
[22] A. N. Frumkin, B. B. Damaskin, and O. A. Petrii, *Elektrokhimiya* **12** (1976) 3.
[23] A. N. Frumkin and O. A. Petrii, *Electrochim. Acta* **15** (1970) 391.
[24] A. N. Frumkin and O. A. Petrii, *Electrochim. Acta* **20** (1975) 347.
[25] J. O'M. Bockris, M. A. V. Devanathan, and K. Müller, *Proc. R. Soc. London Ser. A* **274** (1963) 55.
[26] D. M. Mohilner, in *Electroanalytical Chemistry*, Ed. by A. J. Bard, Dekker, New York, 1966, Vol. 1.
[27] M. J. Spaarnay, *Adv. Colloid Interface Sci.* **1** (1967) 277.
[28] J. O'M. Bockris and M. A. Habib, *J. Electroanal. Chem.* **65** (1975) 473.
[29] J. A. Harrison, J. E. B. Randles, and D. J. Schiffrin, *J. Electroanal. Chem.* **48** (1973) 359.

[30] A. Frumkin, *Ergeb. Exakt. Naturwiss.* **7** (1928) 235.
[31] G. Gouy, *Ann. Phys. (Leipzig)* **7** (1917) 176.
[32] A. N. Frumkin, *Electrocapillary Phenomena and Electrode Potentials*, Odessa, 1919.
[33] A. N. Frumkin, *Philos. Mag.* **40** (1920) 363.
[34] A. Frumkin and A. Gorodetzkaya, *Z. Phys. Chem.* **136** (1928) 451.
[35] A. Frumkin, N. Polianovskaya, N. Grigoryev, and I. Bagotskaya, *Electrochim. Acta* **10** (1965) 793.
[36] A. N. Frumkin, N. B. Grigoryev, and I. A. Bagotskaya, *Dokl. Akad. Nauk SSSR* **157** (1964) 957.
[37] N. B. Grigoryev, V. A. Bulavka, and Yu. M. Loshkarev, *Elektrokhimiya* **11** (1975) 1404.
[38] A. N. Frumkin, *J. Res. Inst. Catal. Hokkaido Univ.* **15** (1967) 61.
[39] S. Trasatti, *Chim. Ind. Milan* **53** (1971) 559.
[40] A. Frumkin and A. Gorodetzkaya, *Z. Phys. Chem.* **136** (1928) 215.
[41] A. Murtazajew and A. Gorodetzkaya, *Acta Physicochim. URSS* **4** (1936) 75.
[42] D. Grahame, *Proceedings of the Fourth Conference on Electrochemistry*, Consultants Bureau, New York, 1961, Vol. 1.
[43] D. C. Grahame, *Anal. Chem.* **30** (1958) 1736.
[44] D. I. Leikis and E. S. Sevastyanov, *Dokl. Akad. Nauk SSSR* **144** (1962) 1320.
[45] A. N. Frumkin, N. S. Polyanovskaya, and N. B. Grigoryev, *Dokl. Akad. Nauk SSSR* **157** (1964) 1455.
[46] I. A. Bagotskaya, A. M. Morozov, and N. B. Grigoryev, *Electrochim. Acta* **13** (1968) 873.
[47] J. N. Butler and M. L. Meehan, *J. Phys. Chem.* **70** (1966) 3582.
[48] J. R. Macdonald, *J. Chem. Phys.* **22** (1954) 1857.
[49] R. J. Watts-Tobin, *Philos. Mag.* **6** (1961) 133.
[50] N. F. Mott and R. J. Watts-Tobin, *Electrochim. Acta* **4** (1961) 79.
[51] B. B. Damaskin, N. V. Nikolaeva-Fedorovich, and A. N. Frumkin, *Dokl. Akad. Nauk SSSR* **121** (1958) 129.
[52] C. A. Barlow, in *Physical Chemistry. An Advanced Treatise*, Ed. by H. Eyring, Academic Press, New York, 1970, Vol. 9A.
[53] A. Frumkin, B. Damaskin, N. Grigoryev, and I. Bagotskaya, *Electrochim. Acta* **19** (1974) 69.
[54] A. Frumkin, I. Bagotskaya, and N. Grigoryev, *Denki Kagaku* **43** (1975) 2.
[55] S. Levine, G. M. Bell, and A. L. Smith, *J. Phys. Chem.* **73** (1969) 3534.
[56] A. N. Frumkin, *Svensk Kem. Tidskr.* **77** (1965) 300.
[57] B. Jakuszewski, *Bull. Acad. Pol. Sci. Ser. Chim.* **9** (1961) 11.
[58] V. Novakovsky, E. Ukshe, and A. Levin, *Zh. Fiz. Khim.* **29** (1955) 1847.
[59] L. I. Antropov, *Ukr. Khim. Zh.* **29** (1963) 555.
[60] R. Vasenin, *Zh. Fiz. Khim.* **27** (1953) 878; **28** (1954) 1672.
[61] A. Frumkin, B. Damaskin, I. Bagotskaya, and N. Grigoryev, *Electrochim. Acta* **19** (1974) 75.
[62] B. Jakuszewski and Z. Kozlowski, *Rocz. Chem.* **36** (1962) 1873.
[63] J. R. Macdonald and C. A. Barlow, *J. Chem. Phys.* **36** (1962) 3062.
[64] D. C. Grahame, *J. Am. Chem. Soc.* **76** (1954) 4819.
[65] D. C. Grahame, *J. Am. Chem. Soc.* **79** (1957) 2093.
[66] D. C. Grahame, *Chem. Rev.* **41** (1947) 441.
[67] A. N. Frumkin, Z. A. Iofa, and M. A. Gerovich, *Zh. Fiz. Khim.* **30** (1956) 1455.
[68] A. Frumkin, *Electrochim. Acta* **2** (1960) 351.

[69] J. O'M. Bockris, M. Green, and D. A. J. Swinkels, *J. Electrochem. Soc.* **111** (1964) 743.
[70] R. Parsons and F. G. R. Zobel, *J. Electroanal. Chem.* **9** (1965) 333.
[71] J. E. B. Randles, *Trans. Faraday Soc.* **52** (1956) 1573.
[72] S. Trasatti, *J. Electroanal. Chem.* **28** (1970) 257.
[73] G. J. Hills and R. Payne, *Trans. Faraday Soc.* **61** (1965) 326.
[74] B. E. Conway and L. G. M. Gordon, *J. Phys. Chem.* **73** (1969) 3609.
[75] G. J. Hills and S. Hsieh, *J. Electroanal. Chem.* **54** (1974) 289.
[76] N. H. Cuong, C. V. D'Alkaine, A. Jenard, and H. D. Hurwitz, *J. Electroanal. Chem.* **51** (1974) 377.
[77] P. Vanel, *C.R. Acad. Sci. Ser. C* **282** (1976) 287.
[78] R. Parsons, *Electrochim. Acta* **21** (1976) 681.
[79] J. O'M. Bockris and M. A. Habib, *J. Electrochem. Soc.* **123** (1976) 24.
[80] E. Lange and K. Mischenko, *Z. Phys. Chem.* **149** (1930) 1.
[81] H. Gerischer, in *Advances in Electrochemistry and Electrochemical Engineering*, Ed. by P. Delahay and C. W. Tobias, Interscience, New York, 1961, Vol. 1.
[82] D. L. Begley, D. A. Bryan, R. W. Alexander, R. J. Bell, and C. A. Goben, *Surface Sci.* **60** (1976) 99.
[83] C. P. Smyth, *Dielectric Behaviour and Structure*, McGraw-Hill, New York, 1955.
[84] J. B. Hasted, *Aqueous Dielectrics*, Chapman and Hall, London, 1973.
[85] J. B. Hasted, in *Structure of Water and Aqueous Solutions*, Ed. by W. A. P. Luck, Verlag Chemie, Berlin, 1974, p. 377.
[86] P. Schuster, in *Structure of Water and Aqueous Solutions*, Ed. by W. A. P. Luck, Verlag Chemie, Berlin, 1974, p. 115.
[87] F. H. Stillinger, in *Advances in Chemical Physics*, Ed. by I. Prigogine and S. Rice, Wiley, New York, 1975, Vol. 31.
[88] L. Antropov, *Kinetics of Electrode Processes and Null Points of Metals*, Council of Scientific and Industrial Research, New Delhi, 1960.
[89] N. B. Grigoryev, V. P. Kuprin, and Yu. M. Loshkarev, *Elektrokhimiya* **9** (1973) 1842.
[90] V. Ya. Bartenev, E. S. Sevastyanov, and D. I. Leikis, *Elektrokhimiya* **4** (1968) 745.
[91] K. V. Rybalka and D. I. Leikis, *Elektrokhimiya* **3** (1967) 383.
[92] N. B. Grigoryev, I. A. Gedvillo, and N. G. Bardina, *Elektrokhimiya* **8** (1972) 409.
[93] S. Trasatti, *J. Electroanal. Chem.* **52** (1974) 313.
[94] A. De Battisti and S. Trasatti, *J. Chim. Phys.* **74** (1977) 60.
[95] B. Ershler, *Usp. Khim.* **21** (1952) 237.
[96] A. Frumkin, *Usp. Khim.* **15** (1946) 385.
[97] J. O'M. Bockris and S. D. Argade, *J. Chem. Phys.* **49** (1968) 5133.
[98] A. N. Frumkin, *Elektrokhimiya* **1** (1965) 394.
[99] N. B. Grigoryev, S. A. Fateev, and I. A. Bagotskaya, *Elektrokhimiya* **8** (1972) 1525.
[100] N. B. Grigoryev, S. A. Fateev, and I. A. Bagotskaya, *Elektrokhimiya* **8** (1972) 1633.
[101] I. A. Bagotskaya, S. A. Fateev, N. B. Grigoryev, and A. N. Frumkin, *Elektrokhimiya* **9** (1973) 1676.
[102] A. M. Kalyuzhnaya, N. B. Grigoryev, and I. A. Bagotskaya, *Elektrokhimiya* **10** (1974) 1717.

103 S. D. Argade and E. Gileadi, in *Electrosorption*, Ed. by E. Gileadi, Plenum Press, New York, 1967.
104 V. A. Kiryanov, V. S. Krylov, and N. B. Grigoryev, *Elektrokhimiya* **4** (1968) 408.
105 L. E. Rybalka and D. I. Leikis, *Elektrokhimiya* **11** (1975) 1619.
106 M. Rosen, D. R. Flinn, and S. Schuldiner, *J. Electrochem. Soc.* **116** (1969) 1112.
107 T. N. Andersen, J. L. Anderson, D. D. Bodè, and H. Eyring, *J. Res. Inst. Catal. Hokkaido Univ.* **16** (1968) 449.
108 A. Frumkin, N. Balashova, and V. Kazarinov, *J. Electrochem. Soc.* **113** (1966) 1011.
109 A. Frumkin, O. Petry, and R. Marvet, *J. Electroanal. Chem.* **12** (1966) 504.
110 A. N. Frumkin and B. B. Damaskin, in *Modern Aspects of Electrochemistry*, Ed. by J. O'M. Bockris and B. E. Conway, Butterworths, London, 1964, Vol. 3.
111 S. Trasatti, *J. Electroanal. Chem.* **53** (1974) 335.
112 K. Müller, *J. Res. Inst. Catal. Hokkaido Univ.* **14** (1966) 224.
113 R. Parsons, *Rev. Pure Appl. Chem.* **7** (1968) 91.
114 A. Frumkin and B. Damaskin, *Pure Appl. Chem.* **15** (1967) 263.
115 B. B. Damaskin and A. N. Frumkin, *J. Electroanal. Chem.* **34** (1972) 191.
116 A. N. Frumkin, B. B. Damaskin, and A. A. Survila, *J. Electroanal. Chem.* **16** (1968) 493.
117 E. Dutkiewicz, J. D. Garnish, and R. Parsons, *J. Electroanal. Chem.* **16** (1968) 505.
118 B. Damaskin, A. Frumkin, and A. Chizhov, *J. Electroanal. Chem.* **28** (1970) 93.
119 B. B. Damaskin, A. A. Survila, and L. E. Rybalka, *Elektrokhimiya* **3** (1967) 146.
120 S. Trasatti, *J. Electroanal. Chem.* **91** (1978) 1.
121 B. B. Damaskin, *Elektrokhimiya* **11** (1975) 428.
122 F. Pulidori, G. Borghesani, R. Pedriali, A. De Battisti, and S. Trasatti, *J. Chem. Soc. Faraday Trans. 1* **74** (1978) 79.
123 J. E. B. Randles, in *Advances in Electrochemistry and Electrochemical Engineering*, Ed. by P. Delahay and C. W. Tobias, Interscience, New York, 1963, Vol. 3.
124 J. E. B. Randles, XVième Conseil International de Chemie, Solvay, June 1972.
125 J. Llopis, in *Modern Aspects of Electrochemistry*, Ed. by J. O'M. Bockris and B. E. Conway, Butterworths, London, 1971, Vol. 6.
126 B. Kamienski, *Electrochim. Acta* **1** (1959) 272.
127 B. Case and R. Parsons, *Trans. Faraday Soc.* **63** (1967) 1224.
128 H. J. M. Nedermeijer-Denessen and C. L. De Ligny, *J. Electroanal. Chem.* **57** (1974) 265.
129 E. J. W. Verwey, *Rec. Trav. Chim.* **62** (1942) 127.
130 N. A. Izmailov, *Dokl. Akad. Nauk SSSR* **149** (1963) 884.
131 M. F. Halliwell and S. C. Nyburg, *Trans. Faraday Soc.* **59** (1963) 1126.
132 I. Eliezer and P. Krindel, *J. Chem. Phys.* **57** (1972) 1884.
133 E. J. W. Vervey, *Rec. Trav. Chim.* **61** (1942) 564.
134 F. H. Stillinger and A. Ben-Naim, *J. Chem. Phys.* **47** (1967) 4431.
135 N. H. Fletcher, *Philos. Mag.* **18** (1968) 1287.
136 J. E. B. Randles and D. J. Schiffrin, *J. Electroanal. Chem.* **10** (1965) 480.
137 D. J. Schiffrin, *Trans. Faraday Soc.* **66** (1970) 2464.
138 J. Mingens and B. A. Pethica, *J. Chem. Soc. Faraday Trans. 1* **69** (1973) 500.

[139] H. J. M. Nedermeijer-Denessen and C. L. De Ligny, *J. Electroanal. Chem.* **59** (1975) 1.
[140] M. Blank and R. H. Ottewill, *J. Phys. Chem.* **68** (1964) 2206.
[141] R. J. Demchak and T. Fort, *J. Colloid Interface Sci.* **46** (1974) 191.
[142] R. I. Kaganovich and A. N. Frumkin, *Elektrokhimiya* **9** (1973) 1338.
[143] N. L. Jarvis and M. A. Scheiman, *J. Phys. Chem.* **72** (1968) 74.
[144] N. N. Kochurova, B. A. Noskov, and A. I. Rusanov, *Dokl. Akad. Nauk SSSR* **227** (1976) 1386.
[145] A. Frumkin and B. Damaskin, *J. Electroanal. Chem.* **66** (1975) 150.
[146] B. E. Conway, H. P. Dhar, and S. Gottesfeld, *J. Colloid Interface Sci.* **43** (1973) 303.
[147] M. E. Nicholas, P. A. Joyner, B. M. Tessem, and M. D. Olson, *J. Phys. Chem.* **65** (1961) 1373.
[148] G. Gouy, *Ann. Phys. (Leipzig)* **6** (1916) 5.
[149] S. Trasatti, *J. Electroanal. Chem.* **65** (1975) 815.
[150] W. V. Kayser, *J. Colloid Interface Sci.* **56** (1976) 622.
[151] T. Yamanaka and K. Tanabe, *J. Phys. Chem.* **80** (1976) 1723.
[152] A. C. Zettlemoyer, F. J. Micale, and K. Klier, in *Water. A Comprehensive Treatise*, Ed. by F. Franks, Plenum Press, New York, 1975, Vol. 5.
[153] E. McCafferty, V. Pravdic, and A. C. Zettlemoyer, *Trans. Faraday Soc.* **66** (1970) 1720.
[154] R. G. Gast, E. R. Landa, and G. W. Meyer, *Clays Clay Miner.* **22** (1974) 31.
[155] A. V. Volkov, A. V. Kiselev, and V. I. Lygin, *Kolloid. Zh.* **38** (1976) 330.
[156] H. Jeziorowski, H. Knözinger, W. Meye, and H. D. Müller, *J. Chem. Soc. Faraday Trans. 1* **69** (1973) 1744.
[157] A. C. Zettlemoyer, *J. Colloid Interface Sci.* **28** (1968) 343.
[158] F. S. Baker and K. S. W. Sing, *J. Colloid Interface Sci.* **55** (1976) 605.
[159] *Natl. Bur. Stand. (U.S.) Circ. No. 500* (1952).
[160] A. I. Vitvitskii and I. Ya. Tyuryaev, *Zh. Prikl. Khim.* **49** (1976) 322.
[161] O. Exner, *Collect. Czech. Chem. Commun.* **37** (1972) 1425; **38** (1973) 799.
[162] A. J. Appleby, *Catal. Rev.* **4** (1970) 221.
[163] B. E. Conway and H. P. Dhar, *Colloid Polym. Sci.* **253** (1975) 11.
[164] T. N. Andersen and J. O'M. Bockris, *Electrochim. Acta* **9** (1964) 347.
[165] S. Cerny and V. Ponec, *Catal. Rev.* **2** (1968) 249.
[166] O. Johnson, *J. Catal.* **28** (1973) 503.
[167] G. Blyholder and R. W. Sheets, *J. Catal.* **39** (1975) 152.
[168] C. F. Melius, *Chem. Phys. Lett.* **39** (1976) 287.
[169] J. L. Cavanau, *Water and Solute-Water Interactions*, Holden-Day, San Francisco, 1964.
[170] J. E. Desnoyers and C. Jolicoeur, in *Modern Aspects of Electrochemistry*, Ed. by J. O'M. Bockris and B. E. Conway, Butterworths, London, 1969, Vol. 5.
[171] D. W. James, R. F. Armishaw, and R. L. Frost, *J. Phys. Chem.* **80** (1976) 1346.
[172] G. Ebert and Ch. Ebert, *Colloid Polym. Sci.* **254** (1976) 25.
[173] W. Drost-Hansen, *Ind. Engin. Chem.* **61** (1969) 10.
[174] B. V. Derjagin, *J. Phys. Chem.* **1** (1932) 29.
[175] A. Hartkopf and B. L. Karger, *Acc. Chem. Res.* **6** (1973) 209.
[176] M. Salve and U. Palm, *Uch. Zap. Tartu. Gos. Univ.* **332** (1974) 71.
[177] M. Haga and V. Past, *Uch. Zap. Tartu. Gos. Univ.* **235** (1969) 47.
[178] A. N. Frumkin, N. B. Grigoryev, and I. A. Bagotskaya, *Elektrokhimiya* **2** (1966) 329.
[179] A. N. Frumkin and N. B. Grigoryev, *Elektrokhimiya* **4** (1968) 533.

[180] M. S. Metsik, V. D. Perevertaev, V. A. Liopo, G. T. Timoshtchenko, and A. B. Kiselev, J. Colloid Interface Sci. **43** (1973) 662.
[181] K. Kaneko, M. Serizawa, T. Ishikawa, and K. Inouye, Bull. Chem. Soc. Japan **48** (1975) 1764.
[182] E. McCafferty and A. C. Zettlemoyer, Trans. Faraday Soc. **66** (1970) 1732.
[183] T. McMullen, Can. J. Chem. **51** (1973) 4038.
[184] R. Martens, H. Nägerl, and F. Freund, Ind. Chim. Belg. **38** (1973) 519.
[185] V. A. Kiryanov and V. S. Krylov, Elektrokhimiya **6** (1970) 1596.
[186] V. A. Kiryanov, Elektrokhimiya **6** (1970) 1175.
[187] L. E. Rybalka, B. B. Damaskin, and D. I. Leikis, Elektrokhimiya **9** (1973) 414.
[188] R. Ya. Pullerits, U. V. Palm, and V. E. Past, Elektrokhimiya **5** (1969) 886.
[189] L. E. Rybalka, D. I. Leikis, and B. B. Damaskin, Elektrokhimiya **9** (1973) 62.
[190] L. E. Rybalka, B. B. Damaskin, and D. I. Leikis, Elektrokhimiya **11** (1975) 9.
[191] L. Antropov, quoted in Ref. 59.
[192] O. Klein and E. Lange, Z. Elektrochem. **43** (1937) 570.
[193] D. J. Schiffrin, J. Phys. Chem. **73** (1969) 1632.
[194] G. F. Dionne, J. Appl. Phys. **44** (1973) 5637.
[195] E. Zaremba, Phys. Lett. **57A** (1976) 156.
[196] T. L. Einstein, Surface Sci. **45** (1974) 713.
[197] T. B. Grimley, Proc. Phys. Soc. **92** (1967) 776.
[198] D. K. Klemperer and J. C. Snaith, Surface Sci. **28** (1971) 209.
[199] B. E. Nieuwenhuys, O. G. Van Aardenne, and W. M. H. Sachtler, Chem. Phys. **5** (1974) 418.
[200] J. Müller, Surface Sci. **45** (1974) 314.
[201] D. F. Klemperer and J. C. Snaith, Surface Sci. **45** (1974) 314.
[202] B. E. Nieuwenhuys, Surface Sci. **49** (1975) 363.
[203] J. Müller, Surface Sci. **49** (1975) 681.
[204] M. A. Chesters, M. Hussain, and J. Pritchard, Surface Sci. **35** (1973) 161.
[205] O. Johnson, J. Res. Inst. Catal. Hokkaido Univ. **19** (1972) 152; **20** (1972) 125.
[206] F. C. Tomkins, in The Solid-Gas Interface, Ed. by E. A. Flood, Dekker, New York, 1967.
[207] P. R. Antoniewicz, Surface Sci. **52** (1975) 703.
[208] W. C. Meixner and P. R. Antoniewicz, Phys. Rev. B **13** (1976) 3276.
[209] B. E. Niewenhuys, Ned. Tijdschr. Vacuumtech. **13** (1975) 41.
[210] B. E. Nieuwenhuys and W. M. H. Sachtler, J. Colloid Interface Sci. **58** (1977) 65.
[211] R. R. Ford and J. Pritchard, Trans. Faraday Soc. **67** (1971) 216.
[212] N. S. Polyanovskaya and A. N. Frumkin, Elektrokhimiya **6** (1970) 246.
[213] J. W. Schultze, Ber. Bunsenges. Phys. Chem. **73** (1969) 483.
[214] J. W. Schultze, Electrochim. Acta **17** (1972) 451.
[215] C. Kemball, Proc. R. Soc. London Ser. A **90** (1947) 117.
[216] R. Suhrmann, J. M. Heras, L. Viscido de Heras, and G. Wedler, Ber. Bunsenges. Phys. Chem. **68** (1964) 511.
[217] R. Suhrmann, J. M. Heras, L. Viscido de Heras, and G. Wedler, Ber. Bunsenges. Phys. Chem. **72** (1968) 854.
[218] C. M. Quinn and R. W. Roberts, Trans. Faraday Soc. **60** (1964) 899.
[219] E. E. Huber and C. T. Kirk, Surface Sci. **5** (1966) 447.
[220] E. V. Osipova, N. A. Shurmovskaya, and R. Kh. Burshtein, Elektrokhimiya **5** (1969) 1139.
[221] A. G. Pshenichnikov, R. Kh. Burshtein, and V. D. Kovalevskaya, Elektrokhimiya **11** (1975) 1465.

[222] L. Ya. Egorov and I. M. Novoselskii, *Elektrokhimiya* **6** (1970) 521.
[223] A. N. Frumkin, *J. Electroanal. Chem.* **64** (1975) 247.
[224] A. J. Sargood, C. W. Jowett, and B. J. Hopkins, *Surface Sci.* **22** (1970) 343.
[225] C. W. Jowett, P. J. Dobson, and B. J. Hopkins, *Surface Sci.* **17** (1969) 474.
[226] B. J. Hopkins, M. Leggett, and G. D. Watts, *Surface Sci.* **28** (1971) 581.
[227] J. C. Fuggle, L. M. Watson, D. J. Fabian, and S. Affrossman, *Surface Sci.* **49** (1975) 61.
[228] J. Kubota and K. Azuma, *J. Res. Inst. Catal. Hokkaido Univ.* **23** (1975) 190.
[229] G. M. Kornacheva, R. Kh. Burshtein, and N. A. Shurmovskaya, *Elektrokhimiya* **9** (1973) 81.
[230] M. A. Leban and A. T. Hubbard, *J. Electroanal. Chem.* **74** (1976) 253.
[231] S. Trasatti, *J. Electroanal. Chem.* **54**, (1974) 19.
[232] R. L. Wells and T. Fort, *Surface Sci.* **32** (1972) 554.
[233] N. A. Surplice and W. Brearly, *Surface Sci.* **52** (1975) 62.
[234] M. A. Chesters and G. A. Somorjai, *Surface Sci.* **52** (1975) 21.
[235] W. M. H. Sachtler, G. J. H. Dorgelo, and A. A. Holscher, *Surface Sci.* **5** (1966) 221.
[236] B. J. Wood, *J. Phys. Chem.* **75** (1971) 2186.
[237] I. G. Murgulescu and M. I. Vass, *Rev. Roum. Chim.* **13** (1968) 373.
[238] J. H. Thomas and S. P. Sharma, *J. Vac. Sci. Technol.* **13** (1976) 549.
[239] A. Cetronio and J. P. Jones, *Surface Sci.* **44** (1974) 109.
[240] D. M. Zehner, B. R. Appleton, T. S. Noggle, J. W. Miller, J. H. Barrett, L. H. Jenkins, and O. E. Schow, *J. Vac. Sci. Technol.* **12** (1975) 454.
[241] R. W. Joyner, *Surface Sci.* **39** (1973) 450.
[242] J. Lecoeur, C. Sella, J.-C. Martin, L. Tertian, and J. Deschamps, *C.R. Acad. Sci. Ser. C* **281** (1975) 71.
[243] N. V. Smith, *Phys. Rev. B* **9** (1974) 1365.
[244] A. N. Frumkin, quoted by A. Matsuda, *J. Res. Inst. Catal. Hokkaido Univ.* **23** (1975) 144.
[245] G. C. Bond, *Gold Bull.* **5** (1971) 11.
[246] G. C. Bond, P. A. Sermon, G. Webb, D. A. Buchanan, and P. B. Wells, *J. Chem. Soc. Chem. Commun.* (1973) 444.
[247] G. C. Bond and P. A. Sermon, *Gold Bull.* **6** (1973) 102.
[248] S. Trasatti, *Z. Phys. Chem. N. F.* **98** (1975) 75.
[249] R. Parsons, *J. Electroanal. Chem.* **5** (1963) 397.
[250] B. B. Damaskin, A. A. Survila, S. A. Vasina, and A. I. Fedorova, *Elektrokhimiya* **3** (1967) 825.
[251] A. Hamelin and G. Valette, *C. R. Acad. Sci. Ser. C* **267** (1968) 127.
[252] A. Hamelin and G. Valette, *C. R. Acad. Sci. Ser. C* **267** (1968) 211.
[253] D. J. Schiffrin, *Trans. Faraday Soc.* **67** (1971) 3318.
[254] G. J. Hills and R. N. Reeves, *J. Electroanal. Chem.* **31** (1971) 269.
[255] T. I. Popova, N. A. Simonova, and L. M. Dubova, *Elektrokhimiya* **8** (1972) 246.
[256] M. P. Pyarnoya, U. V. Palm, and N. B. Grigoryev, *Elektrokhimiya* **11** (1975) 575.
[257] V. A. Panin and K. V. Rybalka, *Elektrokhimiya* **8** (1972) 1202.
[258] B. B. Damaskin, I. M. Ganzhina, and R. V. Ivanova, *Elektrokhimiya* **6** (1970) 1540.
[259] B. B. Damaskin, *Elektrokhimiya* **1** (1965) 1123.
[260] R. Parsons, *Proc. R. Soc. London Ser. A* **261** (1961) 79.
[261] R. Parsons and P. C. Symons, *Trans. Faraday Soc.* **64** (1968) 1077.
[262] U. V. Palm, Yu. I. Erlikh, and T. E. Erlikh, *Elektrokhimiya* **10** (1974) 1180.

263 L. E. Rybalka and B. B. Damaskin, *Elektrokhimiya* **9** (1973) 1062.
264 N. B. Grigoryev, A. M. Kalyuzhnaya, and I. A. Bagotskaya, *Elektrokhimiya* **11** (1975) 1574.
265 N. B. Grigoryev, V. P. Kuprin, and Yu. M. Loshkarev, *Elektrokhimiya* **11** (1975) 638.
266 I. A. Bagotskaya, S. A. Fateev, N. B. Grigoryev, and N. G. Bardina, *Elektrokhimiya* **6** (1970) 369.
267 N. B. Grigoryev and D. N. Machavariani, *Elektrokhimiya* **6** (1970) 89.
268 V. S. Krylov and N. B. Grigoryev, *Elektrokhimiya* **7** (1971) 511.
269 N. B. Grigoryev and V. S. Krylov, *Elektrokhimiya* **4** (1968) 763.
270 N. B. Grigoryev and D. N. Machavariani, *Elektrokhimiya* **5** (1969) 87.
271 K. V. Rybalka, *Elektrokhimiya* **8** (1972) 400.
272 A. Bewick and J. Robinson, *J. Electroanal. Chem.* **60** (1975) 163.
273 A. Bewick and J. Robinson, *J. Electroanal. Chem.* **71** (1976) 131.
274 R. Parsons, R. M. Reeves, and P. N. Taylor, *J. Electroanal. Chem.* **50** (1974) 149.
275 A. Frumkin and B. Damaskin, *J. Electroanal. Chem.* **66** (1975) 150.
276 R. Parsons, *Handbook of Electrochemical Constants*, Butterworths, London, 1959.
277 J. O'M. Bockris and E. C. Potter, *J. Chem. Phys.* **20** (1952) 614.
278 J. O'M. Bockris and M. A. Habib, *J. Electroanal. Chem.* **68** (1976) 367.
279 D. D. Bodè, in *Chemical Dynamics*, Ed. by J. O. Hirschfelder, Wiley, New York, 1971, p. 361.
280 C. L. Gardner, *J. Electroanal. Chem.* **61** (1975) 113.
281 J. W. Schultze and F. D. Koppitz, *Electrochim. Acta* **21** (1976) 327.
282 N. F. Mott, R. Parsons, and R. J. Watts-Tobin, *Philos. Mag.* **7** (1962) 483.
283 J. O'M. Bockris and M. A. Habib, *J. Res. Inst. Catal. Hokkaido Univ.* **23** (1975) 47.
284 K. Robinson and S. Levine, *J. Electroanal. Chem.* **47** (1973) 395.
285 B. A. Abd-El-Nabey and S. Trasatti, *J. Chem. Soc. Faraday Trans. 1* **71** (1975) 1230.
286 A. De Battisti, B. A. Abd-El-Nabey, and S. Trasatti, *J. Chem. Soc. Faraday Trans. 1* **72** (1976) 2076.
287 B. B. Damaskin, *Elektrokhimiya* **11** (1975) 428.
288 J. I. Padova, in *Modern Aspects of Electrochemistry*, Ed. by B. E. Conway and J. O'M. Bockris, Butterworths, London, 1972, Vol. 7.
289 S. Levine, J. Mingins, and G. M. Bell, *J. Electroanal. Chem.* **13** (1967) 280.
290 J. B. Hasted, in *Water. A Comprehensive Treatise*, Ed. by F. Franks, Plenum Press, New York, 1972, Vol. 1.
291 J. E. Chamberlain, M. N. Afsar, J. B. Hasted, M. S. Zafar, and G. J. Davies, *Nature* **225** (1975) 319.
292 H. P. Schwan, R. J. Sheppard, and E. H. Grant, *J. Chem. Phys.* **64** (1976) 2257.
293 K. R. Foster and H. A. Resing, *J. Phys. Chem.* **80** (1976) 1390.
294 B. E. Conway, J. O'M. Bockris, and I. A. Ammar, *Trans. Faraday Soc.* **47** (1951) 756.
295 R. G. Barradas and J. M. Sedlak, *Electrochim. Acta* **17** (1972) 683.
296 R. Pottel, K. Giese, and U. Kaatze, in *Structure of Water and Aqueous Solutions*, Ed. by W. A. P. Luck, Verlag Chemie, Berlin, 1974, p. 391.
297 U. Kaatze, *Ber. Bunsenges. Phys. Chem.* **77** (1973) 447.
298 L. K. Partridge, A. C. Tansley, and A. S. Porter, *Electrochim. Acta* **13** (1968) 2029.

[299] M. A. V. Devanathan, *Proc. R. Soc. London Ser. A* **264** (1961) 133.
[300] C. N. R. Rao, in *Water. A Comprehensive Treatise*, Plenum Press, New York, 1972, Vol. 1.
[301] I. L. Cooper and J. A. Harrison, *J. Electroanal. Chem.* **66** (1975) 85.
[302] R. R. Salem, *Zh. Fiz. Khim.* **48** (1974) 2581.
[303] R. R. Salem, *Zh. Fiz. Khim.* **50** (1976) 1352.
[304] P. Jackson and G. D. Parfitt, *Trans. Faraday Soc.* **67** (1971) 2469.
[305] G. E. Van Gils, *J. Colloid Interface Sci.* **30** (1969) 272.
[306] P. W. Schindler, E. Waelti, and B. Fuerst, *Chimia* **30** (1976) 107.
[307] V. N. Pak, *Zh. Fiz. Khim.* **48** (1974) 2338.
[308] J. A. Schufle, C.-T. Huang, and W. Drost-Hansen, *J. Colloid Interface Sci.* **54** (1976) 184.
[309] P. C. Owzarski and W. E. Ranz, *J. Phys. Chem.* **73** (1969) 3628.
[310] F. W. Röllgen and H. D. Beckey, *Surface Sci.* **26** (1971) 100.
[311] E. P. Mikheeva and N. P. Keier, *Dokl. Akad. Nauk SSSR* **219** (1974) 906.
[312] A. J. Bennet, *Surface Sci.* **50** (1975) 77.
[313] S. Gottsfeld and B. E. Conway, *J. Chem. Soc. Faraday Trans. 1* **70** (1974) 1793.
[314] J. R. Macdonald and C. A. Barlow, *J. Chem. Phys.* **39** (1963) 412.
[315] G. D. Parfitt, in *Progress in Surface and Membrane Science*, Ed. by D. A. Cadenhead and J. F. Danielli, Academic Press, New York, 1976, Vol. 11.
[316] F. Dumont, D. Van Tan, and A. Watillon, *J. Colloid Interface Sci.* **55** (1976) 678.
[317] F. S. Baker, C. Phillips, and K. S. W. Sing, in *Proceedings of the Symposium on Oxide–Electrolyte Interfaces*, The Electrochemical Society, Princeton, 1973, p. 65.
[318] B. E. Conway, D. J. Mackinnon, and B. V. Tilak, *Trans. Faraday Soc.* **66** (1970) 1203.
[319] K. Johansson and J. C. Eriksson, *J. Colloid Interface Sci.* **40** (1972) 398.
[320] V. N. Khabarov, A. I. Rusanov, and N. N. Kochurova, *Kolloidn. Zh.* **38** (1976) 120.
[321] B. E. Conway, *J. Electroanal. Chem.* **65** (1975) 491.
[322] K. W. Allen, D. R. Lewis, and K. G. A. Pankhurst, *J. Chem. Soc. A* (1971) 3028.
[323] H. S. Frank, in *Water. A Comprehensive Treatise*, Ed. by F. Frank, Plenum Press, New York, 1972, Vol. 1.
[324] K. Tempelhoff, *Z. Phys. Chem. (Leipzig)* **257** (1976) 49.
[325] H. S. Frank, in *Structure of Water and Aqueous Solutions*, Ed. by W. A. P. Luck, Verlag Chemie, Berlin, 1974, p. 9.
[326] G. M. Bell and D. W. Salt, *J. Chem. Soc. Faraday Trans. 2* **72** (1976) 76.
[327] G. Nemethy, in *Structure of Water and Aqueous Solutions*, Ed. by W. A. P. Luck, Verlag Chemie, Berlin, 1974, p. 73.
[328] G. M. Bell and H. Sallouta, *Mol. Phys.* **29** (1975) 1621.
[329] M. A. Ryazanov, *Zh. Fiz. Khim.* **49** (1975) 788.
[330] A. V. Karyakin, G. A. Kriventsova, and N. V. Soboleva, *Dokl. Akad. Nauk SSSR* **221** (1975) 1096.
[331] J. W. Perram and S. Levine, *Adv. Mol. Relaxation Processes* **6** (1974) 85.
[332] B. L. Lentz, A. T. Hagler, and H. A. Scheraga, *J. Phys. Chem.* **78** (1974) 1531.
[333] M. Weissmann and L. Blum, *Trans. Faraday Soc.* **64** (1968) 2605.
[334] J. A. Pople, *Proc. R. Soc. London Ser A* **205** (1951) 163.
[335] J. Del Bene and J. A. Pople, *J. Chem. Phys.* **52** (1970) 4858.

[336] O. Ya. Samoilov, *Structure of Aqueous Electrolyte Solutions and the Hydration of Ions*, Consultant Bureau, New York, 1965.
[337] G. Nemethy and H. A. Scheraga, *J. Chem. Phys.* **36** (1962) 3382.
[338] E. Forslind, *Progr. Colloid Polym. Sci.* **56** (1975) 12.
[339] A. H. Narten, M. D. Danford, and H. A. Levy, *Discuss. Faraday Soc.* **43** (1967) 97.
[340] R. W. Hendricks, P. G. Mardon, and L. B. Shaffer, *J. Chem. Phys.* **61** (1974) 319.
[341] B. B. Damaskin, *Elektrokhimiya* **1** (1965) 63.
[342] T. Smith, *Adv. Colloid Interface Sci.* **3** (1972) 161.
[343] S. Levine, G. M. Bell, and M. Calvert, *Can. J. Chem.* **40** (1962) 518.
[344] R. Parsons and J. M. Parry, *J. Electrochem. Soc.* **113** (1966) 992.
[345] H. Dhar, B. E. Conway, and K. M. Joshi, *Electrochim. Acta* **18** (1973) 789.
[346] J. Lawrence and R. Parsons, *J. Phys. Chem.* **73** (1969) 3577.
[347] J. O'M. Bockris and D. A. Swinkels, *J. Electrochem. Soc.* **111** (1964) 736.
[348] B. E. Conway and H. P. Dhar, *Surface Sci.* **44** (1974) 261.
[349] A. Goel, A. S. N. Murthy, and C. N. R. Rao, *Indian J. Chem.* **9** (1971) 56.
[350] J. T. Law, Ph.D. Dissertation, Royal College of Science, London, 1951.
[351] K. V. Rybalka and V. A. Panin, *Elektrokhimiya* **9** (1973) 172.
[352] L. S. Prabhumirashi and C. I. Jose, *J. Chem. Soc. Faraday Trans. 2* **72** (1976) 1721.
[353] G. H. F. Diercksen, W. P. Kraemer, and B. O. Roos, *Theor. Chim. Acta* **36** (1975) 249.
[354] C. Braun and H. Leidecker, *J. Chem. Phys.* **61** (1974) 3104.
[355] M. S. Gordon, D. E. Tallman, C. Monroe, M. Steinbach, and J. Armbrust, *J. Am. Chem. Soc.* **97** (1975) 1326.
[356] T. H. Spurling and I. K. Snook, *Chem. Phys. Lett.* **32** (1975) 159.
[357] F. H. Stillinger and H. L. Lemberg, *J. Chem. Phys.* **62** (1975) 1340.
[358] O. Matsuoka, E. Clementi, and M. Yoshimine, *J. Chem. Phys.* **64** (1976) 1351.
[359] J. H. Keighley and A. J. Whitaker, *Mol. Phys.* **31** (1976) 643.
[360] A. Kitaura and K. Morokuma, *Int. J. Quantum Chem.* **10** (1976) 325.
[361] L. L. Shipman, J. C. Owicki, and H. A. Scheraga, *J. Phys. Chem.* **78** (1974) 2055.
[362] J. O'M. Bockris and B. E. Conway, *J. Chem. Phys.* **28** (1958) 707.
[363] G. G. Susbielles, *J. Electroanal. Chem.* **12** (1966) 230.
[364] C. W. De Kreuk, M. Sluyters-Rehbach, and J. H. Sluyters, *J. Electroanal. Chem.* **35** (1972) 137.
[365] J. O'M. Bockris, E. Gileadi, and K. Müller, *J. Chem. Phys.* **44** (1966) 1445.
[366] S. M. Nelson, H. H. Huang, and L. E. Sutton, *Trans. Faraday Soc.* **65** (1969) 225.
[367] J. O'M. Bockris and M. A. Habib, *Z. Phys. Chem. N. F.* **98** (1975) 43.
[368] J. O'M. Bockris and M. A. Habib, *An. Quim.* **71** (1975) 952.
[369] A. M. Morozov, N. B. Grigoryev, and I. A. Bagotskaya, *Elektrokhimiya* **2** (1966) 1235.
[370] C. V. D'Alkaine, E. R. Gonzalez, and R. Parsons, *J. Electroanal. Chem.* **32** (1971) 57.
[371] A. De Battisti and S. Trasatti, *J. Electroanal. Chem.* **59** (1975) 137.
[372] R. Parsons and R. Payne, *Z. Phys. Chem. N. F.* **98** (1975) 9.
[373] R. Parsons, R. Peat, and R. M. Reeves, *J. Electroanal. Chem.* **62** (1975) 151.
[374] A. De Battisti, V. Faggiano, and S. Trasatti, *J. Electroanal. Chem.* **73** (1976) 327.

[375] J. E. B. Randles and K. S. Whiteley, *Trans. Faraday Soc.* **52** (1956) 1509.
[376] D. C. Grahame, E. M. Coffin, J. I. Cummings, and M. A. Poth, *J. Am. Chem. Soc.* **74** (1952) 1207.
[377] W. Paik, T. N. Andersen, and H. Eyring, *J. Phys. Chem.* **71** (1967) 1891.
[378] Z. Koczorowski and Z. Figaszewski, *Rocz. Chem.* **44** (1970) 191.
[379] R. G. Barradas and J. M. Sedlak, *Electrochim. Acta* **17** (1962) 67.
[380] M. Privat, J. Dupin, and R. Grand, *J. Chim. Phys.* **10** (1972) 1415.
[381] J. A. Harrison, J. E. B. Randles, and D. J. Schiffrin, *J. Electroanal. Chem.* **25** (1970) 197.
[382] S. Trasatti, *J. Electroanal. Chem.*, **82** (1977) 391.
[383] K. V. Rybalka, *Elektrokhimiya* **7** (1971) 242.
[384] V. A. Panin, K. V. Rybalka, and D. I. Leikis, *Elektrokhimiya* **8** (1972) 390.
[385] B. B. Damaskin, *Elektrokhimiya* **1** (1965) 1258.
[386] V. F. Ivanov, B. B. Damaskin, A. N. Frumkin, A. A. Ivashchenko, and N. I. Peshkova, *Elektrokhimiya* **1** (1965) 279.
[387] V. F. Ivanov, B. B. Damaskin, A. A. Ivashchenko, V. F. Khovina, and N. I. Melekhova, *Elektrokhimiya* **8** (1972) 886.
[388] H. D. Hurwitz and C. V. D'Alkaine, *J. Electroanal. Chem.* **42** (1973) 77.
[389] R. R. Salem, *Zh. Fiz. Khim.* **43** (1969) 2876.
[390] F. D. Koppitz and J. W. Schultze, *Electrochim. Acta* **21** (1976) 337.
[391] V. V. Eletskii and Yu. V. Pleskov, *Elektrokhimiya* **10** (1974) 179.
[392] B. E. Conway, J. G. Mathieson, and H. P. Dhar, *J. Phys. Chem.* **78** (1974) 1226.
[393] I. L. Cooper and J. A. Harrison, *J. Electroanal. Chem.* **66** (1975) 85.
[394] D. Schuhmann, P. Vanel, and C. Bertrand, *J. Chim. Phys.* **74** (1977) 643.
[395] B. E. Conway, *Ann. Rev. Phys. Chem.* **17** (1966) 481.
[396] K. Heinziger, *Z. Naturforsch.* **31a** (1976) 1073.
[397] H. G. Hertz and C. Rädle, *Ber. Bunsenges. Phys. Chem.* **77** (1973) 521.
[398] E. Clementi and H. Popkie, *J. Chem. Phys.* **57** (1972) 1077.
[399] B. E. Conway and J. O'M. Bockris, in *Modern Aspects of Electrochemistry*, Ed. by J. O'M. Bockris and B. E. Conway, Butterworths, London, 1954, Vol. 1.
[400] M. J. Jaycock and J. C. R. Waldsax, *J. Chem. Soc. Faraday Trans. 1* **70** (1974) 1501.
[401] P. B. Barraclough and P. G. Hall, *J. Chem. Soc. Faraday Trans. 1* **70** (1975) 2266.
[402] R. W. Rice and G. L. Haller, *Proceedings of the 5th International Congress on Catalysis, 1972*, North-Holland Publ. Co., Amsterdam, 1973, p. 317.
[403] J. Clavilier and N. Van Huong, *J. Electroanal. Chem.* **41** (1973) 193.
[404] K. Palts, R. Pullerits, and V. Past, *Uch. Zap. Tartu. Gos. Univ.* **235** (1969) 64.
[405] R. Payne, in *Advances in Electrochemistry and Electrochemical Engineering*, Ed. by P. Delahay and C. W. Tobias, Wiley, New York, 1970, Vol. 7.
[406] R. Payne, *J. Phys. Chem.* **71** (1967) 1548.
[407] U. V. Palm, M. G. Byaèrtnyu, and E. K. Petyarv, *Elektrokhimiya* **11** (1975) 1849.
[408] B. B. Damaskin, R. V. Ivanova, and A. A. Survila, *Elektrokhimiya* **1** (1965) 767.
[409] S. Minc, J. Jastrzebska, and M. Brzostowska, *J. Electrochem. Soc.* **108** (1961) 1161.
[410] G. H. Nancollas, D. S. Reid, and C. A. Vincent, *J. Phys. Chem.* **70** (1966) 3300.
[411] E. Dutkiewicz and R. Parsons, *J. Electroanal. Chem.* **11** (1966) 196.

412 R. Parsons, J. Electroanal. Chem. **53** (1974) 479.
413 R. Payne, J. Electroanal. Chem. **47** (1973) 265.
414 V. F. Ivanov, B. B. Damaskin, V. F. Khonina, A. A. Ivashchenko, and N. I. Melekhova, Elektrokhimiya **9** (1973) 1852.
415 Z. N. Ushakova and V. F. Ivanov, Elektrokhimiya **8** (1972) 1880.
416 W. R. Fawcett and R. O. Loutfy, J. Electroanal. Chem. **39** (1972) 185.
417 B. B. Damaskin and Yu. M. Povarov, Dokl. Akad. Nauk SSSR **140** (1961) 394.
418 D. C. Grahame, Z. Elektrochem. **59** (1955) 740.
419 R. Parsons and M. A. V. Devanathan, Trans. Faraday Soc. **49** (1953) 673.
420 S. Minc, J. Jastrzebska, and M. Jurkiewicz-Herbich, J. Electroanal. Chem. **65** (1975) 351.
421 J. E. B. Randles, B. Behr, and Z. Borkowska, J. Electroanal. Chem. **65** (1975) 775.
422 J. D. Garnish and R. Parsons, Trans. Faraday Soc. **63** (1967) 1754.
423 E. K. Petyar, K. A. Kolak, and U. V. Palm, Elektrokhimiya **8** (1972) 100.
424 E. K. Petyar and U. V. Palm, Elektrokhimiya **9** (1973) 1343.
425 Z. N. Ushakova and V. F. Ivanov, Elektrokhimiya **12** (1976) 485.
426 V. F. Ivanov, B. B. Damaskin, N. I. Peshkova, A. A. Ivashchenko, and V. F. Balashov, Elektrokhimiya **4** (1968) 851.
427 S. Minc and J. Jastrzebska, J. Electrochem. Soc. **107** (1960) 135.
428 Z. Koczorowski and I. Zagorska, Rocz. Chem. **44** (1970) 911.
429 A. Mellier and M. Leard, J. Chim. Phys. **73** (1976) 379.
430 Z. N. Ushakova, V. F. Ivanov, V. F. Khonina, A. A. Ivashchenko, and N. I. Melekhova, Elektrokhimiya **9** (1973) 1883.
431 R. R. Salem, Zh. Fiz. Khim. **43** (1969) 1839.
432 R. R. Salem and A. F. Sharovarnikov, Elektrokhimiya **9** (1973) 1016.
433 D. I. Dzhaparidze, G. A. Tedoradze, and Sh. S. Dzheparidze, Elektrokhimiya **5** (1969) 955.
434 D. I. Dzhaparidze, A. E. Kakhadze, and V. A. Chagelishvili, Elektrokhimiya **9** (1973) 1318.
435 R. Payne, J. Phys. Chem. **73** (1969) 3598.
436 Ya. Dolido, R. V. Ivanova, and B. B. Damaskin, Elektrokhimiya **4** (1968) 567.
437 E. K. Petyarv and U. V. Palm, Elektrokhimiya **9** (1973) 1836.
438 I. M. Ganzina, B. B. Damaskin, R. I. Kaganovich, and R. V. Ivanova, Elektrokhimiya **7** (1971) 362.
439 I. M. Ganzina, B. B. Damaskin, and R. I. Kaganovich, Electrokhimiya **8** (1972) 93.
440 R. R. Salem, Elektrokhimiya **11** (1975) 1096.
441 R. I. Kaganovich, B. B. Damaskin, and M. K. Kaisheva, Elektrokhimiya **6** (1970) 1359.
442 I. M. Ganzhina and B. B. Damaskin, Elektrokhimiya **6** (1970) 1715.
443 U. V. Palm and E. K. Petyarv, Elektrokhimiya **11** (1975) 139.
444 W. R. Fawcett and R. O. Loutfy, Can. J. Chem. **51** (1973) 230.
445 P. Champion, C. R. Acad. Sci. Ser. C **269** (1969) 1159.
446 A. Frumkin, I. Bagotskaya, and N. Grigoryev, Z. Phys. Chem. **98** (1975) 3.
447 N. B. Grigoryev, A. M. Kalyuzhnaya, and I. A. Bagotskaya, Elektrokhimiya **12** (1976) 418.
448 N. B. Grigoryev and Yu. M. Povarov, Elektrokhimiya **12** (1976) 471.
449 B. Case, N. S. Hush, R. Parsons, and M. E. Peover, J. Electroanal. Chem. **10** (1965) 360.

[450] A. De Battisti and S. Trasatti, *J. Electroanal. Chem.* **48** (1973) 213.
[451] R. Payne, *J. Am. Chem. Soc.* **89** (1967) 489.
[452] J. Lawrence and R. Parsons, *Trans. Faraday Soc.* **64** (1968) 751.
[453] J. Lawrence and R. Parsons, *Trans. Faraday Soc.* **64** (1968) 1656.
[454] W. R. Fawcett and M. D. Mackey, *J. Chem. Soc. Faraday Trans. 1* **69** (1973) 634.
[455] W. R. Fawcett and M. D. Mackey, *J. Chem. Soc. Faraday Trans. 1* **70** (1974) 947.
[456] T. Biegler and R. Parsons, *J. Electroanal. Chem.* **21** (1969) App. 4.
[457] V. A. Kuznetsov, N. G. Vasilkevich, and B. B. Damaskin, *Elektrokhimiya* **6** (1970) 1339.
[458] B. Jakuszewski and Z. Kozlowski, *Soc. Sci. Lodz. Acta Chim.* **10** (1965) 5.
[459] Z. Kozlowski and B. Jakuszewski, *Soc. Sci. Lodz. Acta Chim.* **11** (1966) 5.
[460] B. Jakuszewski, Z. Kozlowski, and W. Granosik, *Soc. Sci. Lodz. Acta Chim.* **13** (1968) 15.
[461] L. I. Antropov, M. A. Gerasimenko, and Yu. S. Gerasimenko, *Elektrokhimiya* **11** (1975) 1571.
[462] S. Minc and I. Zagorska, *Electrochim. Acta* **16** (1971) 1213.
[463] D. Bax, C. L. De Ligny, and A. G. Remijnse, *Rec. Trav. Chim.* **91** (1972) 1225.
[464] H. J. M. Denessen, C. L. De Ligny, and A. G. Remijnse, *J. Electroanal. Chem.* **51** (1974) 215.
[465] I. Zagorska and Z. Koczorowski, *Rocz. Chem.* **44** (1970) 1559.
[466] D. Bax, C. L. De Ligny, and M. Alfenaar, *Rec. Trav. Chim.* **91** (1972) 452.
[467] C. L. De Ligny, H. J. M. Denessen, and M. Alfenaar, *Rec. Trav. Chim.* **90** (1971) 1265.
[468] D. Bax, C. L. De Ligny, and A. G. Remijnse, *Rec. Trav. Chim.* **91** (1972) 965.
[469] B. Jakuszewski, M. Przasnyski, and A. Siekowska, *Bull. Acad. Pol. Sci. Ser. Sci. Chim.* **20** (1972) 43.
[470] B. Jakuszewski, M. Przasnyski, H. Scholl, and A. Siekowska, *Electrochim. Acta* **20** (1975) 119.
[471] H. J. M. Denessen, C. L. De Ligny, and A. G. Remijnse, *J. Electroanal. Chem.* **44** (1973) 153.
[472] B. Jakuszewski and H. Scholl, *Electrochim. Acta* **17** (1972) 1105.
[473] R. Parsons and B. T. Rubin, *J. Chem. Soc. Faraday Trans. 1* **70** (1974) 1636.
[474] U. Palm and V. Past, *Usp. Khim.* **44** (1975) 2035.
[475] A. Frumkin and B. Damaskin, *J. Electroanal. Chem.*, **79** (1977) 259.
[476] A. N. Frumkin and B. B. Damaskin, *Dokl. Akad. Nauk SSSR* **221** (1975) 395.
[477] R. Payne, in *Techniques of Electrochemistry*, Ed. by E. Yeager and A. Salkind, Wiley-Interscience, New York, 1972, Vol. 1, p. 71.
[478] J. Clavilier and N. Van Huong, *C. R. Acad. Sci. Ser. C* **269** (1969) 736.
[479] N. B. Grigoryev and A. M. Kalyuzhnaya, *Elektrokhimiya* **10** (1974) 1287.
[480] L. E. Rybalka and B. B. Damaskin, *Elektrokhimiya* **9** (1973) 1562.
[481] G. A. Dobrenkov and G. D. Shilotkach, *Elektrokhimiya* **6** (1970) 1416.
[482] N. B. Grigoryev and D. N. Machavariani, *Elektrokhimiya* **8** (1972) 406.
[483] A. R. Alumaa and U. V. Palm, *Elektrokhimiya* **8** (1972) 790.
[484] Yu. P. Ipatov, V. V. Batrakov, and V. V. Shalaginov, *Elektrokhimiya* **12** (1976) 286.
[485] Yu. P. Ipatov and V. V. Batrakov, *Elektrokhimiya* **11** (1975) 1282.
[486] N. B. Grigoryev, V. P. Kuprin, Yu. M. Loshkarev, and R. V. Malaya *Elektrokhimiya* **11** (1975) 1400.

[487] B. B. Damaskin, V. M. Gerovich, I. P. Gladkikh, and R. I. Kaganovich, *Zh. Fiz. Khim.* **38** (1964) 2495.
[488] B. B. Damaskin, I. P. Mishutushkina, V. M. Gerovich, and R. I. Kaganovich, *Zh. Fiz. Khim.* **38** (1964) 1797.
[489] V. A. Panin, K. V. Rybalka, and D. I. Leikis, *Elektrokhimiya* **8** (1972) 1507.
[490] V. Bartenev, E. Sevastyanov, and D. I. Leikis, *Elektrokhimiya* **6** (1970) 1868.
[491] Z. Görlich, quoted in Ref. 472.

3

Electrochemical Aspects of Adsorption on Mineral Solids

P. Somasundaran

Henry Krumb School of Mines, Columbia University, New York, New York

E. D. Goddard

Union Carbide Corporation, Tarrytown Technical Center, Tarrytown, New York

I. INTRODUCTION

Adsorption is a phenomenon involving the transfer of material from one phase (most frequently a fluid) to an interface which may be liquid(L)/gas(G), solid(S)/G, L/L, or L/S. In this chapter, emphasis will be placed on mineral solids as the adsorbent, water as the liquid phase, and surfactants as the transfer material or adsorbate. The great practical importance of adsorption is that it is the governing factor in a number of processes such as mineral flotation, flocculation and comminution, tertiary oil recovery using surfactant flooding, and detergency. Adsorption of surfactants from aqueous solution onto minerals is often a very complex process since it can be influenced by the properties of all the components of the system. Thus, it is affected by properties of the solid, such as solubility and interfacial electrochemical potential difference, solution properties such as salinity, pH, and temperature, and, of course, by the properties of the surfactant itself, which are determined mostly by its structural makeup. The manner in which these physicochemical properties affect adsorption of surfactants on minerals is discussed below with the help of typical examples drawn from the literature.

II. BASIC PRINCIPLES

Adsorption of a particular material from solution onto a solid is the result of interactions with species on or in the surface of the solid, which are intrinsic to the solid, or in some cases, with species which are themselves preadsorbed because of a more energetically favorable state than that in which they exist in the bulk solution. As is well known, the interactions that are responsible for the accumulation of the adsorbate at the interface can be either chemical or physical in nature.[1,2] Thus, even differences in the extent of van der Waals' interactions existing in the bulk and at the interface can contribute toward adsorption. If the standard free energy involved in transfer of the adsorbate from the bulk (b) to the interfacial region (i) is $\Delta G°_{bi}$, its concentration C_i in the interfacial region can be expressed as

$$C_i = C_b \exp\left(\frac{-\Delta G°_{bi}}{RT}\right) \quad (1)$$

where C_b is the bulk concentration in mole cm^{-3}, R is the gas constant, and T is the absolute temperature. To express the adsorption density (now designated Γ_1) in terms of mole cm^{-2}, Eq. (1) is multiplied by the thickness of the adsorbed layer, τ, leading to

$$\Gamma_i = \tau C_b \exp\left(\frac{-\Delta G°_{ads}}{RT}\right) \quad (2)$$

$\Delta G°_{ads}$ is equal to $\Delta G°_{bi}$ and is the cumulative result of a number of contributing energy changes such as those due to electrostatic attraction, covalent bonding, hydrogen bonding, or nonpolar bonding between the adsorbate and adsorbent species, lateral interaction between the adsorbed species, and solvation or desolvation of any adsorbing species or species displaced from the interface during adsorption.[3] In addition, differences in physical properties, such as structure and dielectric constant, of the medium at the interface and in the bulk can also contribute toward adsorption. These differences are, however, not precisely known and cannot, therefore, be quantitatively taken into account in interpreting adsorption data. It is also to be noted that one or more of the above factors can play a governing role in determining adsorption, depending on the mineral–solution system. Some of the factors can even contribute a positive change to the free energy of adsorption, thus minimizing the extent of adsorption, but nevertheless allowing the species to adsorb if the total change in free energy is negative. In certain cases, adsorption can occur even if the total charge in free energy, ΔG°_{ads}, under equilibrium conditions is positive, but is negative at some point in time prior to the attainment of equilibrium and, at which stage, adsorption is "irreversible." Recognition of these factors is important lest misleading conclusions be reached regarding the mechanism of adsorption. For example, in the case of oleate adsorption on hematite, mere fluctuation in solution pH during the equilibration can produce abnormal values of the final extent of adsorption. Nevertheless, careful control of variables during adsorption studies has yielded some valuable information on the role of some of the above-mentioned factors in governing the adsorption of surfactants.[4]

III. CHARGE GENERATION

The adsorption that ionic surfactants undergo on minerals is often due to electrostatic attraction since most minerals are charged when in contact with aqueous solutions. In fact, this represents one of the most important types of adsorption in practice and will be illustrated by several examples later. It is therefore pertinent to inquire into the nature of this charge generation in some detail. (A second, compelling reason for this inquiry derives from the fact that minerals in contact with water will always have a finite solubility in water, and the resulting simple or complex ions can have a significant effect on the uptake of any adsorbate by the mineral, as will be discussed later.) The interfacial charge may be the result of preferential adsorption of ions from the solution, preferential dissolution of lattice ions, as in the case of silver iodide, or solvation of the surface species followed by dissociation, as in the case of alumina:[5]

$$-M(H_2O)^+ \overset{H^+}{\leftrightarrows} -MOH_{surface} \overset{OH^-}{\rightleftarrows} -MO^-_{surface} + H_2O \quad (3)$$

The lattice ions are considered as potential-determining ions for AgI-type solids, whereas H^+ and OH^- ions, which are the chemical constituents of the solvent reacting with the solid surface, are generally considered as potential determining for oxide minerals. For salt-type minerals that are sparingly soluble and at the same time reactive toward the solvent, both the solvent constituent ions and the lattice ions, as well as their complexes with the solvent species, are found to be potential determining. As an example, calcite undergoes all the reactions (4) to (13) (see below) and produces the following charged species[6]: HCO_3^-, CO_3^{2-}, Ca^{2+}, $CaHCO_3^+$, and $CaOH^+$.

$$CaCO_3(s) \rightleftarrows CaCO_3(aq), \quad K_1 = 10^{-5.09} \quad (4)$$
$$CaCO_3(aq) \rightleftarrows Ca^{2+} + CO_3^{2-}, \quad K_2 = 10^{-3.25} \quad (5)$$
$$CO_3^{2-} + H_2O \rightleftarrows HCO_3^- + OH^-, \quad K_3 = 10^{-3.67} \quad (6)$$
$$HCO_3^- + H_2O \rightleftarrows H_2CO_3 + OH^-, \quad K_4 = 10^{-7.65} \quad (7)$$
$$H_2CO_3 \rightleftarrows CO_2(g) + H_2O, \quad K_5 = 10^{1.47} \quad (8)$$
$$Ca^{2+} + HCO_3^- \rightleftarrows CaHCO_3^+, \quad K_6 = 10^{0.82} \quad (9)$$
$$CaHCO_3^+ \rightleftarrows H^+ + CaCO_3(aq), \quad K_7 = 10^{-7.90} \quad (10)$$
$$Ca^{2+} + OH^- \rightleftarrows CaOH^+, \quad K_8 = 10^{1.40} \quad (11)$$
$$CaOH^+ + OH^- \rightleftarrows Ca(OH)_2(aq), \quad K_9 = 10^{1.37} \quad (12)$$
$$Ca(OH)_2(aq) \rightleftarrows Ca(OH)_2(s), \quad K_{10} = 10^{2.45} \quad (13)$$

It can be seen from these equations that when calcite approaches equilibrium with water at high pH values, an excess of negative HCO_3^- and CO_3^{2-} ions will exist, whereas at low pH values an excess of positive Ca^{2+}, $CaHCO_3^+$ species will arise. These ionic species may be produced at the solid–solution interface or may form in solution and subsequently adsorb on the mineral in amounts related to their concentrations in solution. In either case, the net result will be a positive charge on the surface at low pH values and a negative charge at high pH values. The potential-determining ions to be considered in this case are, in addition to Ca^{2+} and CO_3^{2-}, OH^-, H^+, and HCO_3^-. Since the concentration of all the species present can be calculated as a function of pH from available thermodynamic data, the point of zero charge (pzc) of the solid can be obtained assuming correspondence between it and the "isoelectric" point of the solution, by estimating the pH at which the total concentration of negative "potential-determining" ions is equal to that of the positive "potential-determining" ions. (The rationale for this approach has been presented in some detail by Parks and de Bruyn.[7] They point out that the isoelectric point corresponds, at least for several oxidic materials involving complex ionic equilibria in aqueous solution, with the point of minimum overall solubility of the solid; the correspondence between this point and the point of zero charge implies equal adsorbability of the negative and positive potential-determining ions. In turn, determination of the minimum solubility of the solid allows confirmation of the actual potential-determining ions from the thermodynamic activity versus pH plot for all relevant species present.) Employing this approach for calcite, a value of 8.2 was obtained for the point of zero charge, in close agreement with that obtained from measurements of solution pH changes caused by the addition of calcite, from solubility measurements, and values estimated from flotation experiments[6] (see Figs. 1a,b). It can be seen from Fig. 1b that, in addition to the constituent ions Ca^{2+} and CO_3^{2-}, other ions such as OH^-, H^+, and HCO_3^- also have a major role as potential-determining ions for calcite. It should be pointed out that the isoelectric point obtained from the zeta potential measurements can differ from the point of zero charge* owning to nonequilibrium conditions that exist even after prolonged contact of the mineral with aqueous solutions. Thus a continuous shift from an initial pH value of 10.8 toward the equilibrium value has been observed for calcite by using streaming potential measurements.

The mechanism of charge generation for minerals that are

---

* In this chapter we use the current IUPAC definition of terms. Thus, point of zero charge (pzc) refers to a particle or surface carrying no fixed charge, and isoelectric point (iep) to a particle showing no electrophoresis or a surface showing no electro-osmosis.

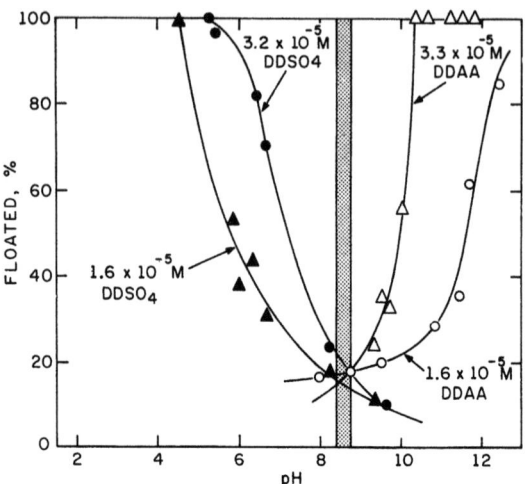

Figure 1a. Determination of point of zero charge from flotation of calcite with dodecylammonium acetate (DDA) and sodium dodecylsulfate (DDSO$_4$) solutions (Somasundaran and Agar[6]).

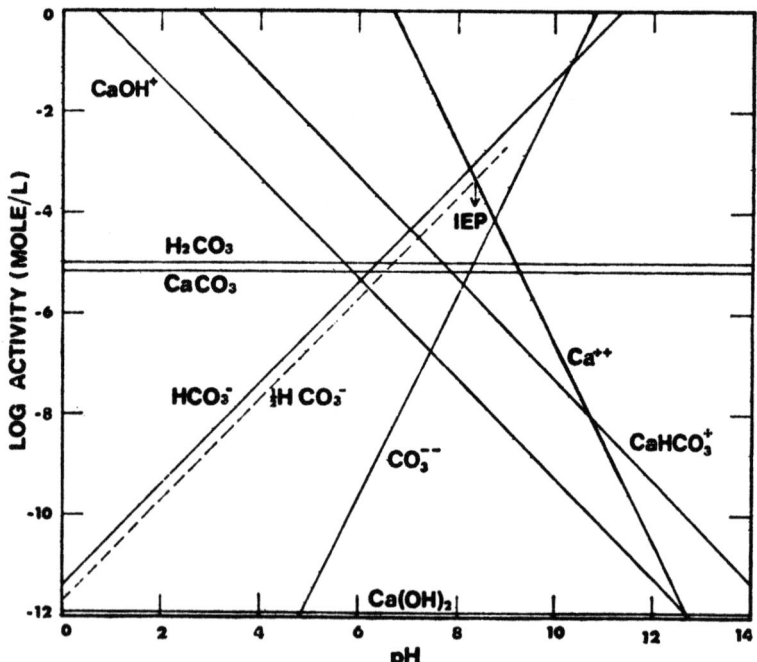

Figure 1b. Determination of point of zero charge of calcite from thermodynamic data (Somasundaran and Agar.[6]).

tri-ionic crystals, such as apatite, is much more complex. As pointed out by Saleeb and de Bruyn,[8] the concentration of any two ions can be varied independently for such tri-ionic solids and one can still obtain a constant solubility product permitting the location of a series of points of zero charge which lie on a line on the equilibrium solution surface. This is obtained by plotting concentrations of the three lattice ions that are potential determining on three mutually perpendicular axes. Substitution of the constituents in the apatite with other species such as F^-, Sr^{2+}, and Ba^{2+} produces significant changes in its electrokinetic characteristics. A summary of values of the solubility and point of zero charge of various ionic solids is given in Table 1.[9] For natural ore-apatite, partially saturated with fluoride, Somasundaran[10] obtained an isoelectric point of pH 4 which, as seen in Fig. 2, shifted toward a final value of pH 6 with equilibration. It is to be noted that the nature of the mineral had a marked effect on the zeta potential.

Measurements of solution pH changes gave a value of 7 for the point of zero charge of fluorapatite. In addition to hydrogen and hydroxyl species, results given in Table 2 suggest that the phosphate species is also potential determining. Zeta potential changes accompanying the addition of calcium and fluoride also are significant. The major effect of calcium appeared to be due to its specific adsorption characteristics. As for the case of calcite, the mechanism of charge development at the apatite surface involves[10] a number of pH-dependent hydrolysis reactions of the surface species:

$$Ca_{10}(PO_4)_6(OH)_2 + 6H_2O$$
$$\leftrightarrows 4Ca_2(HPO_4)(OH)_2 + 2Ca^{2+} + 2HPO_4^{2-} \quad (14)$$
$$Ca_2(HPO_4)(OH)_2 \leftrightarrows 2Ca^{2+} + HPO_4^{2-} + 2OH^- \quad (15)$$

Regardless of whether it is the complex $Ca_2(HPO_4)(OH)_2$ or the original $Ca_{10}(PO_4)_6(OH)_2$ which undergoes dissolution to produce Ca^{2+}, and HPO_4^{2-}, these ions would undergo further hydrolysis in solution and complex formation according to the reactions given below:

$$Ca^{2+} + OH^- \rightleftarrows CaOH^+ \quad (16)$$
$$CaOH^+ + OH^- \rightleftarrows Ca(OH)_2 \quad (17)$$

Table 1
Solubility Constants and pzc and iep Values of Natural and Synthetic Salt-Type Minerals[a]

Salt-type compound	Solubility		Electrochemical properties		
	pK_{SO}	iep	pzc or iep, pH	Method[b]	Conditions[c]
AgCl	10.0	pAg 4.1–4.6	pzc 4.9–6.0	SP	Syn.
AgI	16.0	pAg 5.1–6.2	iep 6.2	SP	Syn.
Fluorite, CaF_2	10.3	pCa 2.6–7.7	iep 2.3–10.7	SP	Nat.
Gypsym, $CaSO_4 \cdot 2H_2O$	4.6		iep 2.3	SP	Nat.
Celestite, $SrSO_4$	6.2				
Barite, $BaSO_4$	9.7	pBa 3.9–7.0	iep 3.4	ME	Nat.
Scheelite, $CaWO_4$	9.0	pCa 4.0–4.8	iep 10.2	ME	Nat., aged 2 hr, NaCl
Calcite, $CaCO_3$	8.4		iep 10.8	SP	Nat., aged hours
Calcite, $CaCO_3$		pCa 3.5, pCO_3 3.0	iep 8–8.5	SP	Nat., aged days
Calcite, $CaCO_3$			pzc 8.2	TD	Nat., aged days
Calcite, $CaCO_3$			iep 5.5–6	SP	Aged 30 min.
Aragonite, $CaCO_3$	8.2		pzc <5	SP	Nat.
Magnesite, $MgCO_3$	4.9		iep 6–6.5	SP	Nat., aged 30 min, KCl
Dolomite, $(Ca, Mg)CO_3$	16.7		iep <7.0	SP	Nat., aged 30 min, KCl
Eggonite, $AlPO_4 \cdot 2H_2O$	21.0		pzc 4.0	DR	Syn.
Strengite, $FePO_4 \cdot 2H_2O$	23.0		pzc 2.8	DR	Syn.
Monazite, $(Ce, La, Th)PO_4$			iep 3.4	ME	Nat.
Fluroapatite, $Ca_{10}(PO_4)_6(F, OH)_2$			iep 5.6	SP	Nat., aged hours, KNO_3
Fluorapatite, $Ca_{10}(PO_4)_6(F, OH)_2$			iep ~4	SP	Nat., aged hours, KNO_3
Fluorapatite, $Ca_{10}(PO_4)_6(F, OH)_2$			iep ~6	SP	Nat., aged days, KNO_3
Fluorapatite, $Ca_{10}(PO_4)_6F_2$	119.1	pCa 4.38, pF 4.63, $pHPO_4$ 5.2	pzc 6.9 ± 0.2	TI	Syn., aged 504 hr, KNO_3
Fluorapatite, $Ca_{10}(PO_4)_6F_2$			iep 7.0	ME	Syn.
Hydroxyapatite, $Ca_{10}(PO_4)_6(OH)_2$	115.3	pCa 4.38, $pHPO_4$ 4.19	pzc 8.5 ± 0.2	TI	Syn., aged 504 hr, KNO_3
Hydroxyapatite, $Ca_{10}(PO_4)_6(OH)_2$			iep 8.5	ME	Syn.
Francolite, $Ca_{10}(C, PO_4)_6(F, OH)_2$			iep 3.8–4.9	SP	Nat. microcryst, aged 30 min, KCl

[a] After Hanna and Somasundaran, Ref. 9.
[b] SP, streaming potential; ME, microelectrophoresis; TD, from thermodynamic data; DR, from pH drift; TI = $(T_H - L_{OH})$ determined by titration.
[c] Nat., natural crystalline; Syn., synthetic.

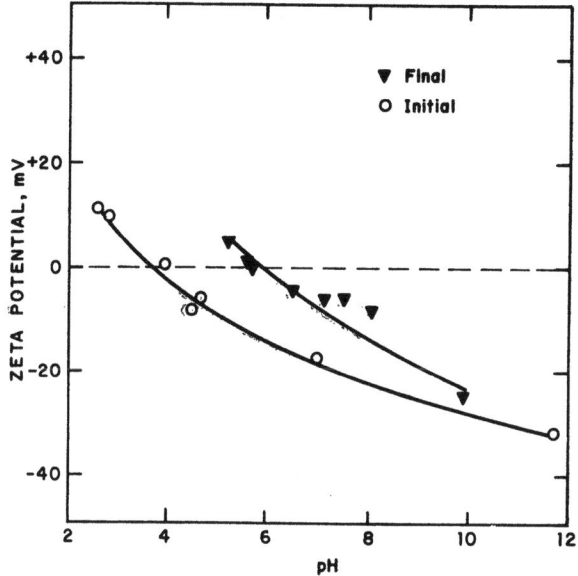

Figure 2. Effect of aging of apatite on its zeta potential as a function of pH (Somasundaran[10]).

Table 2
Isoelectric Point of Alkaline Earth Apatites[a]

Solid	iep		
	I[b]	II[c]	III[d]
CaFA	pF = 4.63 pCa = 4.38 pHPO$_4$ = 5.22		
CaHA	pH = 7.15 pHPO$_4$ = 4.19	pH = 7.00 pHPO$_4$ = 4.48	pCa = 4.38
SrHA	pH = 8.00 pHPO$_4$ = 4.00	pH = 9.80 pHPO$_4$ = 4.40	pSr = 4.19
BaA	pH = 8.57 pHPO$_4$ = 3.43	pH = 11.88	

[a] After Saleeb and de Bruyn, Ref. 8.
[b] Experimentally determined by addition of K_2HPO_4 to suspensions of CaHA, SrHA, and BaA.
[c] Experimentally determined by addition of KOH to suspensions.
[d] By extrapolation.

$$Ca(OH)_2(aq) \rightleftarrows Ca(OH)_2(s) \qquad (18)$$
$$PO_4^{3-} + H_2O \rightleftarrows HPO_4^{2-} + OH^- \qquad (19)$$
$$HPO_4^{2-} + H_2O \rightleftarrows H_2PO_4^- + OH^- \qquad (20)$$
$$H_2PO_4^- + H_2O \rightleftarrows H_3PO_4 + OH^- \qquad (21)$$
$$Ca^{2+} + HPO_4^{2-} \rightleftarrows CaHPO_4 \qquad (22)$$
$$Ca^{2+} + H_2PO_4^- \rightleftarrows CaH_2PO_4^+ \qquad (23)$$

And fluoride ions present in contact with fluorapatite would be involved in the following reactions:

$$F^- + H_2O \rightleftarrows HF + OH^- \qquad (24)$$
$$Ca^{2+} + 2F^- \rightleftarrows 1CaF_2 \qquad (25)$$

Surface species such as $-Ca^+$, $-F$, and $-PO_4^{2-}$ would similarly undergo hydrolysis according to the reactions given above and the extent of such hydrolysis for each species would, together with solution pH, essentially determine the surface charge.

On the basis of the above considerations, there is of course no reason why, for example, silicate species in solution should not be considered as potential-determining ions for silica. Clay minerals, which have layered structures consisting of sheets of SiO_4 tetrahedra and sheets of AlO_6 or MgO_6 octahedra linked with each other by means of shared oxygen ions, are negatively charged under most natural conditions owing mainly to substitution, for example, of Al^{3+} for Si^{4+} in the silica tetrahedra. This charge is internal to the structure and is not dependent on solution concentration.[11] The edges of the clay particles will, on the other hand, exhibit pH-dependent charge characteristics owing to hydroxylation and ionization of the broken Si—O and Al—O bonds at the edges. The point of zero charge of the clay is thus determined by the algebraic sum of face and edge charges. It is to be noted that, at the point of zero charge, both the sides and faces will be charged and thus possess adsorptive properties that other minerals might not possess at their points of zero charge.

In contrast to clays, nonclay silicates, such as chrysotile, possess points of zero charge that are determined by the mineralogical composition of the surface exposed to the solution; this composition can be different from its bulk chemical constitution. Thus, chrysotile possesses a point of zero charge that is more alkaline than that expected on the basis of the bulk chemical composition since the surface exposed to the solution is richer in magnesium.

There is almost no basic information on the interfacial charge characteristics of salt minerals such as sylvite (KCl) and halite (NaCl). This situation derives from the fact that it has been difficult to study these minerals using the conventional electrokinetic techniques owing to the high ionic strength of their saturated solutions.

The surface potential ψ_0 in these systems is given by

$$\psi_0 = \frac{RT}{z_+ F} \ln \frac{a_+}{a_+^0} \quad \text{or} \quad \frac{RT}{z_- F} \ln \frac{a_-}{a_-^0} \tag{26}$$

where a_+ and a_- are the activities of the positive and negative potential-determining ions in the solution with valencies z_+ and z_-, and a_+^0 and a_-^0 are their activities under conditions of zero surface charge on the mineral. On the somewhat questionable assumption that potential differences due to dipole orientation remain constant, the total double-layer potential is considered zero when the surface charge is zero, which, in fact, represents the definition of the point of zero charge. As mentioned earlier, for oxides, hydrogen and hydroxyl ions are potential determining and therefore surfaces of such materials will carry a positive charge in solutions that are more acidic than those corresponding to the pzc, and a negative charge in those that are more alkaline. Since the system as a whole must be electrically neutral, the medium surrounding the particles must contain an equivalent net amount of ions of charge opposite to that on the surface of the particle. Owing to the attraction by the charged sites, these counterions will not be uniformly distributed in the solution phase, but will be adsorbed at the oxide–solution interface in the form of diffuse layer because of thermal agitation. This gives rise to an electrical double layer consisting of a layer of surface charge and another layer of counterions. A schematic representation of this well-known type of electrical double layer is given in Fig. 3.

The point of zero charge of minerals is an important, experimentally accessible interfacial property since it determines the

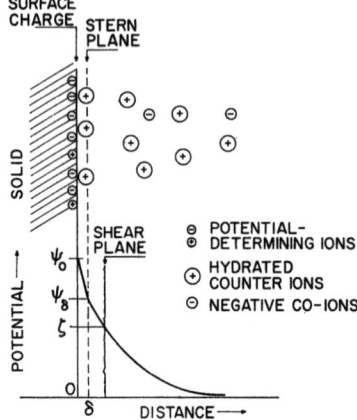

Figure 3. Schematic representation of the double layer according to Stern's model.

adsorption of various ions under the various conditions encountered and, consequently, the behavior of the solid in several interfacial processes. Typical pzc values are given in Table 3.[12] It is important to note that the values obtained are often affected by the presence of impurities, previous history including pretreatments, method of storing and aging, and the extent of aging.[13,14] Variations in the source or method of preparation, including mechanical treatments and washing and drying, and the presence of surface defects and of adsorbed and structural impurities, also produce significant changes in the pzc. The type of acid used for

Table 3
Point of Zero Charge or Isoelectric Point of Various Typical Minerals[a]

Minerals	pzc or iep
Quartz, SiO_2	pH 2–3.7
Rutile, TiO_2	pH 6.0
Corundum, Al_2O_3	pH 9.0
Magnesia, MgO	pH 12.0
Fluorapatite (natural), $Ca_5(PO_4)_3$ (F, OH)	pH 6
Fluorapatite (synthetic)	pCa 4.4, pF 4.6, $pHPO_4$ 5.22
Hydroxyapatite, $Ca_5(PO_4)_3(OH)$, synthetic	pH 7–7.15, $pHPO_4$ 4.19–4.48
Calcite, $CaCO_3$	pH 9.5
Barite, $BaSO_4$	pBa 6.7
Silver iodide, AgI	pAg 5.6
Silver sulfide, Ag_2S	pAg 10.2

[a] After Somasundaran, Ref. 12.

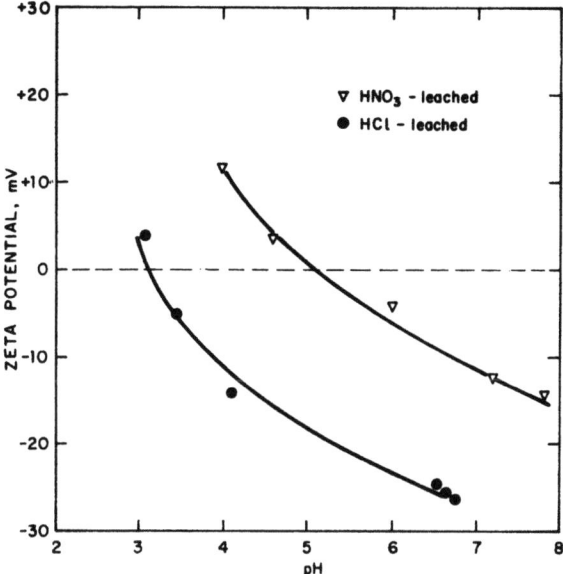

Figure 4. Effect of acid leaching on the isoelectric point of apatite (Somasundaran[13]).

cleaning minerals can also cause considerable effects (see Fig. 4). In fact, it has been found recently that the iep of quartz can be raised from below 2 to as high as 6 by leaching it in hydrofluoric acid solution. Upon aging the HF-leached quartz in water, the iep can be brought to its original value but only over a period of several days (see Fig. 5). Washing the minerals with hot solutions also produces similar long-term effects.[14] The change in these properties during subsequent aging is also found to be governed by the type of treatment used.

IV. ELECTROSTATIC ADSORPTION

The dependence of adsorption on the electrical nature of the interface has been tested for several mineral–surfactant systems. For example, the role of electrostatic forces in the adsorption of alkyl sulfonates on alumina is clearly illustrated in Fig. 6 where adsorption of dodecanesulfonate is plotted as a function of pH at a constant equilibrium bulk concentration of 3×10^{-5} mole liter^{-1}.

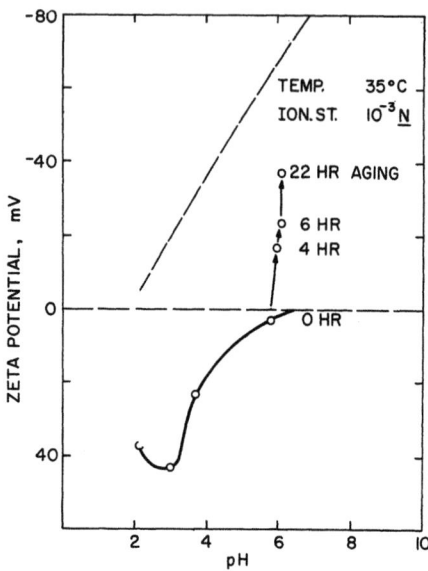

Figure 5. Zeta potential of quartz treated with HF and then with hot NaOH solutions as a function of pH without aging, and at pH = 6 after 4, 6, and 22 hours of aging; dashed curve is for equilibrium values obtained for HNO_3-treated quartz, for comparison (Kulkarni and Somasundaran[15]).

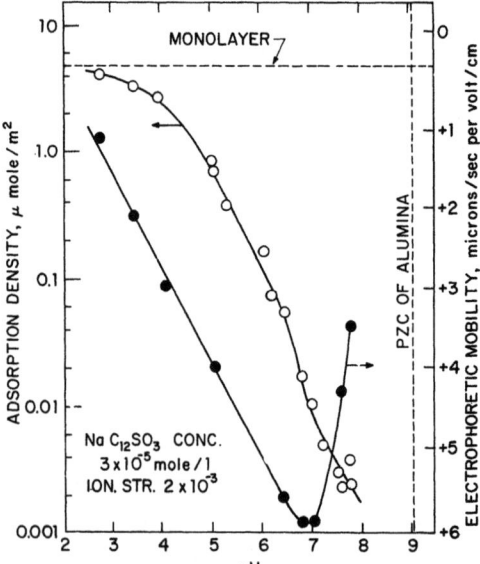

Figure 6. Adsorption of dodecanesulfonate on alumina as function of pH (Somasundaran and Fuerstenau[16]).

The point of zero charge of alumina is at pH 9.1. It is then to be noted that only below pH 9.1, where the alumina is positively charged, does the anionic sulfonate adsorb in significant amounts.[16] A similar correlation has also been obtained for calcite in alkanesulfonate and amine solutions.[6] In fact, there is abundant information in the literature that simple electrical considerations as outlined above, viz., adsorption of a negatively charged surfactant on a positively charged solid, and vice versa, apply to a large number of systems investigated.[17,18]

The correlation obtained for the adsorption of quaternary ammonium salts on tricalcium phosphate and calcium carbonate with the zeta potential of these solids is particularly interesting. It can be seen from Fig. 7 that the adsorption density of cetyltrimethylammonium bromide (CTAB) on tricalciumphosphate is about four times that on calcium carbonate. A close-packed monolayer (with a molecular area of 51 Å^2/CTAB molecule) was found for the case of $Ca_3(PO_4)_2$ but not for $CaCO_3$. This is

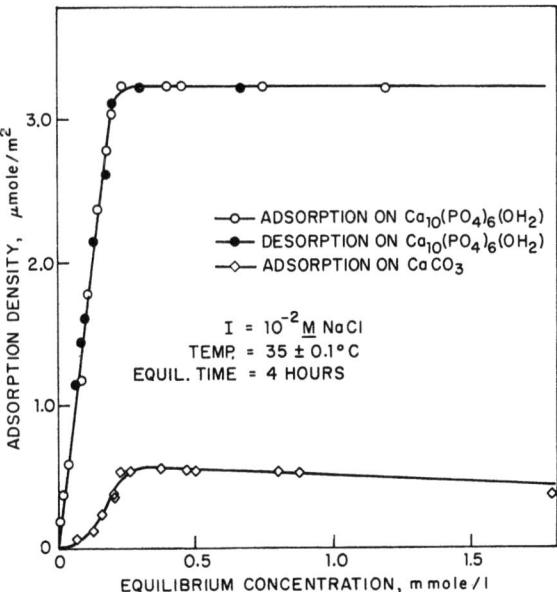

Figure 7. Adsorption isotherm of cetyltrimethylammonium bromide (CTAB) on tricalciumphosphate and calcium carbonate (Hanna et al.[19]).

attributed to the smaller number of negative sites on $CaCO_3$ than on $Ca_3(PO_4)_2$. Moreover, it can be seen from Fig. 7 that the adsorption and desorption isotherms coincide with each other, indicating total reversibility of adsorption. These observations support the above interpretation based on the hypothesis that adsorption for these systems is mainly due to electrostatic interaction. It has repeatedly been demonstrated[18,20,21] that there is a direct relationship between such adsorption and flotation of the mineral, i.e., a direct relationship between electrical interaction and mineral surface modification exists.[22]

In systems similar to those referred to above, where electrostatic forces play a major role, the presence of other charged species in solution can influence adsorption owing to competition for adsorption sites by ions that are charged similarly to the adsorbate species. For example, addition of potassium nitrate can depress the flotation of quartz using dodecylammonium ions owing to the reduction of adsorption of the amine cation by the competing potassium ions. This causes a compression of the electrical double layer and reduces the interfacial potential that is responsible for the adsorption of the dodecylammonium ion. If the added salt contains multivalent ions, this effect can be much stronger and can even cause a reversal of the zeta potential. The more marked depression that has been obtained in the past for quartz flotation using calcium[23] and barium[24] salts results from the stronger tendency of bivalent ions over monovalent ions to adsorb and compete with the surfactant ions. These effects can be investigated further by measuring the relative effect of the monovalent and bivalent ions in depressing the surfactant adsorption and thereby the flotation, and comparing the values obtained with those calculated on the basis of a simple electrical model. Towards this end, Eq. (2) can be rewritten in the following manner for the adsorption of ions of bulk concentration c in the Stern plane δ potential ψ_δ

$$\Gamma_\delta = 2rc \exp\left(\frac{-zF\psi_\delta + \phi}{RT}\right) \quad (27)$$

where r is the radius of the adsorbate ion, z its valency, F the Faraday constant, ψ_δ the potential at plane δ, and ϕ the specific adsorption energy. This equation can now be used for estimating the effect of the adsorption of inorganic ions on flotation. The approach is to measure the relative concentrations of the monovalent and bivalent ions required to depress the flotation by the cationic surfactant by a certain fraction and to compare these ratios with those determined by calculations involving measurements of the zeta potentials corresponding to these conditions. All such estimations involve certain assumptions and, in order to proceed with a model calculation, we assume the following:

1. ψ_δ is equal to the measured zeta potential.
2. Adsorption values for divalent ions under constant zeta potential conditions are half those for the monovalent ions.
3. The adsorption density of the collector ions at the solid–liquid interface is constant for a given degree of flotation.
4. The association energy of the hydrocarbon chains is independent of the ionic strength of the solution.

The assumption that ψ_δ is equal to the zeta potential is safe under conditions of low adsorption in the Stern plane. At high adsorption densities, however, the large number of counterions between the δ plane and the shear plane can make the zeta potential significantly lower than ψ_δ and therefore, under these conditions, the first assumption may not be valid. Changes in viscosity at the interface and the resultant possibility of a shift in the shear plane make the calculations very complex. The problem of ψ_δ not being equal to zeta is, however, immaterial for the present purpose. This derives from the fact that it is uncertain whether or not it is the actual adsorption of the collector ions in the Stern plane that should be calculated. Roy and Fuerstenau[25] have suggested that the adsorption of the surfactant takes place, possibly, on a layer of adsorbed water molecules. Barium on the other hand, could conceivably approach the solid surface very closely. This particular point will be taken into account in the calculations. One can then write an equation for the ratio c_{Na}/c_{Ba} for equivalent flotation in the following manner[23]:

$$\frac{\Gamma_{Na}}{\Gamma_{Ba}} = \frac{r_{Na}}{r_{Ba}} \frac{c_{Na}}{c_{Ba}} \exp\left(\frac{-\phi_{Ba} + 2F\zeta_{Ba} - F\zeta_{Na}}{RT}\right) \qquad (28)$$

Γ_{Na} and Γ_{Ba} are the adsorption densities of sodium and barium counterions, respectively, under conditions of equivalent flotation and constant concentration of the collector; c_{Na} and c_{Ba} are the corresponding bulk concentrations of sodium and barium ions, and ζ_{Na} and ζ_{Ba} the measured zeta potentials. The parameters r_{Na} and r_{Ba} represent the radii of the ions and ϕ_{Ba} is the specific adsorption energy, per one mole of barium ions, which may include some energy term associated with dehydration. The adsorption of hydrated sodium ion on oxide minerals is nonspecific, since its presence is known not to change the point of zero charge of these minerals in water. Using the value of $3RT$ for ϕ_{Ba}[24] we obtain from Eq. (28)

$$\frac{\Gamma_{Na}}{\Gamma_{Ba}} = \frac{r_{Na}}{r_{Ba}} \frac{c_{Na}}{c_{Ba}} \exp\left[3 - \frac{F(2\zeta_{Ba} - \zeta_{Na})}{RT}\right] \quad (29)$$

Equation (29) can be used, with the help of data in literature, to calculate the relative effectiveness of sodium and barium salts in depressing amine cation flotation of quartz. The values for the ratio c_{Na}/c_{Ba} are given in Table 4. It can be seen that the calculated values are in fair agreement with the experimental values. Even though the agreement is reasonable, the calculations must be regarded as only illustrative; the assumptions used in the calculations need to be checked in more detail for validity. For example, it was found necessary to employ the radius of the dehydrated Ba ion, as well as the assigned ϕ_{Ba} value, to obtain the above-mentioned agreement.

Table 4

Calculated and Experimental Values for the Relative Effectiveness of Sodium Ions and Barium Ions in Depressing Quartz Flotation at Natural pH Using 10^{-4} M Dodecylammonium Acetate[a]

Floated (%)	c_{Na} (mole/liter)	c_{Ba} (mole/liter)	ζ_{Na} (mV)	ζ_{Ba} (mV)	Calculated c_{Na}/c_{Ba}	Experimental c_{Na}/c_{Ba}
90	1×10^{-2}	2×10^{-5}	-63	-50	64	50
70	6×10^{-3}	1.5×10^{-4}	-49	-35	33	40
50	1.5×10^{-2}	8×10^{-4}	-40	-22	18	20
30	5×10^{-2}	5×10^{-3}	-32	-10	10	10

[a] After Somasundaran.[23]

Many other examples of depression of surfactant adsorption, and hence of flotation, accompanying the addition of a polyvalent inorganic gegenion can be found in the literature. In contrast to the above, ions that are charged oppositely to that of the mineral can provide a means of enhancing the adsorption of a surfactant bearing the same sign of charge as that originally held by the solid.[20] This clearly implies a reversal of the sign of charge of the mineral by the polyvalent inorganic ion. In fact, this approach has been successfully employed in flotation, at least in model systems. Also, it is interesting to note in this regard the observation of Saleeb and Hanna[26,27] that the introduction of SO_4^{2-} ions increases the adsorption of a cationic surfactant, cetyltrimethylammonium bromide, on $CaCO_3$, but not on $Ca_3(PO_4)_2$. This was attributed to an increase in the number of negative sites on the $CaCO_3$ surface owing to sorption of the bivalent sulfate ions. Such an increase could not be obtained on $Ca_3(PO_4)_2$ since this mineral is apparently saturated with negative sites even in the absence of sulfonate. This postulate is supported by the observation that introduction of Mg^{2+} into the solution reduces the adsorption of the cationic surfactant on both $CaCO_3$ and $Ca_3(PO_4)_2$. It should be mentioned here that alkaline earth ions such as calcium have been related to function most effectively in the pH range where they are in hydrolyzed soluble form.[28-30] The high adsorption affinity of partially hydrolyzed ions onto charged surfaces was recognized by Wolstenholme and Schulman[31] and this point was adopted extensively in connection with coagulation studies on model suspensions by Matijevic and co-workers.[32]

V. LATERAL INTERACTION BETWEEN ADSORBATES

An examination of the data given in Fig. 6 for the dependence on pH of electrophoretic mobility of alumina in alkane sulfonate solutions suggests that there must be another primary force responsible for adsorption in addition to electrostatic attraction below about pH 7. Adsorption isotherms (as well as settling data) obtained for the alumina–sulfonate systems also suggest such a possibility (see Fig. 8). Both the above properties are found to undergo a sharp increase at a given adsorption density. This was attributed to lateral association of the surfactant species in the interfacial region above a critical adsorption density. Such association to form two-dimensional aggregates, called hemimicelles, is analogous to micelle formation in solution and results from the favorable energetics of (partial) removal of the alkyl chains from the aqueous environment. The energy gained from the process of hemimicellization has been estimated in a number of ways. For example, the data given in Fig. 9 for the zeta potential of quartz as a function of the concentration of amines of different chain lengths can be treated in the following manner to obtain the standard free energy of hemimicellization.[35] Equation (2) is written with $\Delta G°_{ads}$ divided into the two terms, one corresponding to electrostatic interaction and the other to the energy of hemimicellar association:

$$\Gamma_i = rC_b \exp\left(\frac{-\Delta G°_{el} - \Delta G°_{hm}}{RT}\right) \tag{30}$$

Concentrations corresponding to hemimicelle formation (C_{HMC}) were obtained from the data given in Fig. 9 by graphically

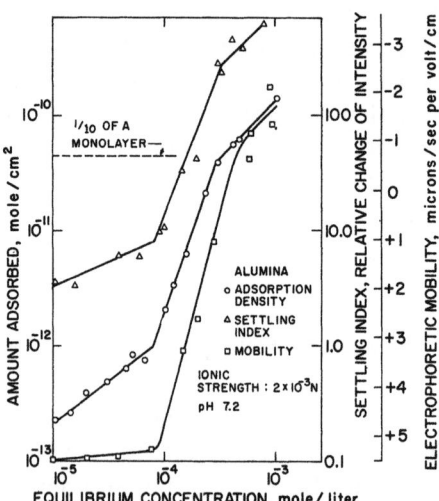

Figure 8. Adsorption density of dodecanesulfonate, electrophoretic mobility, and settling rate of alumina–sodium-dodecanesulfonate system as a function of the concentration of sodium dodecanesulfonate (Somasundaran[33]).

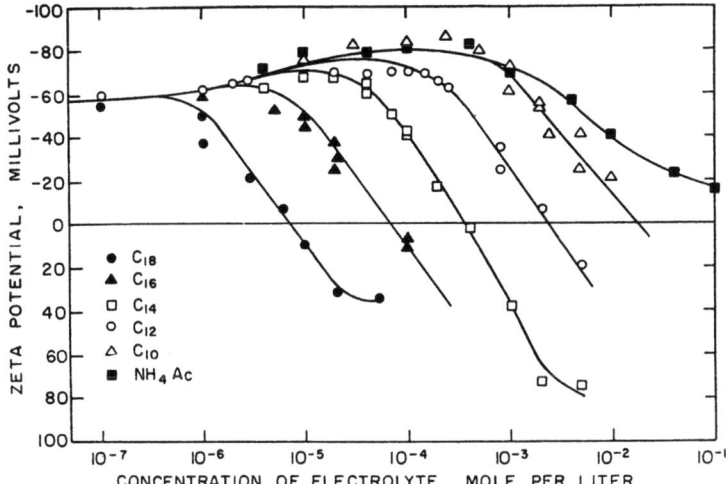

Figure 9. Effect of hydrocarbon chain length on the ζ potential of quartz in solutions of alkylammonium acetates and in solutions of ammonium acetate (Somasundaran et al.[34]).

locating the concentrations at which the extrapolations of the straight-line portions of the streaming potential curves intersect the curve for ammonium acetate. Substitution of C_{HMC} for C_b and $n\phi_{\text{h(sl)}}$ for ΔG_{lu} yields

$$\Gamma_i = \tau C_{\text{HMC}} \exp\left(\frac{-\Delta G^\circ_{\text{el}} - n\phi_{\text{h(sl)}}}{RT}\right) \quad (31)$$

where n is the number of CH_2 and CH_3 groups in the chain and $\phi_{\text{h(sl)}}$ the average free energy of transfer of 1 mole of CH_2 groups from an aqueous environment into a hemimicelle environment. Rearrangement of Eq. (31) leads to

$$\log C_{\text{HMC}} = \frac{n\phi_{\text{h(sl)}}}{RT} + \frac{\Delta G^\circ_{\text{el}}}{RT} + \log\frac{\Gamma_i}{\tau} \quad (32)$$

A least-squares plot of $\log C_{\text{HMC}}$ as a function of number of carbon atoms in the alkyl chain is given in Fig. 10. The value of $\phi_{\text{h(sl)}}$ obtained from this line is $-0.95kT$, comparable to that obtained for the free energy of micellization[36] (see Fig. 11). Heat and standard entropy changes associated with various adsorption processes can be calculated by considering the adsorption of the

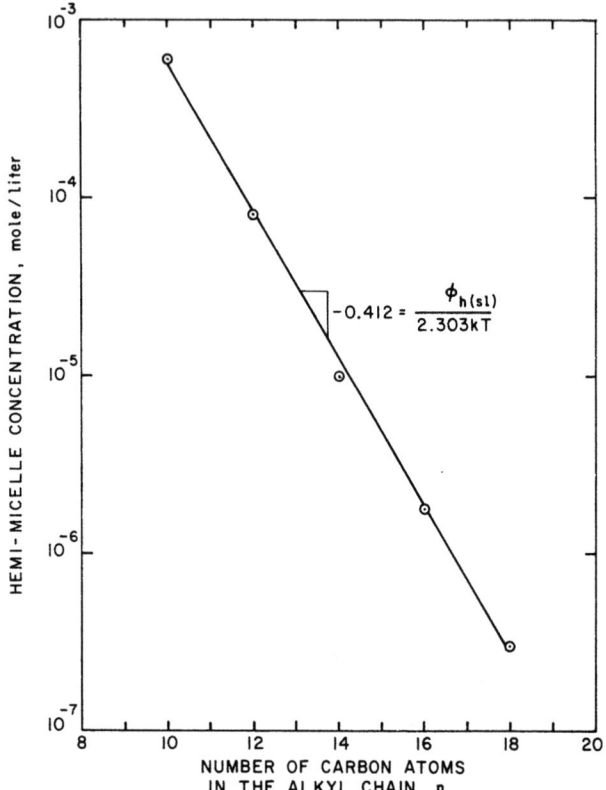

Figure 10. Least-squares plot of $\log C_{\text{HMC}}$ of alkylammonium acetates from streaming potential data as a function of the number of carbon atoms in the chain (Lin and Somasundaran[35]).

long-chain ions, X, as follows.[37] Consider the process of transfer of solute from solution to the surface in two steps:

1. $X(\text{soln.}, a = 1) \to X(\text{soln.}, a = C_1)$, for which we represent the free-energy change as $(\Delta \bar{G}^\circ)_{T_1} = RT_1 \ln C_1$, and
2. $X(\text{soln.}, a = C_1) \leftrightarrows X(\text{surf.}, \Gamma)$, for which $\Delta G = 0$. Hence, for the overall process $X(\text{soln.}, a = 1) \to X(\text{surf.}, \Gamma)$,

$$(\Delta \bar{G}^\circ_{X_\sigma})_{T_1} = RT_1 \ln C_1 \tag{33}$$

Here a is the activity of the alkyl salt in solution and C_1 its concentration in mole/liter corresponding to an adsorption density Γ mole/cm^2 at temperature T_1, and $(\Delta \bar{G}^\circ_{X_\sigma})_{T_1}$ is the relative standard partial molar free energy of the absorbed ions at this temperature. The standard state for X in the solution is such that the activity coefficient approaches unity as the concentration tends to zero. Similarly, at temperature T_2, the relative standard partial molar free energy of adsorbed ions will be

$$(\Delta \bar{G}^\circ_{X_\sigma})_{T_2} = RT_2 \ln C_2. \qquad (34)$$

Knowing C_1 and C_2 for the same adsorption density, it is possible to evaluate $\Delta \bar{G}^\circ_{X_\sigma}$ at both temperatures, from which the standard heat and entropy terms involved, $\Delta \bar{H}^\circ_{X_\sigma}$ and $\Delta \bar{S}^\circ_{X_\sigma}$, can be calculated. These represent the changes corresponding to the transfer of solute, at unit activity, from solution to the surface, at a

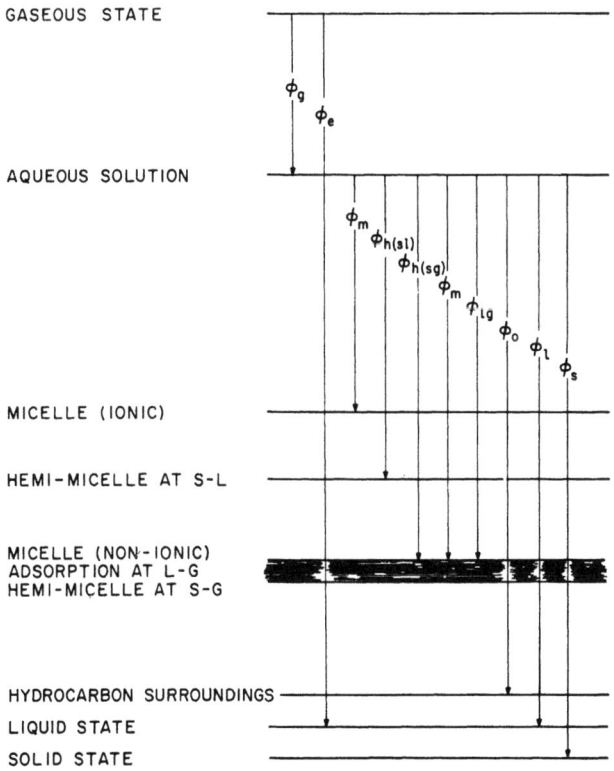

Figure 11. Schematic diagram for free energy of transfer of —CH_2— groups from aqueous solution to various environments (Lin and Somasundaran[35]).

density Γ:

$$\Delta \bar{H}_{X_\sigma} = \bar{H}_{X_\sigma} - \bar{H}^\circ_{X_1} \approx \Delta H_{X_{ad}} \tag{35}$$

where $\Delta \bar{H}_{X_{ad}}$ is the change in partial molar heat content of the alkyl ion on adsorption. Correspondingly, the standard entropy of adsorption can be obtained using the following equation, which includes a contribution from the entropy of dilution, ΔS_{dil}, to the total entropy change, $\Delta \bar{S}^\circ_{X_\sigma}$:

$$\Delta \bar{S}^\circ_{X_\sigma} = \Delta S_{dil} + \Delta \bar{S}^\circ_{X_{ad}}$$
$$= -R \ln C + \bar{S}_{X_\sigma} - S^\circ_{X_1} \tag{36}$$

From Eqs. (33) to (35)

$$\Delta \bar{H}^\circ_{X_{ad}} = \frac{R(\ln C_2 - \ln C_1)}{1/T_2 - 1/T_1} \tag{37}$$

Similarly,

$$\Delta \bar{S}^\circ_{X_\sigma} = \frac{R(T_1 \ln C_1 - T_2 \ln C_2)}{T_2 - T_1} \tag{38}$$

and from Eq. (36)

$$\Delta \bar{S}^\circ_{X_{ad}} = \frac{R(T_1 \ln C_1 - T_2 \ln C_2)}{T_2 - T_1} - \Delta S_{dil} \tag{39}$$

By evaluating $\Delta \bar{H}^\circ_{X_{ad}}$ and $\Delta \bar{S}^\circ_{X_{ad}}$ as a function of the concentration, it is possible to obtain the changes in them due to such processes as hemimicellization. A word of caution is, however, in order: adsorption in these systems is complicated by interaction with other components present. For example, when sulfonate ions are adsorbed on alumina, in a system at constant ionic strength, they will be displacing anions of the supporting electrolyte from the double layer and possibly water molecules bound to the surface. For concentrations below the hemimicelle concentration, adsorption of sulfonate ions takes place by simple ion exchange.[16] Since neutral 1:1 electrolyte is generally added during these experiments to maintain constant ionic strength in the system, adsorption of sulfonate ions below the hemimicelle concentration is accompanied by displacement of chloride ions from the surface.

Alkane sulfonate ions are considerably larger than chloride ions, and hence displacement of a relatively large number of water molecules in a structured state from the surface may also take place when the surfactant is adsorbed. The measured changes in thermodynamic quantities on adsorption will embrace all these effects and can therefore be expected to be different from the changes in the corresponding quantities for the adsorbed sulfonate ions only, according to the simple picture outlined above.

It is also possible to evaluate the thermodynamic quantities associated with hemimicelle formation by approximating it to a phase change and using the Clausius–Clapeyron-type equation:

$$\frac{d \ln (\mathrm{HMA})}{dT} = \frac{-\Delta H_{\mathrm{HM}}}{RT^2} \qquad (40)$$

where (HMA) is the critical concentration and ΔH_{HM} is the heat of hemimicelle formation at temperature T. In Eq. (40) either the bulk concentration or the adsorption density is employed. Choice of the latter has certain advantages in interpreting hemimicelle formation as a two-dimensional association process. In passing we note that, in the case of micelles in bulk, a number of workers have considered the micelle to be a different phase from the aqueous solution one and, on this basis, have treated micelle formation as a phase change.[38] Recent experiments by Mysels and co-workers[39] show that, with proper experimental precautions, the activity of the surfactant will be observed to increase above the critical micelle concentration (CMC) and that this increase can be explained on the mass action model. It should, however, be noted that the increase in activity above the CMC is very small and treatments based on a pseudophase separation model, even though not strictly correct, can yield useful information. Measurements using specific ion electrodes actually indicate a decrease of single surfactant ion activity above the CMC and substantiate the implied increase in counterion activity.[40] In any event, for long-alkyl-chain surfactants, there is fair agreement between the mass action and phase separation models.

The heat and entropy of adsorption calculated in the manner outlined above for dodecane sulfonate adsorbed on alumina particles[37] show marked changes at particular concentrations and

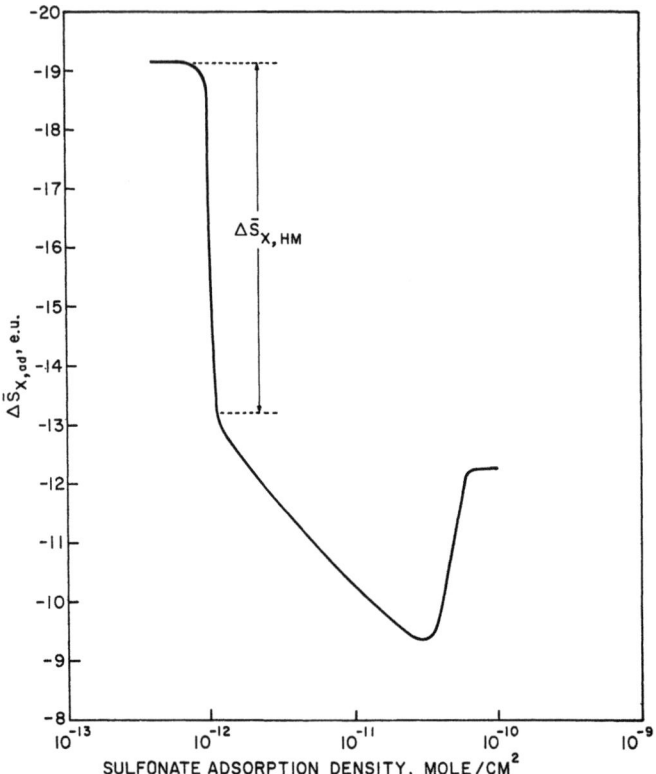

Figure 12. Partial molar entropy of adsorption of sulfonate as a function of its adsorption density (Somasundaran and Fuerstenau[37]).

are in general agreement with the postulate of interaction of surfactant ions to form two-dimensional aggregates. Most interestingly, the association was found to produce a net increase in the entropy of the system, suggesting a decrease, upon aggregation, in the ordering the water molecules that were originally surrounding isolated surfactant chains (see Fig. 12). Because of the necessity to initiate the surfactant adsorption on the solid by some primary adsorption force such as electrostatic attraction and of the requirement to pack the molecules in an orderly manner, the extent of aggregation and the concentration at which the aggregation becomes significant are dependent, among other things, on such system properties as pH, temperature, ionic strength, and the chemical structure of the surfactant.

VI. CHEMICAL FORCES

On the basis of infrared results obtained for adsorbed surfactant layers on certain minerals, it has been proposed that fatty acids, as well as sulfonates, can also adsorb in consequence of covalent bonding[41,42] In addition, the observation that maximum flotation of certain minerals with fatty acids is obtained at the point of zero charge (where the electrostatic attraction should be at a minimum) has led to the postulate of mechanisms depending on covalent bonding between the surfactant species and the mineral surface species.

Both the above observations cannot, however, be considered to provide sufficient evidence for chemisorption. Identification of a particular bonding type in an IR spectrum does not necessarily involve the existence of such bonding on minerals in surfactant solutions because of the possible alterations in chemical state of the adsorbed surfactant during the preparation of the sample for the spectroscopic examination. On the other hand, the dependence of a flotation maximum on pH can be attributed, for a number of systems such as hematite–oleate, to a change in chemical state of the oleate species in solution with pH. This is further described in the next section. Chemisorption, it has been proposed, can take place by stoichiometric ion exchange. Thus, in the case of adsorption of oleate on fluorite, the surfactant ions were found to release an equivalent amount of fluoride ions to form a surface layer of calcium oleate[43,44] Hanna[27,45] also observed salt formation to be the mechanism of adsorption for the first layer of Aerosol OT,* dodecylbenzenesulfonate, and dodecanesulfonate on $CaCO_3$ and $Ca_3(PO_4)_2$. A second layer, with the ionic heads exposed to the bulk solution, was considered to be formed in these cases owing to interchain cohesion. Such a mechanism is supported by the observation that the contact angle and flotation response obtained were at a maximum at concentrations corresponding to the completion of the first layer.

VII. CHEMICAL STATE OF THE ADSORBATE

In addition to micellization and precipitation, hydrolyzable surfactants such as fatty acids and amines can undergo various

* Dioctylsulfosuccinate, sodium salt (American Cyanamid).

associative interactions in aqueous solutions giving rise to dimer, acid soaps, trimers, etc.[38] The important role of such associated species[46-48] in determining adsorption has been neglected in the past. Depending upon solution properties such as pH, ionic strength, temperature, etc., some of the these species can be present in amounts sufficient to play a governing role in adsorption. Owing to the possibility of marked differences in their surface activities, one can obtain considerable variation in adsorption with changes in a parameter as simple as solution pH.

Consider the case, for example, of the best known of all surfactants, viz., fatty acid soaps. The major chemical equilibra in a system consisting of this anionic surfactant (R^-Na^+) are

$$R^- + H^+ \rightleftarrows RH, \quad K_a^{-1} \quad (41)$$

$$RH(aq) \rightleftarrows RH(1), \quad K_{sp}^{-1} \quad (42)$$

$$RH + R^- \rightleftarrows R_2H^-, \quad K_{AD} \quad (43)$$

$$2R^- \rightleftarrows R_2^{2-}, \text{ etc.}, \quad K_D \quad (44)$$

$$Na^+ + R_2H^- \rightleftarrows NaHR_2, \quad K \quad (45)$$

Among the various species listed above, the acid–soap complex may be more surface active than other forms of the surfactant complexes because of its large molecular size and low intrinsic solubility.

Species formed between various ionic and molecular forms of the surfactant are termed ionomolecular complexes or IM complexes.[49] Association interactions to form IM complexes have been considered to be the reason for the deviations in anticipated behavior of surfactant solutions in respect of such properties as conductivity, transport number, partial molar volume, osmosis, etc. It is reasonable to expect such interactions to affect adsorption at different interfaces also.

As an example, a most widely used hydrolyzable surfactant, namely oleic acid, is considered in further detail and constants for the various equilibria represented by Eqs. (41)–(45), for this system, viz., $pK_a = 4.95$; $pK_{sp} = 7.6$; $pK_{AD} = -4.7$; $pK_D = -4.0$; $pK = -9.35$ (Somasundaran et al.,[50,51] Jung[52]), lead to the equilibrium diagram showing the activities of various species at 3×10^{-5} mole liter^{-1} oleate as shown in Fig. 13. The precipitation of oleic acid takes place at pH 7.78. Maximum acid–soap complex formation also takes place at this pH. Even though the numerical

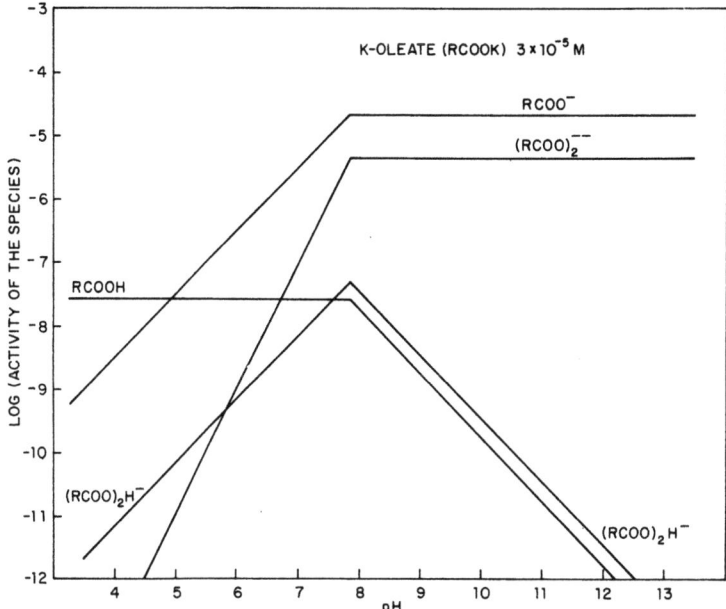

Figure 13. Equilibrium diagram showing the activities of various species of oleate at a concentration of 3×10^{-5} mole/liter (Anathapadmanabhan et al.[51]).

value of the concentration of acid-soap species may be low, it must be remembered that this species has a potentially high surface activity since it represents an entity with the total hydrocarbon chain length of its parent soap and acid but possessing only a single anionic charge. Recent studies on the surface tension of oleic acid solutions do, in fact, suggest the presence of highly surface active species around neutral pH.[53] As seen in Fig. 14, the surface pressure is found to exhibit a maximum at this pH. It is of interest that the surface tension decay rate, $-(d\gamma/dt)$, also exhibits a peak around the same pH. It should be pointed out at this stage that a number of minerals exhibit maximum hydrophobicity, as measured by the flotation technique, in this pH range when oleic acid is used as collector (see Fig. 15). Correlation of flotation with the species distribution diagram suggests that the role of acid–soap dimer, soap dimer, and precipitated oleic acid can be significant in controlling adsorption behavior. In addition, the formation of various ferric oleate

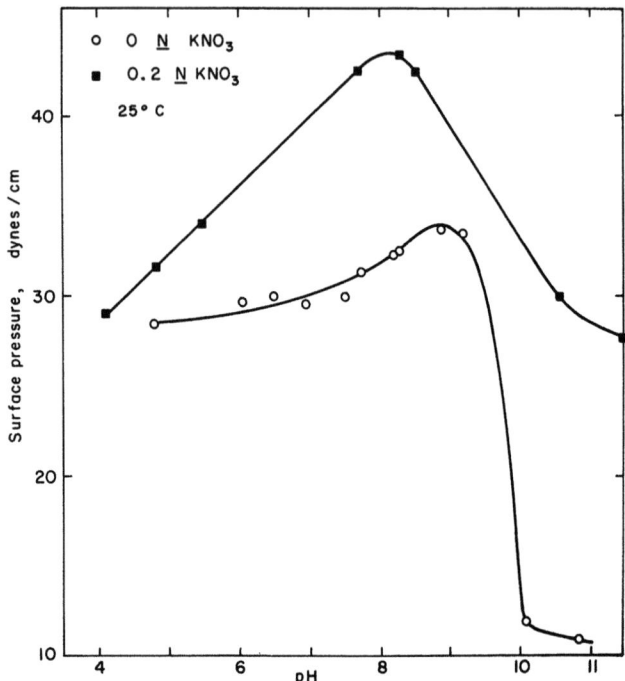

Figure 14. Equilibrium surface pressure of 3×10^{-5} mole/liter potassium oleate solution in water and 2×10^{-1} N KNO_3 (Kulkarni and Somasundaran[53]).

complexes also certainly has to be taken into account in developing a proper understanding of adsorption phenomena in these complex systems.

In the case of the amine–quartz system, the pH of maximum flotation is above about 10.[49,55] Again the amine solutions are found to exhibit maximum surface pressures around this pH. Upon calculation of the distribution of various species, the above pH is also found to coincide with that of maximal ionomolecular complex concentration and with the pH of precipitation of amine. Some early work carried out on alkyl-sulfate–alkyl-alcohol collectors appears to fall into the IM category.[56]

VIII. MISCELLANEOUS

Other factors that can play major roles in certain systems include hydrogen bonding, solvation or desolvation of species, and hydrophobic bonding. Hydrogen bonding can be significant particularly for the adsorption of surfactants containing phenolic groups[11] and possibly of those containing hydroxyl, carboxylic, and amine groups. So-called hydrophobic bonding is important for adsorption on solids that possess a fully or partially hydrophobic surface. In this case surfactant molecules can adsorb in flat configurations on the hydrophobic sites of the solid, as was invoked many years ago to describe the adsorption of dodecyl sulfate on graphitized carbon.[57] A feature of this type of adsorption, overall, is that a second plateau in the isotherm is frequently observed at higher concentrations as the adsorbed layer changes from flat to a more upright orientation. The energy involved in

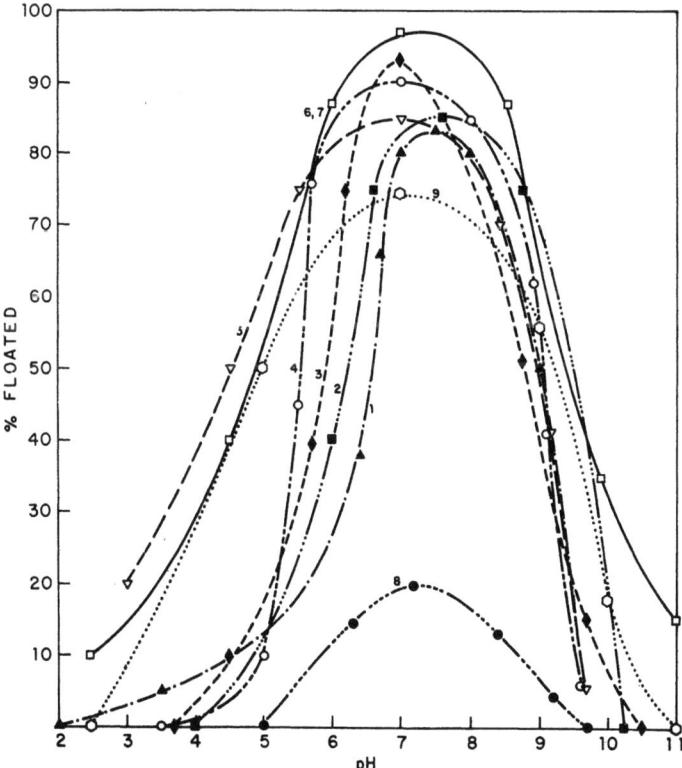

Figure 15. Flotation recovery of various minerals with oleic acid (dose rate 0.1%) as a function of pH: 1, columbite; 2, zircon, 3, tantalite; 4, ilmenite; 5, rutile; 6, garnet; 7, tourmaline; 8, albite; 9, perovskite (Polkin and Najfonow[54]).

this case is additive and, as in the case of hemimicellization, adsorption due to it will increase with the size of the surfactant.

The presence of polymers in solutions has been known to affect adsorption of surfactants on minerals. In some cases, the adsorption is enhanced but in some other cases it is found to be depressed.[58,59]

Much of the work covered so far has been concerned with adsorption of surfactants from very dilute solutions under conditions, in the main, which reflect those prevailing in mineral flotation. We mention briefly here the subject of "micellar" adsorption, that is, the adsorption behavior over a surfactant concentration range which includes the critical micelle concentration (CMC). Over the years, this has been a subject of interest and controversy, in view of the existence of adsorption maxima which are frequently observed as one passes through the CMC. This is, however, by no means always the case as can be seen from the data in Fig. 7 for cetyl trimethylammonium bromide and in Fig. 16 for dodecylamine, both well-characterized surfactants.

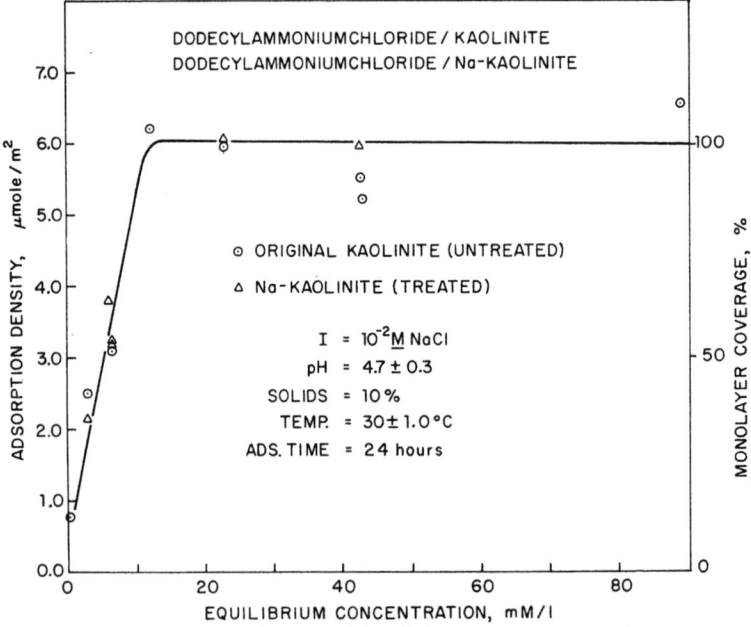

Figure 16. Adsorption isotherms of dodecylammonium chloride on prewetted kaolinites (untreated and NaCl treated) under natural pH conditions.[59]

In both cases, the leveling of adsorption corresponds to the CMC. Instances involving maxima in adsorption of surfactants onto minerals will be discussed in Section XI. The earliest reports of adsorption maxima were concerned with simple adsorption at the air/water interface. This phenomenon, evidently at odds with rational predictions based on the Gibbs adsorption equation for a one-component solute, was clarified by the recognition that most surfactants, unless very carefully purified, can contain highly surface-active impurities.[60] On this basis, many of the earlier reports of adsorption maxima involving solid absorbents may also be explained.[61-63] However, alternate explantions have been attempted, involving, for example, desorption of surface micelles[64] or, more recently, the exclusion of micelles from surfactant-charged surfaces[65] especially in pores. An additional mechanism which is necessary to consider with minerals as adsorbents is that ions, leached from the mineral, especially if polyvalent or hydrolyzed, can promote the adsorption affinity of the surfactant; at higher (supramicellar) concentrations of the surfactant the more surface-active species can be solubilized into the regular micelles and, in this way, a lower level of adsorption will be registered. Instances of this phenomenon are discussed briefly in Section XI.

IX. APPLICATION OF BOUNDARY TENSION AND CONTACT ANGLE MEASUREMENTS

As pointed out earlier, adsorption of surfactant collector is a vital part of the mineral flotation process inasmuch as it decreases the L/S contact angle and facilitates attachment to the bubble. In consequence, there have been several studies of the wetting angle of solutions of collector surfactant on mineral solids, and, in general, a correspondence exists between flotation efficiency and increased contact angle.[18,21] It should be pointed out that a very modest increase in L/S contact angle is apparently sufficient to significantly enhance floatability.

Quite clearly the configuration at the three-phase boundary, G/L/S, is governed by the three interfacial free energies involved. Unfortunately, only one of these free energies (G/L) is experimentally accessible: there is no simple experimental procedure known or available to measure solid boundary tensions. Partial information can, however, be obtained concerning the solid boundary tensions from Young's equation and the experimentally accessible parameters γ_{LG} and the contact angle θ:

$$\gamma_{SG} - \gamma_{LS} = \gamma_{LG} \cos \theta = \gamma_t \qquad (46)$$

This quantity, γ_t, is known as the adhesion tension of the liquid to the solid; it is intuitively obvious that bubble/particle attachment will become more favorable, the lower the value of the adhesion tension, and such correlations have indeed been demonstrated.[66]

If one applies the Gibbs adsorption equation in order to calculate surface excess concentrations Γ at the relevant interfaces, Eq. (46) yields

$$\Gamma_{LS} - \Gamma_{SG} = \frac{1}{2} RT \frac{d(\gamma_{LG} \cos \theta)}{d \ln a} \qquad (47)$$

(for a 1:1 surfactant electrolyte). This is a potentially very useful equation but has enjoyed rather little use.[66,67] It has been shown that the left-hand side of the equation is frequently negative, i.e., adsorption at the solid/gas interface often predominates over that at the solid/liquid interface; indeed, the necessity for this has been pointed out[68,69] and is clearly a consequence of the interfacial requirements for successful flotation. A further point of utility of the equation is that it allows an estimate of the adsorption of surfactant at the S/G interface if the (more customarily determined) adsorption at the L/S interface has been measured.

X. ADSORPTION KINETICS

While the rate of adsorption of surfactants is of prime concern in a number of practical systems, this aspect is considered to be outside the scope of this article. Suffice it is to say that the nature of adsorption kinetics and reversibility are dependent to a large extent on the mechanisms involved. For example, in Fig. 17 adsorption density of sodium dodecanesulfonate is given as a function of time.[33] Adsorption takes place relatively fast, as expected for a case of physical adsorption. Oleate adsorption on hematite is also found to occur rapidly at pH 8[70] (Fig. 18).

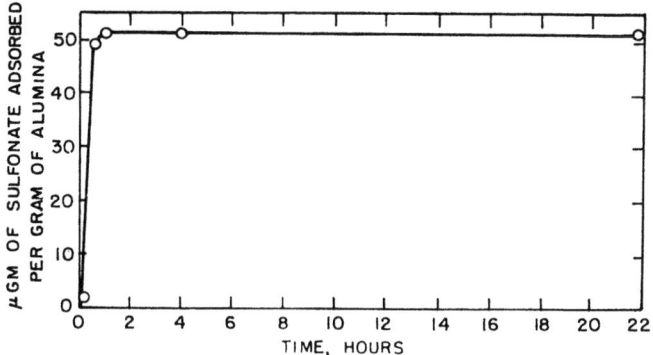

Figure 17. Adsorption of sodium dodecanesulfonate on alumina as a function of time (Somasundaran[33]).

However, the same system is found to exhibit a markedly different type of kinetics at a pH of 4.8 (Fig. 19). Equilibrium is not attained in this case, even in several hours. Evidently adsorption of oleic acid on hematite involves mechanisms that are different from that of oleate or acid soap. The implication of this observation, as to the inherent danger in the common practice of selecting equilibration time for adsorption experiments from tests conducted under only one set of conditions, is to be noted.

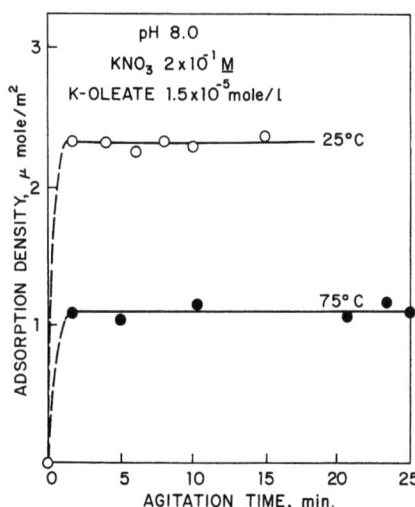

Figure 18. Adsorption of oleate on hematite as a function of time at pH 8 (Kulkarni and Somasundaran[70]).

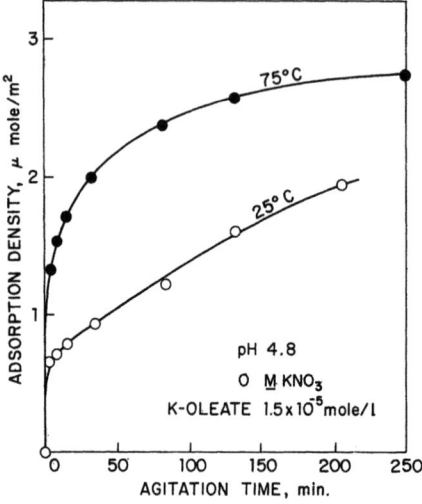

Figure 19. Adsorption of oleate on hematite as a function of time at pH 4.8 (Kulkarni and Somasundaran[70]).

XI. CURRENT RESEARCH TRENDS

Recent emphasis on enhanced (so-called tertiary) oil recovery, from spent oil fields by surfactant flooding, has kindled high interest in surfactant adsorption onto types of minerals likely to be encountered in oil-bearing strata, viz., sandstone, limestone, and clays. Loss of surfactants, generally of the petroleum sulfonate type, by adsorption would constitute a serious materials loss, so that knowledge of their adsorption characteristics has become a very important consideration. It is noteworthy that most of the natural oil fields contain connate water of appreciable salinity. Hence the practical situation is one of considerable complexity, viz., involving natural mineral strata, necessarily cheap and hence impure surfactants, and water of variable salinity and composition.

Current research work, accordingly, spans a range from ideal model systems, on the one hand, to surfactant types and mineral "cores" reflecting actual field usage, on the other; hence a range of adsorption results is to be expected. With these considerations in mind we refer to some very recent data from our own laboratory. Studies on sandstone (Berea)/mahogany sulfonate (a widely used petroleum-based surfactant), limestone/mahogany sulfonate, and kaolinite/dodecylbenzene sulfonate have shown the definite presence of adsorption maxima (see above) under certain conditions of salinity.[71] The nature of the adsorption isotherm obtained was found to be dependent to a great extent on the type of sulfonate used and on the solid. While adsorption maxima were obtained for mahogany sulfonate–AA on sandstone (Berea variety), isotherms of dodecanesulfonate exhibited only positive slopes, but concentrations studied in this case were restricted to the premicellar range owing to solubility problems (see Figs. 20 and 21). An increase in ionic strength, on addition of salts, increased the adsorption in all cases. Furthermore, the adsorption maximum was sensitive to the amount of salt added. At lower NaCl concentrations, the maximum existed for the mahogany sulfonate/Berea sandstone system; at intermediate concentrations, the isotherm exhibited a maximum followed by a shallow minimum; at still higher concentrations the adsorption maximum was not present.

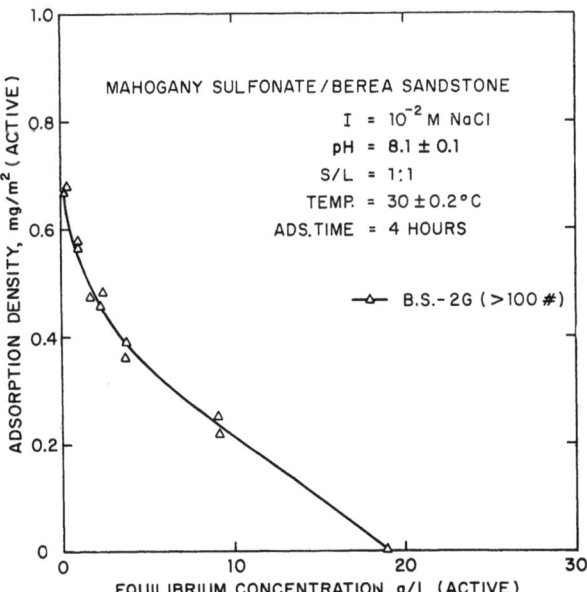

Figure 20. Adsorption isotherms of Mahogany sulfonate–AA (a petroleum sulfonate) on Berea sandstone (Hanna and Somasundaran[71]).

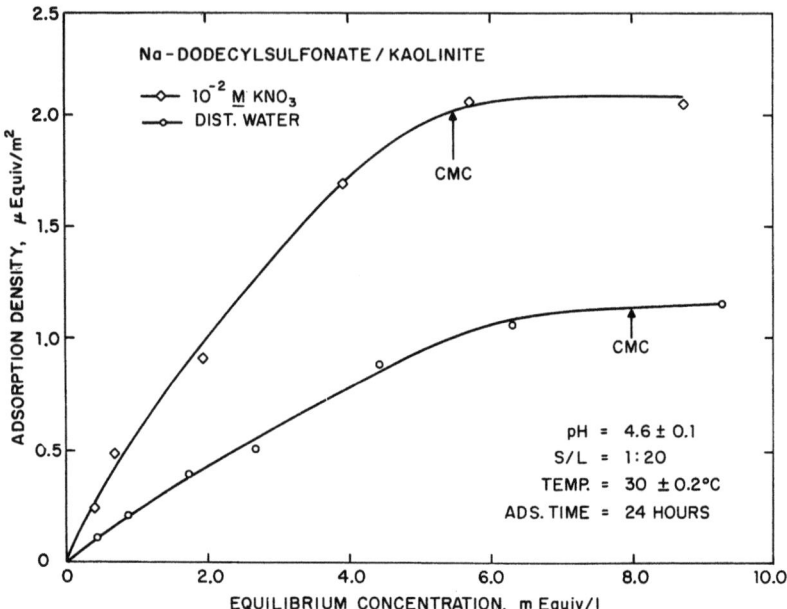

Figure 21. Adsorption isotherms of dodecanesulfonate on kaolinite (Hanna and Somasundaran[72]).

Mineralogical and morphological characteristics of the adsorbent are found to produce significant effects on adsorption. Agricultural limestone (a porous variety) possesses a relatively low adsorption density under natural pH conditions, possibly owing to the nonwetting of pores and the effect of the presence of pores on the interfacial volume from which micelles, which are charged similarly to the solid, are excluded (see Fig. 22). Agricultural limestone also exhibited an unexpected dependence of adsorption on pH (see Fig. 23). This we attribute to the silicates that were found to be present in the sample. Silicates that are concentrated on the surface of the particles, following dissolution of exposed calcite or magnetite, can be expected to be activated by calcium or magnesium species above neutral pH and thereby cause increased adsorption of sulfonate on the mineral as the pH is increased. Bedford limestone which does not contain such silicate exhibits decreased adsorption of sulfonate with increase in pH owing to a concurrent decrease in the positive potential of the mineral particles.

Adsorption studies using clay yielded some interesting new information that proved helpful in understanding the mechanisms involved.[59,72] Both equilibration of dry kaolinite with water and adsorption of sulfonate onto it were found to involve a fast and slow step, the latter possibly involving slow dissolution of aluminum species from kaolinite during prolonged contact of it with water and consequent activation of sulfonate adsorption. Past studies using kaolinite appear not to have considered the possibility of an intermediate metastable condition. The effect of an increase in salinity was in general to increase the adsorption of sulfonate mainly due to the less energetically favorable environment for the surfactant chains in a saline solution. The effect of the inorganic electrolyte's addition was most interesting in that the shape of the isotherm was dependent on both the cation and anion used (see Fig. 24). The adsorption capacity was found to increase in the order $NaCl < NH_4COOCH_3 < NH_4Cl <$

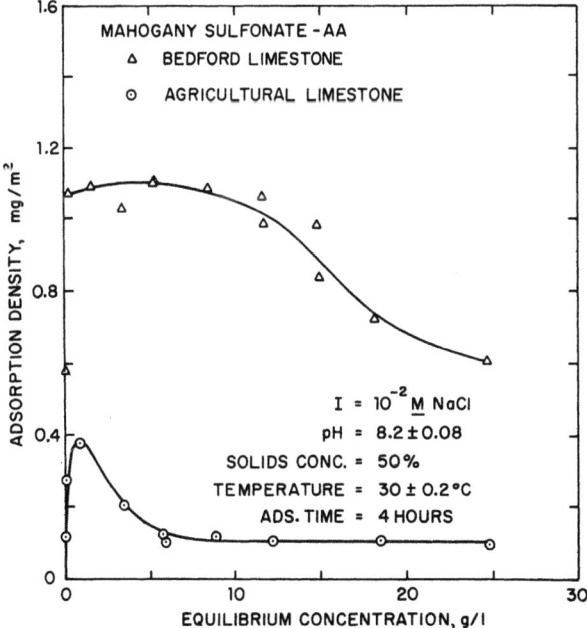

Figure 22. Adsorption of Mahogany sulfonate–AA on Agricultural limestone and Bedford limestone (Hanna and Somasundaran[19]).

$NH_4NO_3 < KCl$, in agreement with the order of decreasing cationic size and hydration capacity of cations. Of more importance was the observation that an adsorption maximum is obtained in inorganic electrolyte solutions containing the so-called "structure-making" ions. No such maximum is obtained when there is a predominance of "structure-breaking" ions.

These brief illustrations of current research activity on mineral sorption suggest that fruitful lines of research for the future might include studies of:

(a) the state of water at interfaces and the effect thereon of inorganic salts,
(b) the kinetics of adsorption of surfactants,

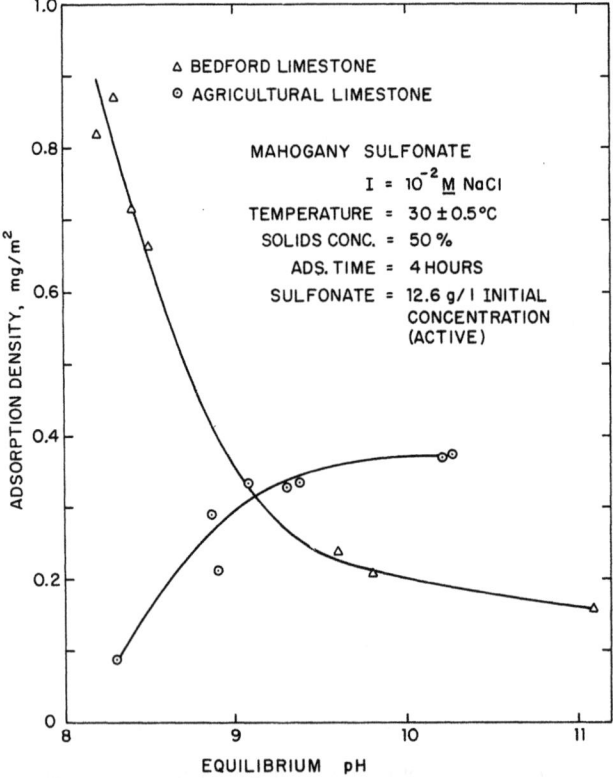

Figure 23. Data for adsorption of Mahogany sulfonate–AA on Agricultural limestone and Bedford limestone as a function of pH (Hanna, Goyal, and Somasundaran[19]).

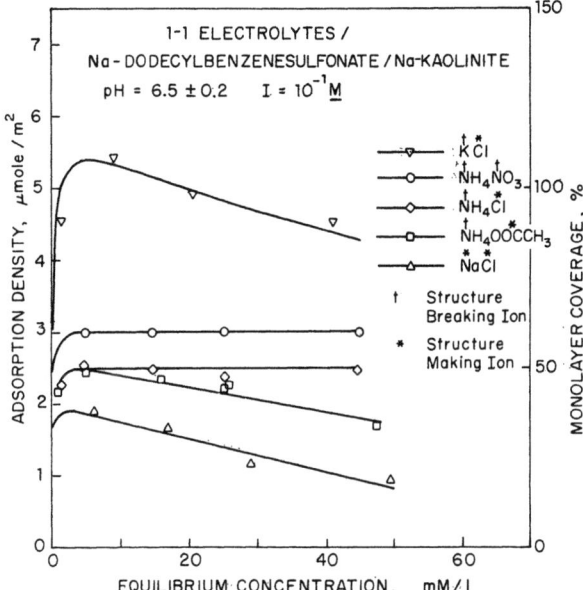

Figure 24. Effect of certain 1-1 electrolytes (10^{-1} mole/liter) on the adsorption of dodecylbenzenesulfonate on kaolinite at pH 6.5 (Hanna and Somasundaran[72]).

(c) the leaching of ionic materials from minerals by water and their effect on dissolved surfactants and their adsorption,
(d) maxima in adsorption, means of offsetting them, and determination of the role of micelles in this phenomenon, and
(e) competitive adsorption of surfactants and polymers.

ACKNOWLEDGMENT

One of us (P.S.) wishes to offer acknowledgment to the National Science Foundation (ENG-76-08756), Amoco Production Company, Chevron Oil Field Company, Exxon Research and Engineering Company, Gulf Research and Development Company, Marathon Oil Company, Mobil Research and Development Company, Shell Development Company, Texaco, Inc., and Union Oil Company of California for support of part of this work.

REFERENCES

[1] P. Somasundaran and H. S. Hanna, "Physico-Chemical Aspects of Adsorption at Solid/Liquid Interfaces, I," in *Basic Principles in Improved Oil Recovery by Surfactant and Polymer Flooding*, Ed. by D. O. Shah and R. S. Schechter, Academic Press, New York, 1977.

[2] E. D. Goddard and P. Somasundaran, *Croatica Chem. Acta* **48** (1976) 451.

[3] D. W. Fuerstenau, "The Adsorption of Surfactants at Solid/Water Interfaces," in *The Chemistry of Biosurfaces*, Ed. by M. L. Hair, Marcel Dekker, Inc., New York, 1971 Vol. 1, p. 143.

[4] R. D. Kulkarni, thesis, "Flotation Properties of Hematite/Oleate System and their Dependence on the Interfacial Adsorption," Columbia University, New York, 1975.

[5] G. A. Parks, *Chem. Rev.* **65** (1965) 177.

[6] P. Somasundaran and G. E. Agar, *J. Colloid Interface Sci.* **24** (1967) 433.

[7] G. A. Parks and P. L. de Bruyn, *J. Phys. Chem.* **66** (1962) 967.

[8] F. Z. Saleeb and P. C. de Bruyn, *Electroanalyt. Chem. Interfacial Electrochem.* **37** (972) 99.

[9] H. S. Hanna and P. Somasundaran, "Flotation of Salt-Type Minerals," in *Flotation*—A.M. Gaudin Memorial Volume, Ed. by M. C. Fuerstenau, American Institute of Mining, Metallurgical, and Petroleum Engineers, Inc., New York, 1976 p. 197.

[10] P. Somasundaran, *J. Colloid Interface Sci.* **27** (1968) 659.

[11] G. A. Parks, "Adsorption in the Marine Environment," in *Chemical Oceanography*, 2nd ed., Ed. by S. P. Riley and G. Skirrow, Academic Press, New York, 1975, p. 241.

[12] P. Somasundaran, "Interfacial Chemistry of Particulate Flotation," in *Advances in Interfacial Phenomena of Particulate Solution/Gas Systems*, Ed. by P. Somasundaran and R. B. Grieves, AIChE Sym. Ser. **71** (1975) 1.

[13] P. Somasundaran, "Pretreatment of Mineral Surfaces and Its Effect on their Properties," in *Clean Surfaces, Their Preparation and Characterization for Interfacial Studies*, Marcel Dekker, New York, 1972, p. 285.

[14] R. D. Kulkarni and P. Somasundaran, in *Oxide Electrolyte Interfaces*, American Electrochemical Society, Inc., Princeton, N. J. 1972, Vol. 31.

[15] R. D. Kulkarni and P. Somasundaran, *Int. J. Mineral Processing* **4** (1977) 89.

[16] P. Somasundaran and D. W. Fuerstenau, *J. Phys. Chem.* **70** (1966) 90.

[17] I. Iwasaki, S. R. B. Cooke and Y. K. Kim, "Some Surface Properties and Flotation Characteristics of Magnetite," *Trans. AIME* **224** (1962) 113.

[18] I. Iwasaki, S. R. B. Cooke and A. F. Colombo, "Flotation Characteristics of Goethite," U.S. Dept. of the Interior, Bureau of Mines, Report 5593, 1960.

[19] H. S. Hanna, A. Goyal and P. Somasundaran, "Surface Active Properties of Certain Micellar Systems for Tertiary Oil Recovery," 7th International Congress on Surface Active Substances, Moscow, 1976.

[20] H. J. Modi and D. W. Fuerstenau, "Flotation of Corundum, An Electrochemical Interpretation," *Trans. AIME* **217** (1960) 381.

[21] D. W. Fuerstenau, T. W. Healy, and P. Somasundaran, *Trans. AIME* **229** (1964) 321.

[22] F. F. Aplan and D. W. Fuerstenau, "Principles of Non-metallic Mineral Flotation," in *Froth Flotation*, Ed. by D. W. Fuerstenau, AIME 50th Ann. Vol., 1962, 170.

[23] P. Somasundaran, *Trans. AIME* **255** (1974) 64.

[24] G. Y. Onoda and D. W. Fuerstenau, in *7th International Mineral Processing Congress*, Gordon and Breach, New York, 1964, Vol. 1, p. 301.
[25] P. Roy and D. W. Fuerstenau, *J. Colloid Interf. Sci.* **26** (1968) 102.
[26] F. Z. Saleeb and H. S. Hanna, *J. Chem. U.A.R.* **12** (1969) 229.
[27] H. S. Hanna, "Relation Between Crystal Lattice Structure and the Adsorption Behavior of Some Salt-Type Minerals," in 7th International Congress on Surface Active Substances, Moscow, 1976.
[28] M. C. Fuerstenau, C. C. Martin and R. B. Bhapu, *Trans. AIME* **226** (1963) 449.
[29] M. C. Fuerstenau, D. A. Rice, P. Somasundaran, and D. W. Fuerstenau, *Trans. I.M.M. (London)* **73** (1965) 381.
[30] T. W. Healy, O. R. James, and R. Cooper, "The Adsorption of Aqueous Co(II) at Silica/Water Interface," in *Adsorption from Aqueous Solution*, Advances in Chemistry Series, No. 79, American Chemical Society, Washington, D.C., 1968, p. 62.
[31] G. A. Wolstenholme and J. H. Schulman, *Trans. Faraday Soc.* **46** (1950) 475.
[32] E. Matijevic, *J. Colloid Interface Sci.* **43** (1973) 217.
[33] P. Somasundaran, "The Effect of van der Waals' Interaction between Hydrocarbon Chains on Solid-Liquid Interfacial Properties," Ph.D. thesis, University of California, Berkeley 1964.
[34] P. Somasundaran, T. W. Healy and D. W. Fuerstenau, *J. Phys. Chem.* **68** (1964) 3562.
[35] I. J. Lin and P. Somasundaran, *J. Colloid Interface Sci.* **37** (1971) 731.
[36] T. W. Healy, *J. Macromol. Sci. Chem.* **A8** (1974) 603.
[37] P. Somasundaran and D. W. Fuerstenau, *Trans. AIME* **252** (1972) 275.
[38] P. Mukerjee, *Adv. Colloid Interface Sci.* **1** (1967) 241.
[39] P. H. Elworthy and K. J. Mysels, *J. Colloid Interface Sci.* **21** (1966) 331.
[40] T. Gilanyi, 51st Colloid and Surface Science Symposium Preprints, Grand Island, New York, 1977 p. 45.
[41] R. O. French, M. E. Wadsworth, M. A. Cook, and I. B. Cutler, *J. Phys. Chem.* **58** (1954) 805.
[42] A. S. Peck and M. E. Wadsworth, "Infrared Studies of the Effect of Fluoride, Sulfate and Chloride on Chemisorption of Oleate on Fluorite and Barite," *Proceedings of the 7th International Mineral Processing Congress*, Ed. by N. Arbiter, Gordon and Breach, New York, 1965, p. 259.
[43] A. Bahr, M. Clement and H. Surmatz, "On the Effect of Inorganic and Organic Substances on the Flotation of Some Non-Sulfide Minerals by Using Fatty Acid-Type Collectors," 8th International Mineral Processing Congress, Leningrad, paper S-11, 1968.
[44] U. Bilsing, "The Mutual Interaction of the Minerals During Flotation For Example the Flotation of CaF_2 and $BaSO_4$," Dissertation, Bergakademie, Freiberg, 1969.
[45] H. S. Hanna, "Adsorption of Anionic Surfactants on Precipitated $CaCO_3$ and Calcite," paper presented at 4th Arab Chemistry Conference, National Research Center, Cairo, 1975.
[46] E. D. Goddard and H. C. Kung, *J. Phys. Chem.* **67** (1963) 1965.
[47] E. D. Goddard and H. C. Kung, *J. Colloid Interface Sci.* **29** (1969) 242.
[48] E. D. Goddard and H. C. Kung, *Kolloid Z. Z. Polym.* **232** (1969) 812.
[49] P. Somasundaran, *Int. J. Mineral Processing* **3**, (1976) 35.
[50] P. Somasundaran, K. P. Anathapadmanabhan and R. D. Kulkarni, "Flotation Mechanism Based on Ionomolecular Complexes," XIIth International Mineral Processing Congress (Sao Paulo), Iron, in Vol. 2, 1977, p. 80.

[51] K. P. Anathapadmanabhan, P. Somasundaran and T. W. Healy, "The Chemistry of Oleate and Amine Solutions in Relation to Flotation," 78-B-67, 107th AIME Annual Meeting, Denver, February 1978.

[52] R. F. Jung, "Oleic Acid Adsorption at the Geothite-Water Interface," M.S. thesis, University of Melbourne, Australia, 1976.

[53] R. D. Kulkarni and P. Somasundaran, "Kinetics of Oleate Adsorption at the Liquid/Air Interface and its Role in Hematite Flotation," in *Advances in Interfacial Phenomena*, Ed. by P. Somasundaran and R. B. Grieves, AIChE Symp. Ser., **150** (1975) 124.

[54] S. I. Polkin and T. V. Najfonow, "Concerning the Mechanism of Collector and Regulator Interaction in the Flotation of Silicate and Oxide Minerals," in 7th International Mineral Processing Congress, Ed. by N. Arbiter, Gordon and Breach, New York, 1965, p. 307.

[55] A. Bleier, E. D. Goddard and R. Kulkarni, *J. Colloid Interface Sci.* **59** (1977) 490.

[56] D. W. Fuerstenau and B. J. Yamada, *Trans. Soc. Min. Eng.* (1962) 50-52.

[57] A. C. Zettlemoyer, *J. Colloid Interface Sci.* **28** (1968) 343. 343.

[58] P. Somasundaran, *J. Colloid Interface Sci.* **31** (1969) 557.

[59] H. S. Hanna and P. Somasundaran, unpublished results, 1977.

[60] G. D. Miles and L. Shedlovsky, *J. Phys. Chem.* **48** (1944) 60.

[61] A. Fava and H. Eyring, *J. Phys. Chem.* **60** (1956) 890.

[62] F. H. Sexsmith and H. J. White, *J. Colloid Sci.* **14** (1959) 630.

[63] F. Z. Saleeb and J. A. Kitchener, *J. Chem. Soc.* (1965) 911.

[64] R. D. Vold and A. K. Phansalkar, *Rec. Trav. Chim.* **74** (1955) 41.

[65] P. Mukerjee and A. Anavil, "Adsorption of Ionic Surfactants to Porous Glass: The Exclusion of Micelles and Other Solutes from Adsorbed Layers and the Problem of Adsorption Maxima," in *Adsorption at Interfaces*, Ed. by K. L. Mittal, ACS. Symp. Ser. **8** (1975) 107.

[66] P. Somasundaran, *Trans. Soc. Min. Eng.* **241** (1968) 105.

[67] W. J. Murphy, M. W. Roberts, and J. R. H. Ross, *Trans. Faraday Soc.* **58** (1972) 1190.

[68] P. L. De Bruyn, J. Th. G. Overbeek and R. Schuhmann, *Mining Eng.* **6** (1954) 519.

[69] C. A. Smolders, *Rec. Trav. Chim.* **80** (1961) 650.

[70] R. D. Kulkarni and P. Somasundaran, "Oleate Adsorption at Hematite/Solution Interface and its Role in Flotation," AIME Annual Meeting, New York, February 1975.

[71] H. S. Hanna and P. Somasundaran, "Physico-Chemical Aspects of Adsorption at Solid/Liquid Interfaces, II. Mahogany Sulfonate/Berea Sandstone, Kaolinite," in *Improved Oil Recovery by Surfactant and Polymer Flooding*, Ed. by D. O. Shah and R. S. Schechter, Academic Press, New York, 1977.

[72] P. Somasundaran and H. S. Hanna, "Adsorption of Sulfonates on Reservoir Rocks," paper S.P.E. 7059, S.P.E.–AIME Symposium on Improved Oil Recovery, February 1978, p. 241.

4

Application of Auger and Photoelectron Spectroscopy to Electrochemical Problems*

J. Augustynski and Lucette Balsenc

Chemistry Department, University of Geneva, Geneva, Switzerland

I. INTRODUCTION

1. Methods

Auger electron spectroscopy (AES) and X-ray photoelectron spectroscopy (XPS), known also as electron spectroscopy for chemical analysis (ESCA), were used originally in surface physics and theoretical chemistry but their range of application has expanded rapidly.

AES became generally accepted as an analytical technique only after Harris[2] demonstrated that the contrast of Auger structures could be increased to a practical extent by differentiating the electron spectrum with respect to energy; it became a popular technique when Weber and Peria[3] showed how the low-energy-electron diffraction (LEED) systems already in existence could be adapted simply to AES.

---

* The present chapter is devoted to the application of AES and XPS to electrochemistry and several aspects of the two techniques relevant to that subject will be treated here. For a review on basic principles, instrumentation, and applications prior to 1973, the reader should refer to the article by Baker published in this series.[1]

The approach of XPS application to surface problems has been more cautious. It has been recognized early that XPS yielded measurements on the surface or the near-surface layers but it was believed that photoelectrons originated at 50 Å or more from the surface; actually, the escape depths of Auger electrons and photoelectrons seem to be of the same order of magnitude, between two and five atomic layers, and for the last three or four years XPS has been used to study surface *per se*.

The possibility that the chemical nature of surface species could be identified and their concentrations determined, makes electron spectroscopies (AES and XPS) very useful complementary techniques for the study of a wide category of stable intermediates and products of electrode reactions. These techniques have also proved to be of value for the characterization of electrode surfaces themselves. After a selective and necessary introductory review on the recent developments of AES and XPS in relation to metallic surface problems, the principal applications to electrochemistry will be discussed. In this context, the potentiality of electron spectroscopy will be compared with that of regular and recently developed electrochemical methods[4] for studying surface processes.

2. Surface Analysis

The surface region of a solid may be defined as consisting of the outermost atomic layers, including any foreign atoms adsorbed onto or absorbed into them either substitutionally or interstitially;[5] in the usual definition, the surface region includes no more than the outermost two atomic layers. A thorough understanding of the properties of solid surfaces must be based on a knowledge of the chemical identity of the atoms present, the geometrical arrangement of these atoms and the distribution of electrons surrounding them, both in energy and in space.

During the last few years a number of techniques have been developed which make surface concepts amenable to direct investigation. We examine here two by now well-established techniques which allow solid surfaces to be characterized in the monolayer range: Auger electron spectroscopy (AES) and X-ray-excited photoelectron spectroscopy (XPS). Both techniques have in common the creation of a hole in an inner atomic level as the first step in the individual physical process. They provide, with different elemental sensitivities, the compositional analysis of the surface.[6] The composition can be deduced either from the energies required to create vacancies in the electronic shells or from the energies of the electronic transitions involved in the decay of the vacancies[7]; in both cases, the binding of the initially ionized level is reflected in the kinetic energy of the electron analyzed.

II. AUGER ELECTRON SPECTROSCOPY

1. Physical Process

In AES an incident electron of energy E_p impinges at the specimen surface; if E_p is greater than the binding energy of a core-level electron, this electron may absorb the excess energy and be excited into an allowed state above the Fermi level; the core vacancy left behind will be filled by an electron from a higher level. Energy is conserved in the decay transition by the emission of another electron (Auger electron) and/or X-ray photons. In the range of interest (<2 keV), the Auger transition is highly favored.[8]

The Auger process thus involves three electrons; if, for instance, the core-level electron originates in the K shell, the second electron in the L shell, and the Auger electron in the valence band, the transition is conventionally noted KLV; the kinetic energy measured by the analyzer (see Section II.3) is given by

$$E_{kin} = E_K - E_L - E_V - \phi_{anal}$$

where E_K, E_L, and E_V are the energies of the K, L, and V levels, respectively, and ϕ_{anal} is the work function of the analyzer.

In transitions involving the valence band the electrons coming from the band can originate anywhere within it; AES thus provides information about the density of occupied states.

The energy of the Auger electron is independent of the energy of the impinging radiation and of the mode of excitation. Although high-energy electrons are usually used as a primary exciting beam, the Auger process occurs whenever a hole is created; for example, Auger electrons are always present in XPS studies.

2. The Auger Spectrum

The Auger spectrum is recorded as the energy distribution of the ejected electrons. Beside the Auger signals a certain number of peaks are usually observed: elastic peak, plasmon loss[9-13] and gain[14,15] peaks, and "true secondary" peak; they are produced by reflection, scattering, or cascade processes[16] involving electrons within the solid. (See Fig. 1.) These features can be found at every energy from E_P to zero, whereas Auger lines are always found at fixed kinetic energies. In addition, whenever the Auger process occurs in a doubly ionized atom, satellite peaks situated on the high-energy side of the Auger transition lines can be observed.[17,18]

Since the Auger process involves three electrons, the Auger

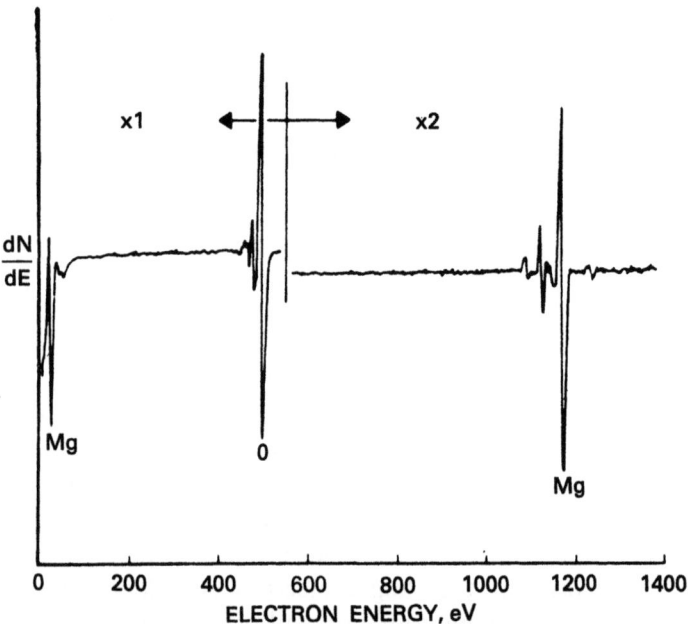

Figure 1. Auger electron spectrum from MgO (from Ref. 82).

peaks are usually broader than the lines observed in the one-electron emission (XPS, UPS). The energy spread depends on the lifetime of the excited state[19] (ca. 10^{-16} sec); for very fast Coster–Kronig transitions, the energy broadening is large, up to 10 eV. The Auger peak width depends also on the atomic levels involved in the process, for instance, the L_1 level is wide, while the L_2 and L_3 levels are narrower. In transitions involving the valence band, the Auger peak could be twice the band width;[20] a classical example is the L_3VV transition in magnesium where the band width is spread over 16 eV. However, as the Auger peak width reflects variation in transition probability across the valence band, it is not unusual to observe narrow peaks in this region. The energy losses, suffered by electrons during escape from the solid or due to interband transitions, bring a further contribution to peak broadening[21,22] and are usually responsible for the characteristic tailing at the low-energy side of the nondifferentiated peak.

Besides these inherent effects, purely instrumental factors, such as limited resolution of the spectrometer and amplitude of the ac modulation voltage, may contribute to a further spread in energy of an Auger line.

3. Instrumentation

Auger analyzers currently in use are of two types[1]: the retarding field analyzer (RFA) based on low-energy electron diffraction (LEED) optics[23,24] and the cylindrical mirror analyzer (CMA), which is a dispersive system.[25–28]

In the RFA all electrons with an energy greater than the energy of interest are transmitted to the collector. The energy distribution is obtained by superimposing a small potential modulation on the analyzer pass energy and detecting the collected electron current with a phase-sensitive detector synchronously tuned to the harmonic of interest. At the fundamental harmonic frequency, the direct energy distribution $N(E)$ is obtained; at the second harmonic frequency the derivative $dN(E)/dE$ is obtained.

In the CMA only electrons with discriminated energy E_i reach the detector; the signal-to-noise ratio is thus several orders of magnitude higher than that of a typical RFA (see Fig. 2); even

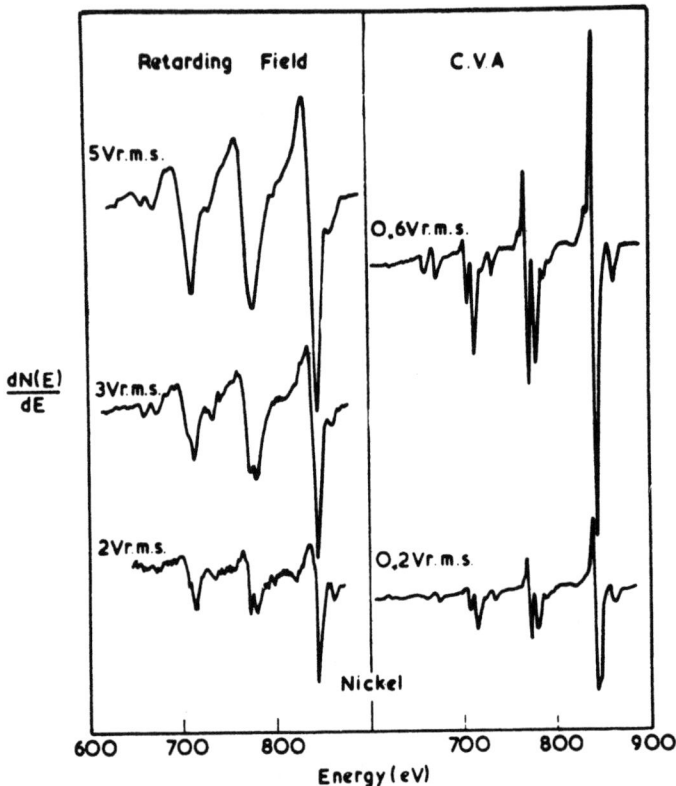

Figure 2. Comparison of part of the nickel differential Auger spectrum taken with an RFA (left-hand side) and a CMA (right-hand side) (from Ref. 25).

if the transmittance is a factor about 10 lower, the overall sensitivity is considerably improved over the RF system. The sensitivity depends critically on the resolution of the analyzer and is optimum when the resolution is approximatively equal to the half-width of the peak to be examined;[25] if very high resolution is required, it may be achieved at the expense of the sensitivity. The resolving power itself is dependent on the specimen position; peak height changes and energy shifts associated with changes in specimen position as small as $\Delta x = 0.1$ mm have been observed;[29] ideally the analyzed area should be at the analyzer focal point. The spot size of the exciting source is also critical.[30,31] It should be mentioned, however, that at low energies (≤ 80 eV) the high dispersion of the system makes it difficult to obtain quantitative data. In addition, the CMA has the advantage of being able to operate at very fast scan times (as high as $1000\,\text{eV}\,\text{sec}^{-1}$ compared with $2\,\text{eV}\,\text{sec}^{-1}$ for the RFA) and with much lower beam currents than are possible for the RFA, with the consequence of reducing considerably beam-induced effects.

In summary, the CMA is to be preferred over the RFA whenever rapid data acquisition, low beam current, and detailed studies of high-energy spectra are needed. On the other hand, the RFA presents two important advantages: firstly, the transmission of the analyzer is energy independent, so such an instrument offers superior transmissions at low-energy voltages (<100 eV), and secondly, the analyzer can be used as a LEED detector as well as an RFA.

4. Qualitative Analysis

Auger electrons are ejected with well-defined energies that are characteristic of the original ion and may serve to identify it.[32,33] Though a large number of Auger transitions are allowed (thousands of them for heavy elements), the Auger spectrum exhibits only a small number of characteristic peaks. As the Auger lines are large, numerous, and close in energy, the observed peaks frequently consist in the superposition of several individual peaks and cannot be identified without ambiguities. One can say, however, that most of the observed peaks originate in the outer shells; even for light elements such as phosphorus, sulphur, or chlorine, the major peaks observed originate in the $L_{2,3}$ level (unresolved) and peaks arising from L_1 transitions are normally difficult to observe.

Because of the complexities mentioned before, elemental analysis is generally based on the matching of spectra against standard plots taken from samples of known composition. For a determined element, the various transition lines are identified by comparing the observed Auger energies with listed values obtained from standards; this procedure seems preferable to calculations, though a number of authors still rely, with comparative success, on calculated values of Auger energies using either semiempirical relationships[34,35] or more refined computations taking into account relaxation effects.[36,37]

AES has proved to be a very sensitive technique, and surface quantities as small as $\sim 10^{-3}$ monolayer (10^{12} atoms) have been identified.

5. Quantitative Analysis

While AES has become a well-established technique for qualitative analysis, its acceptance as a quantitative technique has, however, been less widespread due to the difficulty of establishing suitable calibration standards and to the lack of a comprehensive quantitative theoretical analysis.[38] For quantitative analysis one must relate the observed Auger current measured as peak intensity to the concentration of the element producing that particular peak. The intensity of a particular Auger transition is given by

$$I = v \frac{P_{Ai}}{\sum P_{Ai}} \quad (1)$$

where v is the velocity for hole formation, P_{Ai} the probability for a particular Auger transition, and $\sum P_{Ai}$ the total probability of deexcitation of a hole. Hence intensity is proportional to the number of holes created and, in the first approximation, to the number of atoms present.[39]

(i) Relative Surface Quantities

The determination of relative surface quantities is apparently fairly simple because the emission current of Auger electrons is proportional, in the first approximation, to the number of excited atoms, n_i.

In the direct distribution curve $N(E)$, the area delimitated by the Auger peak, is proportional to n_i and, if we assume that the peak is a Gaussian, the deflection peak-to-peak in the $dN(E)/dE$ distribution is also proportional to n_i.[40,41] The relative amount of an element on the surface can then be determined by simple measurement of the Auger peak height or area, followed by comparison with a standard. Clearly this method is only applicable when all instrumental parameters are held constant and the use of a standard having excitation and escape probabilities very close to those of the specimen under investigation is a necessary requirement. Another prerequisite for quantitative analysis by comparison with standards is the absence of matrix effects: Auger peak shapes and energies must be identical for both the specimen and the standard. Two good examples of how such restrictions should be dealt with are given by Ueda and Shimizu[42] in estimating various impurity concentrations on a silicon surface and by Shell and Rivière,[43] who derived an empirical proportionality factor for determining the relative concentrations of iron and phosphorus in an alloy.

A method of internal calibration by fracturing an alloy *in situ* under ultrahigh vacuum and analyzing the freshly fractured surface was proposed several years ago[44] and applied to the platinum/tin system.[45] More recently, an external calibration method was proposed to interpret quantitatively surface enrichment in alloys.[46,47] The external calibration involves measurements of Auger intensities from the pure elements under identical experimental conditions.

A slightly different approach exists which is widely used in alloy analyses. In this method the unknown sample is first qualitatively analyzed, then a set of standards is measured with all the constituents present; for instance, in binary alloys, a set of standards whose composition varies in the range of the unknown is prepared. Standard manufacturing is tedious and time consuming; nevertheless, whenever accuracy is of the essence this approach is to be preferred.

An attempt has been made to estimate a sensitivity factor (relative Auger yield) without standards.[48] For the present, this method is applicable to transitions that have similar Auger peak shapes and transition probabilities in the range 100–1000 eV. Another formalism based on the equation for the normalized Auger signal has been devised by Chang,[49] who determined relative concentrations with an accuracy as high as ±5%.

(ii) Absolute Surface Quantities

If the determination of relative surface quantities does not present too many difficulties, the determination of absolute surface concentrations (number of atoms per cm^2) is much more complex due to the failure of commonly used detectors to measure absolute Auger currents; hence, reference data obtained from independent calibrations become necessary. Among such methods ellipsometry has the advantage that information comes from the same surface area as the Auger electrons[50]; it was used by Meyer and Vrakking[51] in determining sulfur and oxygen coverage of silicon and germanium substrates. Determination of the amount of sulfur on a nickel surface by counting the ^{35}S radioisotope used as a marker was also proposed by Perdereau.[52] Calibrations have been performed by quartz crystal measurements of vapor-deposited material[53,54]; the use of LEED patterns[55] and work function measurements[56] were also proposed. Independently, Palmberg and Rhodin[57] and Florio and Robertson[58] correlated LEED intensity data at various coverages with Auger signal intensities. Coverages have been determined by ion counting,[3,59] flash desorption,[60] ion scattering,[61] quantitative X-ray fluorescence,[62] and electron-microprobe measurements.[63] Elemental ratios in alloys too, have been extensively used for calibration of Auger signal strength.[64–67] In every case, a linear relationship has been found between Auger peak-to-peak amp-

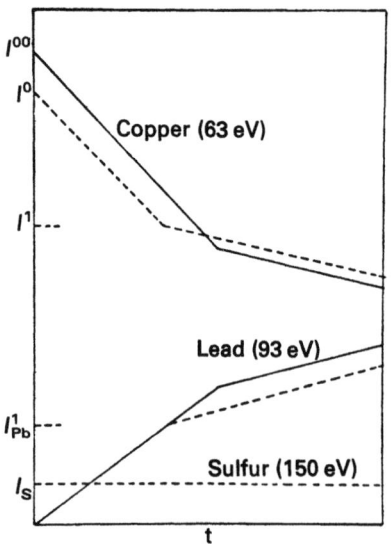

Figure 3. Absolute surface quantities. Schematic plot of the Auger signal intensities during deposition as a function of duration of deposition, t. Full lines for an intially clean substrate. Broken lines for a sulfur-contaminated substrate (from Ref. 69).

litudes and measured parameters, when the coverage was inferior or equivalent to a monolayer.

Argile and Rhead[68] devised an elegant method for measuring submonolayer quantities of an adsorbate on single-crystalline material. The method consists in depositing a second adsorbate (typically an easily deposited metal) first on the uncovered substrate and then on top of the first monolayer. By careful monotoring of the AES signals during deposition it is possible to detect the formation of a complete binary monolayer[69] (see Fig. 3). Experiments at different levels of contamination lead to calibration for both adsorbates. Some information on adsorbate structure may be necessary for a reliable calibration, for instance, confirmation that the metal layer forms a close-packed arrangement at high coverage, but this information is often available from LEED.[70] Such a calibration is valid for a particular adsorbate and for a particular instrument; it should also be mentioned that large quantitative differences have been observed in the Auger response from the same adsorbate on different substrates.

(iii) Quantification without Standards

A semiquantitative analysis can be achieved either by calculating the atom fraction of the component of interest when the Auger current is known, or by determining an absolute sensitivity factor which has the advantage of being unaffected by surface topography and insensitive to instrumental variations.

Until today most attempts to quantify AES have been empirical. The reason for this has been the lack of accurate knowledge concerning the various parameters that govern the magnitude of the Auger current. In some cases, however, when all the factors are known with sufficient accuracy, it is possible to calculate[71] the current resulting from an Auger transition XYZ in an element i by substituting those factors in the expression

$$I_i(XYZ) = I_p W(XYZ) T \phi_i(E_p, E_c) RN_i(A_i^s) \lambda i(A_i^s) r_i(E_p, A_i^s) A_i^s \quad (2)$$

where I_p is the primary beam, $W(XYZ)$ the XYZ Auger transition probability, T the analyzer transmission, $\phi_i(E_p, E_c)$ the ionization cross section of X level depending on the primary beam energy and the critical energy for the X level[72–74] corrected for

Coster–Kronig transitions,[75–77] R the surface roughness factor, N_i the atom density, λ_i the escape depth, r_i the backscattering coefficient,[8,78–81] and A_i^s the atom fraction of the ith component in the volume detected at the surface.

It has often been suggested that quantitative analysis could be performed through elemental sensitivity factors derived empirically from data on elemental standards. Unfortunately, large errors result when the matrix of the specimen is substantially different from that of the standard. Such a difficulty can be overcome by using instead of the pure element, a compound with well-defined chemical composition. It is thus possible to assign to every element an absolute sensitivity factor based on peak-to-peak amplitude, provided that relative escape depths, relative ionization cross sections, and backscattering factors are known.[82] The principal difficulty encountered with this method is the accurate determination of the absolute Auger currents.

Recently, Chang proposed a simple approximate formalism[49] to calculate absolute concentrations with a precision varying from ±20 to ±50%. This interesting method allows layered structures as well as homogenous surfaces to be treated.

From the various points discussed above, AES emerges as a semiquantitative analytical technique. When measuring absolute quantities, the limited accuracy of only 30–50% has to be expected even in favorable cases; if, however, relative concentrations are determined, variations of the order of ±20% are usually obtained. Statistical reproducibility has been checked over long periods of time and variations of the order of ±20 to ±30% have been observed.[38] Such a precision may seem desultory to many analytical chemists: one should, however, bear in mind that such results are obtained in determining exceedingly small quantities, typically of the order of a fraction of a monolayer (10^{13}–10^{15} atoms).

The great sensitivity of AES is partly due to the backscattered electrons which are generated in the bulk by the primary beam; some of them, passing back through the surface, will give rise to additional ionization and so enhance the Auger yield. In some cases, the backscattering contribution which is more pronounced for heavier elements and lower transitions[83] may attain up to 40% of the Auger yield.

6. Chemical Effects

Changes in the chemical environment of an atom may induce four possible changes in the Auger spectrum:

(1) The formation of a chemical bond causes a variation of the energy of a core level which in turn shifts the energy of the Auger transitions involving this level ("chemical shift").

(2) In the valence band, the formation of a chemical bond causes a variation in the density of states which leads to alterations in the peak shapes of the Auger transitions involving this level.

(3) Variations in the peak ratios for transitions of the given same element are observed.

(4) The structure on the low-energy side of the peaks in the $N(E)$ curve ("tailing") is modified due to changes in loss mechanism.

(i) Chemical Shifts

The elucidation of chemical states of atoms in molecules and on surfaces of solids can be performed to a limited extent by AES; the chemical shift is often very small with respect to the Auger peak width and hence not easily detected; if detected, the chemical shift is difficult to interpret because it results from a many-electron process (see Fig. 4).

Though various tentative experiments have been performed, very little systematic work has been done on electron-excited Auger transitions.

Among the authors who have reported methodical experiments, Szalkowski and Somorjai[84,85] compared the peak heights and energy positions of the $L_3M_{2,3}M_{2,3}$ and $L_3M_{2,3}V$ lines from vanadium metal with the same transitions from vanadium in the oxides $VO_{0.92}$, V_2O_3, and VO_2. The energy of the $L_3M_{2,3}M_{2,3}$ transition was found to vary smoothly as a function of oxidation state of vanadium and an average shift of 0.6 eV per oxidation number was reported. In another experiment, the same authors examined a series of vanadium compounds[86] and correlated shifts with the Philips–van Vechten ionicity for the vanadium atom[87] (see Fig. 5). The authors consider that in VSi_2, VC, VN, and VO only the 4s electrons are involved in bonding. The sharp change

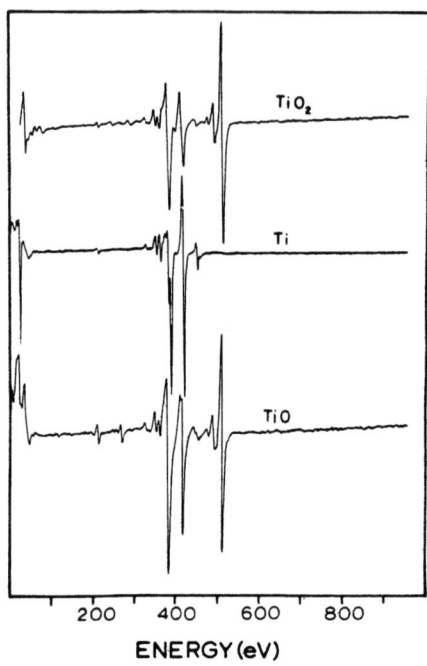

Figure 4. Auger derivative spectra of TiO, TiO_2, and Ti (chemical shift) (from Ref. 91).

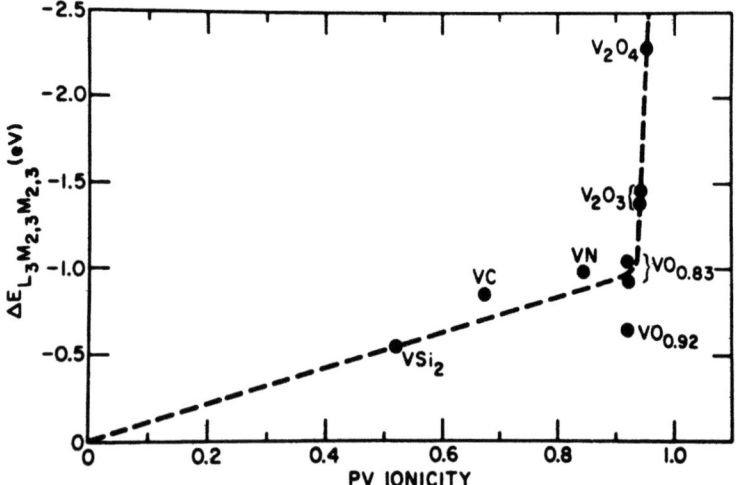

Figure 5. The magnitude of the observed chemical shift of the vanadium $L_3M_{2,3}M_{2,3}$ Auger transition in the vanadium compounds relative to vanadium metal plotted vs. the calculated Phillips–van Vechten ionicity values for those compounds (from Ref. 86).

in slope observed for higher oxides is attributed to involvement of the d electrons in bonding. Measured values have been corrected for electron interactions and static relaxation energy.

In a study of the spectra of Na_2SO_4 and $Na_2S_2O_5$, Farrell[88] showed that the largest Auger peak observed corresponds to a process resulting in at least one vacancy in the nonbonding orbitals in the final state and that the energy spacings of the three sulfur Auger peaks from sodium sulfate were comparable with those found for the different classes of molecular orbitals in Manne's calculations.[89] Two other interesting examples of chemical shifts for the silicon peak in silicon (100) and in a glass Na_3OSiO_2),[90] and for Ti, TiO, and TiO_2,[91] are worth mentioning but such experiments are rare. Interpretation of the Auger energy shifts is made difficult by several factors: firstly, as these shifts result from a two-electron process, a certain ambiguity arises in identifying the original and final levels; secondly, external factors can induce energy shifts; for example, if the sample is an insulator, shifts due to electrostatic charging of the specimen are noticed[92]; changes in modulation amplitudes are also responsible for displacements of the peak. Finally, chemical shifts observed for a reconstructed layer and three-dimensional compound are usually different, which is not surprising as a strong overlayer–substrate interaction takes place in reconstruction, while there is a much weaker interaction in corrosion;[93] such differences may be observed in varying the incidence angle of the impinging beam.[94]

(ii) Auger Line Shape

When, in an Auger transition, one or both of the final-state holes lie in the valence band of a solid, the spectrum observed is simply the self-convolution of the valence band density of states (DOS). Thus, desired information on the valence band should be contained in the shape of a core-valence-valence (core-VV) AES profile.

Variations in chemical environment (chemisorption) induce modifications of the local DOS at the surface that lead to changes in the line shapes of the ejected electrons[95–98] and such variations may be interpreted in terms of molecular orbitals.[99] (see Fig. 6).

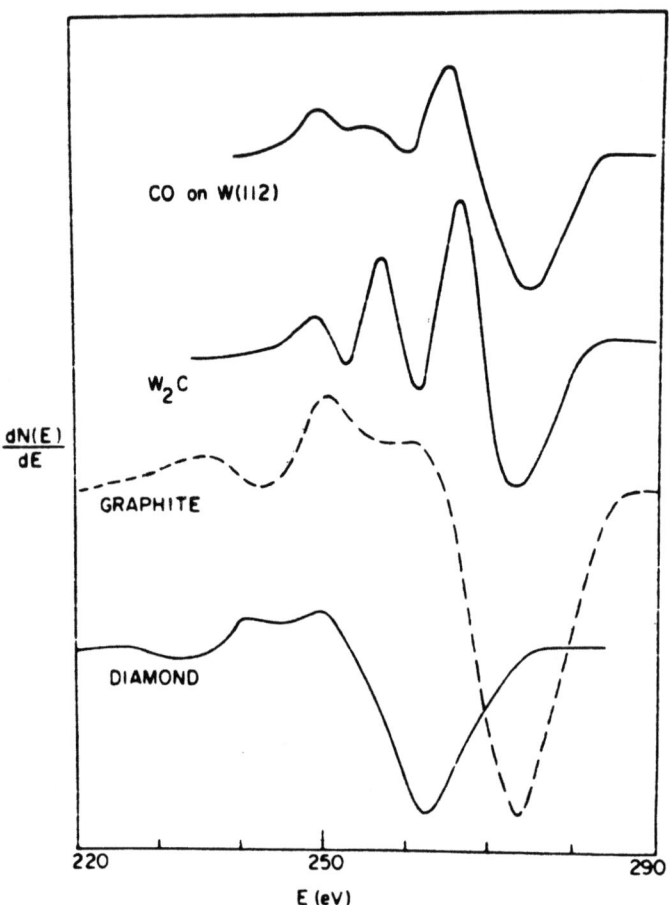

Figure 6. Auger peak shape of carbon in various environments (From Ref. 537).

In studies of sulfur adsorption on polycrystalline nickel, Coad and Rivière[100] detected new adsorbate levels that they correlated with those observed by Becker and Hagstrum[101] in ion-neutralization spectroscopy; virtually the same model had been used by Sickafus and Steinrisser to explain their observed spectra of energy losses, at low primary energies, from sulfur-covered nickel.[102] Such correlations, however, have been the subject of controversy[103] and the final state of the core-VV Auger transition itself has been questioned,[104–106] some authors describing a quasiatomlike final state, other works presenting a bandlike state (in the sense proposed by Lander[107]). Avery[108] has proposed an explanation for the differing behavior of the final state in core-VV emission: if the initial-state core level exerts a strong influence over the valence electrons so that they no longer occupy delocalized valence band states, they will instead occupy more tightly bound localized atomlike states. In some transition metals (group IB, with the exception of Au, and IIB elements) this corresponds to the filling of the d band where the presence of a core-hole induces static increases in the binding energies of the valence electrons (typically 6–9 eV); the $L_{2,3}VV$ spectrum of copper $(d^{10}s^1)$ and zinc $(d^{10}s^2)$[105] and the $M_{4,5}VV$ spectrum of silver[106] can be adequately accounted for by multiplet coupling of the unpaired spins in an atomlike final state. On the other hand, when the influence of the initial core hole on valence electrons is not very strong (as occurs in d^1–d^8 transition metals) the final state retains, at least partially, its bandlike properties during Auger emission. Such a behavior has been actually observed with nickel and in the N_7VV spectra of both clean and adsorbate-covered tungsten.[109]

In compounds with strong ionic character, such as MgO,[110] part of the structure originates by interatomic transitions in which one or two final holes are located in an atom different from that in which the inner hole is created.[111] Two recently published studies[112,113] showing how to identify, deconvolute, and interpret interatomic processes may be consulted with profit.

Auger peak shape studies may also be used to identify the chemical state of the adsorbed species.[114] For example, in the case of CO_2 adsorption on W (100), dissociative adsorption has been postulated by comparison of the carbon Auger peak shape of the adsorbate with those of a carbide, CO_2, and molecular CO.[115]

(iii) Changes in Peak Ratios

Changes in peak ratio for transitions of the same atom are also of interest in studying chemical effects. Such changes are of general occurrence and can be observed whenever adsorption takes place on a clean surface. A predictable effect of adsorption at the surface is the reduction in size of high-energy Auger lines and the disappearance of lower-energy peaks which are much more sensitive to the presence of adsorbates on the surface; very often the extinction of the peaks is simply due to overlaying by the adsorbate as a result of limited escape depth.

Changes in peak ratios in the Auger triplet of titanium as a result of oxidation of the metal surface have been reported.[116] Ueda and Shimizu,[117] by use of photoelectron work function measurements combined with AES, followed changes in the ratio of two low-energy iron peaks during stepwise oxidation of the latter metal and as a result of sulfur segregation at its surface. Similar changes have been studied by Shih et al.[118] who observed different kinds of behavior for the various low-energy peaks of iron when the metal was exposed to sulfur hydride vapors.

(iv) Changes in Loss Spectrum

Interpretation of the Auger line is complicated by inelastic scattering of Auger electrons on their way out the sample; these

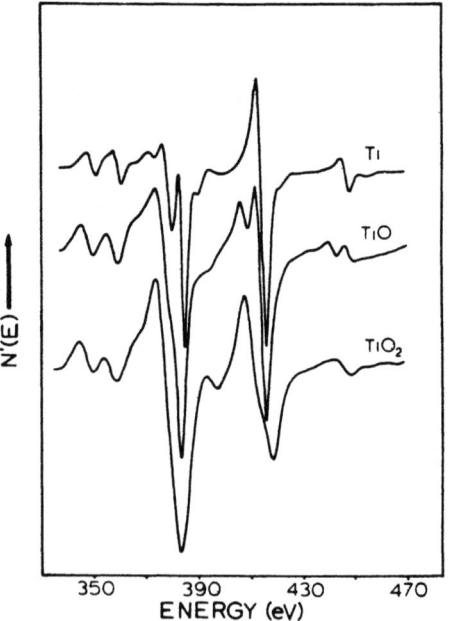

Figure 7. Ti $L_{2,3}$ MM Auger derivative spectra of the pure metal, TiO and TiO_2 (from Ref. 91).

electrons produce a steplike function with an increasing low-energy tail. The function, when not very different in energy of the main peak, overlaps partially with it and is responsible for the broadening and the asymmetry on the low-energy side of the peak. When the energy loss is large, the loss peak may overlap with an Auger line situated at lower energy; such a situation is often encountered in the characteristic Auger triplet of transition metal spectra. In the case of titanium, for instance, the spectrum of the metal contains a shoulder on the high-energy side of the $L_3M_{2,3}M_{2,3}$ line (382 eV), while this feature is noticeably absent in the oxide spectra[91] (Fig. 7). This shoulder peak, which has been reported to be either the first plasmon loss from the $L_{2,3}M_{2,3}M_{4,5}$ peak at 416 eV[116] or a shake-up peak associated with the same transition,[119] is shifted in energy in the oxide spectra (larger loss); it becomes unresolved by overlapping with the Auger peak and so contributes to the distortion and broadening of the Auger features.

7. Background Subtraction and Deconvolution

To extract the discrete Auger signals from a continuous background of unwanted emission, Auger spectra are usually recorded as the first derivative of the energy distribution since differentiation suppresses a large part of the perturbating secondary-electron background; in most cases, this background is actually so large that visualization and positioning of Auger peaks could be performed only on the derivative distribution curve. However, meaningful comparative studies, such as chemical shift measurements, should of necessity be performed on the $N(E)/dE$ distribution; thus, the extremum of the Auger peak does not indicate the true peak position but the energy of the maximum negative slope. From a quantitative point of view as well, the $N(E)$ curve, for which the peak area is proportional to the number of atoms present, should be preferred for rigorous measurements since peak-to-peak amplitude, in the derivative mode, is dependent on the shape of the Auger peak as well as its size. To improve the information received, a number of methods based on data-processing techniques have been developed.

(i) Background Subtraction

The aims of these techniques are to remove the secondary-electron background obscuring the Auger distribution, to generate the direct distribution from the derivative, to retrieve the true Auger line features by elimination of intrinsic and extrinsic aberrations, and finally to relate directly these true feature intensities to the Auger current. Some of the techniques are purely mathematical, the other ones are based on instrumental considerations.

(a) *Mathematical techniques.* With these techniques, the usual approach is to simulate a background and subtract it from the measured spectrum. Sickafus[120] has electronically subtracted an analog of the secondary distribution (excepting Auger electrons) which was approximated by

$$N(E)_{\text{second}} = A(E_0 + E)^{-m} \quad (3)$$

where A, E_0 and m were adjustable parameters determined in a separate experiment. It has also been proposed that the background be approximated by a curve obtained by orthogonal polynomial regression, using from eight to ten suitably chosen data points estimated to be in the background of the directly recorded distribution curve.[121] Spline polynomials have also been used as an approximation for the secondary-electron background[122] and the scheme has been developed to retrieve small experimental structures;[123] a gain in sensitivity of eight has been reported and, in a standard analyzer, the ratio between the strongest and the weakest retrieved line was a thousand.

It has been demonstrated that, if the secondary-electron background is adequately suppressed by differentiation, it is possible to retrieve the characteristic signals themselves by simple integration of derivative data over the region of interest.[124] This scheme, which is referred to as dynamic background subtraction (DBS), corrects the principal disadvantages of differentiation, viz. generation of complex structures difficult to interpret and the fact that contrast is enhanced only at the expense of the signal-to-noise[125] ratio.

DBS takes advantage of the functional differences between

the characteristic Auger features, which have small convergence radii and approach zero rapidly above the high-energy threshold, and the slowly converging Taylor series which give an approximation to background functions[126,127] so that the latter functions can be suppressed to an arbitrary extent by differentiations of a sufficient order.

The normal derived Auger data are very sensitive, both in magnitude and in shape, to the value of the modulation voltage applied to the energy analyzer; in such a distribution, the peak-to-peak heights bear no simple relationship to the amplitude. In the integrated spectrum, however, the peak heights vary almost linearly with the amplitude; it has also been shown that peak areas determined by double integration of heavily overmodulated derivative spectra scale linearly with the modulation amplitude.[128] With DBS it is therefore possible to use large modulation amplitudes to improve the signal-to-noise ratio[129] (thereby increasing the detectability limit for an element) and yet still obtain an exact value for the Auger current as integrated signal strengths increase while the Auger current values remain unchanged.[130] It should be noted that broadening of the peaks by the large modulation potential reduces resolution and spectral contrast. It has also been shown that distortion, resulting from potential modulation differentiation, may be characterized by an instrument response formalism (IRF)[131] and that such a distortion can be exactly corrected in derived data corresponding to the area under the characteristic peak ("double integral"). Thus, DBS allows direct quantitative comparisons to be made between the Auger currents from a standard and a sample of the same element in a different chemical environment.[129]

When such a technique is used for quantification, it should be remembered that if integration does attenuate the high-frequency noise content of derivative spectra[132] it also increases the sensitivity to low-frequency noise, which is often difficult to identify and may degrade peak height and area measurements without the experimenter's knowledge.[133] Moreover, the proportionality between peak height and modulation voltage has not always been found satisfactory[38] and when complex changes and structure formation in the background, due to backscattered electrons, are observed such a scheme is difficult to apply.[134]

Practically, integration can be carried out either by digitizing the derivative form data and integrating them the appropriate number of times by means of a computer or by applying analog integration methods.[135,136]

(b) *Instrumental techniques.* The tailored modulation technique (TMT) has been developed for measuring, in the direct distribution mode, Auger peak heights and areas. Results are obtained by modulating the pass energy with sinusoidal or square waves of large amplitude (and directly detecting the phase-lock-amplifier output signal[137]). The technique is implemented by "tailoring" the modulation waveform to the instrument response function[131] and dynamic background subtraction constraints. The electron energy distributions are obtained over energy ranges equal to the modulation amplitude measured down from the high-energy threshold of each set of peaks.[138]

For a given peak-to-peak modulation voltage, a larger signal-to-noise ratio is obtained by using square rather than sinusoidal waves.[139] With this modulation waveform, the measured electron energy distribution shifts up in energy relative to the normal spectrometer calibration by half the modulation amplitude but the peak intensities remain unaffected. It has been demonstrated that either very small or very large modulation amplitudes must be used to characterize unambiguously the signals either as simple derivatives or integrals. Extension of these waveforms to obtain higher-order derivatives in the background may be carried out; however, it seems preferable to limit the waveforms to the second degree as each successive subtraction results in further narrowing of the integration limits and reduces the signal strength by a factor 2^n in the RFA and by a factor 2^{n-1} in the CMA (n is the order of background subtraction[138]). In this case, measuring the same function over the same spectral region requires a considerable increase in modulation amplitude requiring, in turn, more featureless background ahead of the peak.

Apart from subtracting background, TMT offers two desirable characteristics: improved resolution by compensating for broadening due to the analyzer itself and the possibility of maximizing the signal size.[140] Thus, TMT allows the maximum possible resolution and sensitivity to be simultaneously obtained.

(ii) Deconvolution

Measurements of relative concentrations may be inhibited when two sets of Auger transitions having similar energies and peak shapes overlap. The two more widely used methods for deconvoluting the overlapping peaks are the spectrum subtraction technique (SST) and the differential Auger spectroscopy (DAS) method.

In the first technique a normalized suitable standard spectrum of the perturbing element is subtracted from the spectrum of interest[141] (see Fig. 8). This method allows not only the limit of detection of small concentrations to be enhanced but also the retrieval of the individual Auger line shapes when spectral overlap occurs.

With the second technique[142] one can retrieve the signal for small quantities of an adsorbate whose peaks overlap with those of the host element. The basic idea is to deflect the electron beam

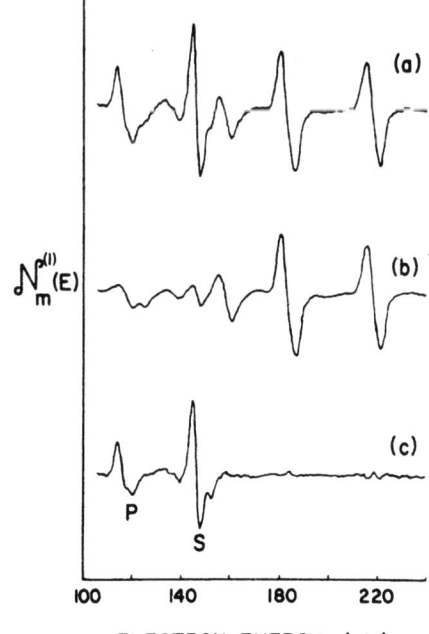

Figure 8. Illustration of the application of SST to retrieve Auger line shapes: (a) part of a high-resolution Auger spectrum of contaminated molybdenum; (b) the corresponding Auger spectrum from clean molybdenum; (c) the result obtained after subtracting part (b) from part (a), showing the Auger line shapes of phosphorus and sulfur in part (a) (from Ref. 141).

alternatively between a standard and the specimen; the primary beam is switched at a frequency of 30 Hz between the two samples and the signal from the standard subtracted electronically from that of the specimen; the difference of the two spectra is obtained as the final output.

Another deconvolution scheme based on an iterative method (van Cittert successive substitution method[143]) has been used to correct the Auger peak distortions induced by electron energy losses.[144] Results of model calculations have been presented for solving the convolution equation governing the relation between the true spectrum, the measured spectrum, and the instrument response function.

The disadvantage of this technique is that the van Cittert deconvolution scheme converges very slowly if the response function features are of greater breadth than those of the true function. In many practical applications, however, convergence requiring a large number of iterations is not possible, since the statistical noise that is present on the measured signal increases with iteration number; in such cases slight improvements might be obtained by smoothing techniques.

Many commercial spectrometers are provided with an automatic deconvolution system. Quantitative results obtained in such a way are, however, not always satisfactory. This apparently results from the fact that a standard FWMH value, the same for any transition of any element, is programmed and fed into the computer. Distortions are particularly severe for elements, such as carbon or boron, which present asymmetric Auger peaks. In consequence, "ready made" computer programs should be carefully revised before use to avoid subsequent disagreements.

8. Electron Beam Effects

During surface investigations several disturbing effects may arise from interactions of the electron beam with the adsorbate and lead to modifications in the surface composition of the specimen. For example, the electron beam has been shown to affect radically the interaction of gases with the surface; carbon monoxide decomposes on both nickel[145,146] and platinum,[147-149] and the decomposition which takes place in two steps results in the buildup of a carbon layer on the surface.[150] A number of authors have reported on surface instability in an electron beam: alkali halides dissociate rapidly,[151,152] surface oxides may be reduced,[153,154] while other substrates are oxidized;[155] such changes in the chemical composition of the surface induce unexpected shifts of the Auger peaks.[156–158] Localized diffusion of elements[159] to[160] and from[161] the surface has also been reported; in addition, the electron beam can desorb material from many virgin surfaces [electron-induced desorption (EID), electron-stimulated desorption (ESD)].[162,163]

An interesting example has been mentioned by Ranke and Jacobi,[164] who have observed arsenic depletion in an oxygen-covered gallium arsenide crystal as a subsequent effect of beam interaction: electron irradiation causes dissociation of the molecular oxygen; reaction between atomic oxygen and arsenic results in the formation of a compact oxide layer which soon becomes depleted in arsenic as a result of sublimation of As_4O_6.

In many cases, therefore, the analysis obtained by AES does not give an accurate view of the "as received" surface because of electron beam effects which can occur almost instantaneously (see Fig. 9). The decrease in the initial surface concentration N_0 as a function of time may be calculated from the relation

$$N = N_0 \exp\left(-\frac{IQt}{e}\right) \quad (4)$$

where Q is the cross section for ESD, e the charge of the electron, and I the current density.[165] If Q is taken to be 10^{-18} cm^2 and I 10 μA/mm^2, it can be calculated that 70% of the adsorbate is desorbed in 200 sec; with a cross section of the order of 10^{-20} cm^2, the desorption becomes negligible.

ESD has been extensively studied not only as a disturbing effect during analysis[166] but also as an independent technique suitable for investigating the adsorbate states on the surface.[167–169] Theoretical models have been developed to explain, at least qualitatively, ESD. The model presented by Menzel,[170] based on a two-step process (Franck–Condon transition from the ground state of the adsorbate system to a repulsive neutral or ionic state of the complex, followed by a recapture process or electron

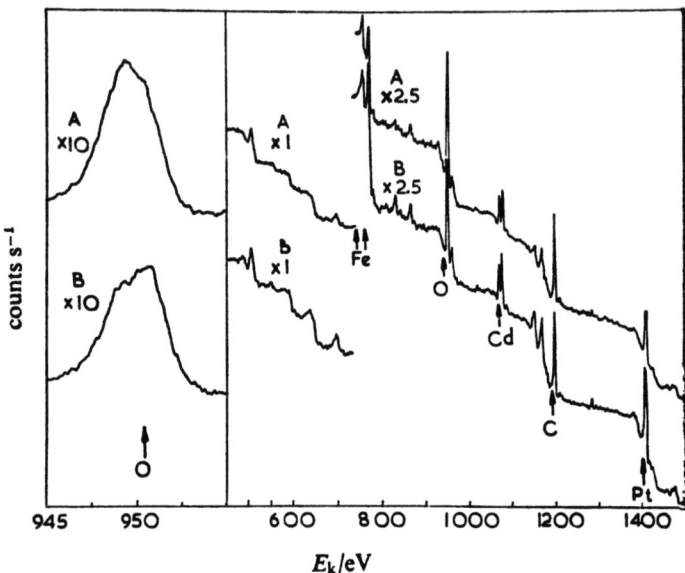

Figure 9. XPS spectra from a contaminated Pt electrode before (traces labeled A) and after (B) electron irradiation. On the right, complete spectra from 450 to 1500 eV recorded with 100 eV pass energy; on the left, detailed scans of the oxygen peak recorded at 50-eV pass energy. Al K_α radiation 1486.6 eV (from Ref. 173).

tunneling) explains quite satisfactorily the various experimental observations. Quantum mechanical treatment of ionic ESD, considering the quantum effects in the motion of the escaping particle, has also been proposed.[171]

Gettings and Coad[172] have investigated experimentally the various beam effects on oxidized or heavily contaminated materials. They have observed discrepancies between analytical results obtained by photoelectron spectroscopy and those obtained by techniques using the electron beam as an exciting source. Different types of materials (metals, glasses) have been analyzed by XPS and SIMS before and after rastering large surface areas with an electron beam.[173]

In practice, the obvious means for minimizing electron beam effects is to reduce drastically both beam intensity[174,175] and duration of the analysis experiment; in any case, the results of Auger analyses should be interpreted with caution.

III. X-RAY-EXCITED PHOTOELECTRON SPECTROSCOPY

1. Physical Process

In the photoelectron spectroscopic technique the specimen is bombarded by soft X-ray photons of well-defined energy, currently the $K\alpha$ radiation of aluminum (1486.6 eV) or the $K\alpha$ radiation of magnesium (1283.6 eV). If an electron absorbs the energy $h\nu$ of an incident photon, it can be ejected into the vacuum where its kinetic energy is analyzed and the remnant electron system goes into a final state of definite energy and of the proper symmetry.

Since the photoemission process involves the excitation of a single electron for each absorbed photon, the kinetic energy measured by the analyzer is

$$E_{kin} = h\nu - E_B - \phi_{anal} \qquad (5)$$

where E_B is the binding energy of the electron in the level considered and ϕ_{anal} the effective work function of the analyzer.

The electron may originate either in one of the core levels or in the valence band which is accessible up to the Fermi level. XPS is therefore a filled-band probe and it has been used in several cases to map the density of states of the valence band in various materials.

2. The Photoelectron Spectrum

The photoelectron spectrum is recorded as the kinetic energy or binding energy distribution of the ejected electrons. In this spectrum all the core levels which have binding energies smaller than the energy of the exciting radiation give rise to a peak. Due to spin–orbit coupling, each of the p, d, and f levels is split in the usual way into several components representing the various sublevels; for example, in the mercury spectrum, peaks corresponding to the $4d_{3/2}$ and $4d_{5/2}$ sublevels can be observed. Figure 10 shows such a distribution where sharp peaks are superimposed on a secondary electron background due to inelastic electron–

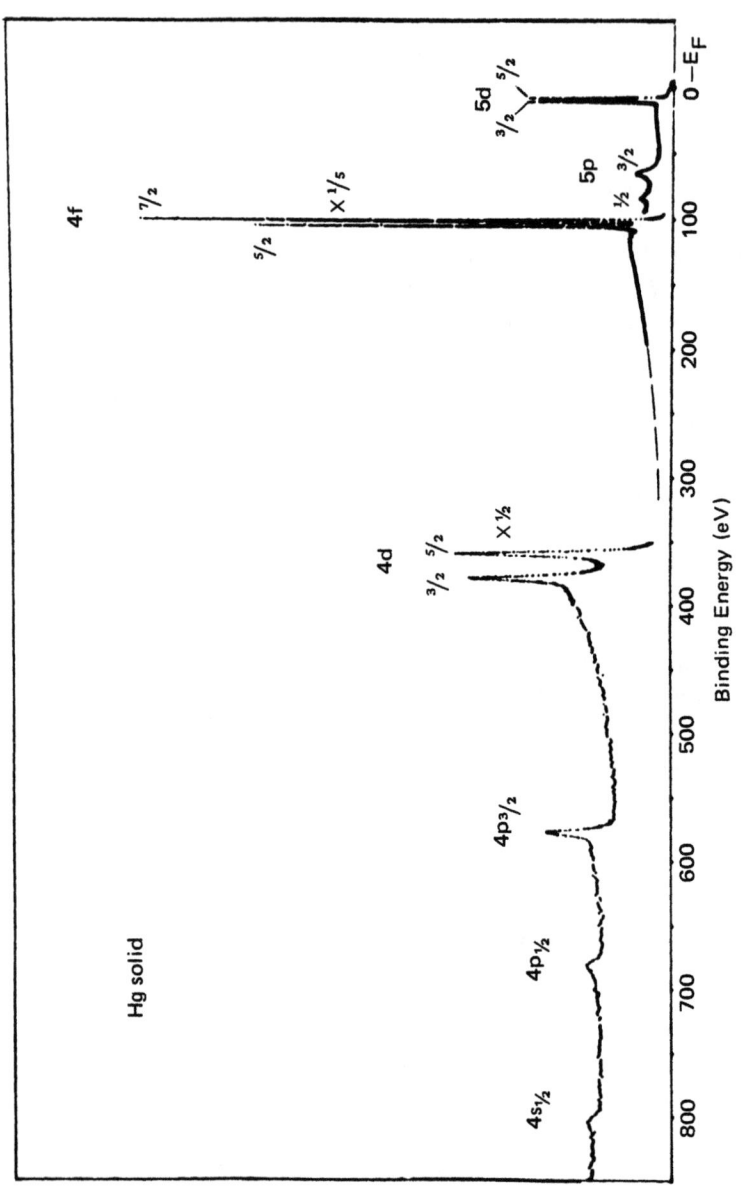

Figure 10. Total XPS spectrum of metallic mercury excited by monochromatized Al K_α X rays (from Ref. 293).

electron scattering; only the electrons giving rise to the sharp peaks have escaped without loss of energy and must therefore have originated very near the surface. If no perturbing effects occur the photoelectron peak is symmetrical around its maximum value and can be approximated by a Lorentzian distribution. However, photoelectron signals are often broadened and asymmetrical peaks are observed; moreover, additional signals due to distinct energy levels of a single electronic configuration or adjacent configuration (satellite peaks) or, arising from inelastic electron scattering (loss peaks), are frequently observed.

Identification of the nature and origin of these secondary signals is essential to thorough understanding and meaningful interpretation of the surface properties. For this purpose a general theoretical framework for treating intensity satellite lines and facilitating interpretation of photoelectron spectra has been developed.[176]

(i) Satellites

Normally the ion which has been created by photoejection of an electron from the core shell is left in the ground state. Frequently, however, excited states are also formed; in such cases the photoelectron is emitted with a kinetic energy reduced by the amount of the excitation and a structure is observed on the low-energy side of the main peak (satellite structure). Two important processes, which can give rise to satellite structure, are electron shake-up and multiplet splitting.[177]

(a) *Electron shake-up.* Electron shake-up occurs when a sudden change in central potential, caused by the removal of a core electron that had been shielding the valence orbital electrons from the nuclear charge, gives rise to monopole excitation of the outer-shell electrons.[177] In other terms this phenomenon can be understood as a nonadiabatic relaxation of the electron cloud upon photoionization of one of the electrons.

The probability of observing a shake-up structure depends on the location of the inner-shell vacancy. This probability is essentially constant for photoionization in the $1s$, $2s$, $2p_{1/2}$, and

$2p_{3/2}$ shells[178] and the energy separation similar. If the vacancy is located in the $3s$ shell or a $3p$ subshell, the probability of exciting an electron in a $3d$ orbital is about one-third of that for a $2p$ subshell. It seems that, for the elements in the first transition series, electron shake-up is the dominant process producing satellite structure.[179,180]

In the case of closed-shell ions, such as scandium(III), titanium(IV), or lanthanum(III), the origin of the satellite structure observed has been attributed to an electron transfer from the filled orbitals of the neighbor atoms to an empty shell of the metal ion.[181-187] The idea has been developed by Jørgensen and Berthou,[188,189] who coined the term "transfer satellite."

The shake-up satellites in the $3d$ group compounds have been extensively discussed.[190] Several comprehensive reviews have been recently published[191-194] and selection rules for shake-up transitions have been tabulated.[195] From these various studies a few relevant characteristics of shake-up effects in the $2p$ spectra of the first series transition metals may be summarized:

(1) The element in the metallic form does not exhibit strong satellite lines; such lines are mainly observed in the compounds.

(2) For transition metal compounds having a strong satellite structure in the $2p$ shell spectrum, there is usually one peak associated with each subshell; this strong peak results from the convolution of several smaller peaks whose energies are closely spaced[196,197] (see Fig. 11).

(3) Satellite structures for the transition metal complexes are more intense for paramagnetic compounds and for weak-field ligand compounds (with the exception of the case of closed shells).

(4) The energy separation between the $2p_{1/2}$ and $2p_{3/2}$ peaks is more important for high-spin ions than for less paramagnetic or diamagnetic compounds of the same metal.[198,199]

It should be noted that for the second- and third-row transition metal compounds, shake-up satellites very seldom arise; this lack of secondary effects is probably due to the fact that such elements have larger crystal-field splittings.

(b) *Multiplet splitting.* Multiplet splittings occur in atoms containing partly filled shells in their ground state. In this process,

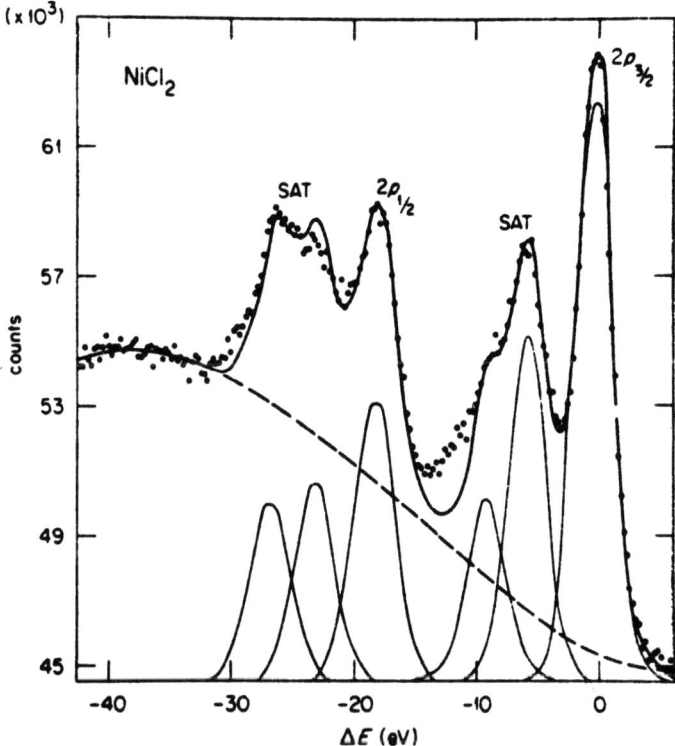

Figure 11. Photoelectron spectra of NiCl ionized in the 2p shell of Ni with Al K_α X rays. The kinetic energies of the photoelectrons are related to that of the main peak. SAT represents satellite structure. The satellite structure has been deconvoluted into two peaks. The approximate background, including characteristic energy losses, is given by the dotted line (from Ref. 194).

the creation by photoejection of a core vacancy leaves a shell with an unpaired spin that can couple with the valence shell. As underlined by Carlson,[177] multiplet splitting is really the consequence of a series of possible spin states rather than the result of a specific excitation.

Such a phenomenon is readily understood if we take as an example the 3s terms in the Mn(II) ion.[200] The ground state of the ion is $3d^5(^6S)$, all spins parallel. Upon emission of a 3s electron two possibilities exist: the remaining electron aligns its spin parallel or antiparallel to the $3d$ spins, giving rise to two final states $3s3d^5(^5S)$ and $3s3d^5(^7S)$. As the spin exchange term in the Hamiltonian gives rise to nonzero matrix elements for parallel spins only, the ^7S state induces a lower binding energy than the ^5S state; it can be calculated that the theoretical energy separation is about 13 eV.[17] The relative intensity ratio can be approximated by $(S+1)/S$, where S is the total spin quantum number.

Clearly, this example is a very simple one; the structure of the $2p$ and $3p$ regions of Mn(II) is more complex and Jørgensen[201] has pointed out that such a structure is partly due to the $2p^53d^5$ and $3p^53d^5$ levels that are accessible according to the rules enunciated by Cox.[202] Numerous examples of multiplet splitting can be observed in the spectra of lanthanide compounds and metals[203] where many J levels are possible final states of the ionized configuration $4f^{q-1}$ (Ref. 204) and the configurations $4s4f^q$ and $5s4f^q$ clearly show such a splitting.[205,206] Considerable structure, probably due to the many levels of $4d^94f^q$ have also been observed in the $4d$ region.[207,208]

Clearly, the subject of multiplet interactions is a complex one and a theoretical discussion would be beyond the scope of this chapter. However, though absolute generalizations should be avoided, a few useful rules can be enunciated.

Regarding the occurrence of energy splitting, Carlson et al.[209] have calculated that the splitting was substantial for vacancies created in both the L and M shells, ranging from 3 to 12 eV for a $2p$ hole; for a K vacancy, multiplet splitting has been found to be negligible (~ 0.1 eV). For the $2p$ photoelectron spectrum, Gupta and Sen[210] have predicted a large number of final states but these states, rather than forming two well-defined satellite peaks in the spectrum will asymmetrically broaden the two main normal peaks. The peak associated with the $2p_{3/2}$ subshell is broadened toward lower kinetic energies, while the $2p_{1/2}$ peak is broadened on the higher-kinetic-energy side and the energy separation between them is increased.[211,212] In the MF_2 spectrum, Kowalczyk et al.[213] have found a small additional satellite peak between the two main peaks, which had been predicted by calculation. In the case of MF_2, however, recent evidence seems to point to a very complex process rather than to simple multiplet splitting.[214]

It has also been found that the 3s splitting decreases as a consequence of the decrease in ionicity of the compound.[215–217] Decreasing ionicity means lower degree of localization of the 3d electrons and hence weaker exchange to the 3s electrons; an additional effect is the diminution of the effective number of unpaired spins leading also to decrease in the splitting. Finally, it has been noticed that multiplet splitting effects were absent in the 2p spectrum of diamagnetic compounds such as $[Fe(CN)_6]K_4$.

(c) *Energy losses.* Characteristic energy losses are always present in photoelectron spectra. The electron, passing through a solid, undergoes inelastic collisions; as a result of such processes, a weak broad peak, whose shape and intensity are characteristic of the material the electron is passing through, can be observed. This peak is characterized by a long low-energy tail and a kinetic energy smaller by 10–25 eV than that of the main peak.[218,219] In fact, in insulators, energy loss is a complex process involving a coupling of both plasmon excitation and interband transitions.

(d) *Auger electrons.* Another extrinsic energy loss process following photoelectron emission may take place by Auger deexcitation. Auger electron emission may occur whenever a hole is created. Filling of the vacancies left by photoelectron emission via the Auger process gives rise to emission of electrons whose energies are determined by the energy levels in the core-ionized atoms.

X-ray-excited Auger electron emission, which has been for a long time considered a perturbing secondary effect, has now gained full recognition as an important source of information. We shall examine a few recent developments of this technique in Section III.9.

3. Instrumentation

In XPS systems the photoelectrons emitted by the specimen are focused on an electron energy analyzer where the energy distribution is measured; the XPS spectrum is recorded as count rate versus kinetic energy (or binding energy).

XPS instruments commonly in use may be grouped into three categories based on the type of analyzer which is the key component of the spectrometer, as its resolution crucially affects the quality of the spectrum. The three groups of instruments are the retarding field analyzer (a nondispersive system), the electrostatic spectrometer, and the magnetic instrument (both dispersive); the comparative advantages and disadvantages of such systems have been reviewed at length by several authors.[220-225]

During the last few years modifications were reported in the construction of spectrometers to improve the performance of the analyzers both in resolution and in sensitivity. Though electrostatic analyzers,[226-228] especially the CMA,[229-232] have been the more popular, magnetic instruments still receive attention[233] and new retarding field analyzers have been developed.[234-237]

XPS peaks are limited in width only by the inherent linewidth of the exciting radiation, typically 1–1.2 eV for aluminum and magnesium $K\alpha$ lines; resolution is improved in modern instruments by the use of a monochromator.[238-240]

Various methods have been proposed to determine the collecting efficiency and the transmission of the analyzer[241-242] and the way to achieve simultaneously the maximum attainable resolution and transmission of an energy analyzer has been examined.[223]

4. Elemental Analysis

Each element in the periodic system has its characteristic binding energies in the core levels which can be used to identify it.[1] XPS allows identification of all the elements to be made except hydrogen. Elemental analysis is performed by comparison with standard spectra obtained from known samples or with tabulated core-level energy values. XPS is not a very sensitive technique and the signal-to-noise ratio is usually poor. As an order of magnitude, one can say that one atom percent of an element is easily detected in a mixture, the mass of the sample being typically 10^{-6} g. Therefore, the chief interest of the technique resides in analysis of the chemical state of the element rather than simply in the elemental determination.

5. Quantitative Analysis

Photoelectron spectroscopy can be used as a quantitative tool for the determination of chemical composition of the surface region.

Quantitative analysis by XPS requires knowledge of the relative photoelectron signal intensities for the various elements. Photoline intensities are a function of orbital degeneracy and of orbital cross sections, where the latter can be thought of as the probability of ionization occurring at a particular excitation energy[243-245]; therefore, determination of the photoionization cross section is of major concern in any quantitative application of XPS. Cross sections can be calculated from a Hartree–Slater model[246-248] or obtained directly from experimental measurements.

The proportionality factor varies with the element, the particular orbital of the element, the photon energy, and the design of the instrument; thus the relation between line intensity and photoionization cross section is not always obvious. The various factors on which intensity depends have been reviewed by Leckey,[249] who mentions the geometry of the spectrometer,[222,223] the degree of attenuation by inelastic scattering,[250] and the asymmetry of emissions originating from shells other than s orbitals.[251] Chemical environment too, appears to influence the atomic sensitivities[252]; for example, Wyatt et al.[253] have reported that, for a series of lead salts, the proportionality factor varied with the anion; these variations have been attributed to the different crystalline structures which may affect the average escape depth (Section IV.1). Thus it has been observed that, for a given core level, the photoionization cross section is unaffected by bonding between the metal and its anion.[254]

The various intrinsic (shake-up, shake-off) and extrinsic (Auger, loss peaks) loss channels may reduce significantly the main photopeak intensities; these secondary peaks must be included into intensity calculations since the total photoemission intensity is shared between all of them.

Several compilations of experimentally determined photoelectron line relative sensitivities have appeared in the literature. The pioneer work in this domain is due to Wagner,[255] who produced relative intensity data for a wide range of elements by recording the spectra of either fluorine- or sodium-containing stoichiometric compounds; the fluorine $1s$ line has been used as a standard; similar studies have been performed by Jørgensen and Berthou.[256,257] All the data thus produced are only valid for one type of instrument and a particular excitation source; it should not be said too often that such data are not directly transferable. Nefedov et al.[258,259] have determined experimentally relative intensities for a large number of elements with $3 \leq Z \leq 56$ and calculated the photoionization cross sections for inner levels of the same elements.

Ideally, the most satisfactory procedure for measurement of relative photoionization cross sections and line intensities would be to cleave in the spectrometer, under UHV conditions, a sample of the element, of an alloy of known composition or a monocrystalline compound of ascertained stoichiometry and then to examine the surface thus produced. A similar procedure was used by Brillson and Ceasar,[260] who measured the relative photoemission intensities from subshells of 51 elements and simple compounds; good correlation ($r = 0.96$) between theory and experiment has been observed over two orders of magnitude for most of the elements studied. It has also been suggested to condense a series of molecular species on an inert surface as thin films and thus obtain the relative sensitivities of the element in the molecular species.[49]

It has often been discussed whether peak height or peak area would be a more accurate representation of the photoelectron line intensity. Wagner[255] has demonstrated a good correspondence (averaging ±15%) between peak height and peak area sensitivities over a wide range of elements. However, discrepancies between measurements of peak heights and peak areas become significant for the $4d$ lines of elements with $Z = 70$ to 90. Generally speaking, peak height measurements are not appropriate when the observed linewidths are appreciably larger than 2 eV; it seems, in any case, that a rigorous quantitative comparison requires the measurement of peak and satellite areas rather than peak heights.

From the studies mentioned above, XPS appears as a semiquantitative technique with relative errors of the order of ±50%. When comparing surfaces of chemically similar samples, viz., samples where the same mechanisms are responsible for inelastic electron scattering and where the escape depths are similar, a reproducibility of 10–15% may be attained.

6. Chemical Effects

Since, in XPS, the core-level transitions do not involve states near the Fermi level, the shapes of the peaks are relatively unaffected by chemical environment; the emphasis has therefore been on the measurement of binding energy shifts. Actually the ability to provide information on chemical bonding or, at least, to distinguish between nonequivalent atoms of the same element by the use of chemical shifts is fundamental to the technique.

(i) Calibration

As a consequence of the interest in chemical shifts, one of the major problems of XPS has been to find a suitable procedure for energy calibration of the photoelectron lines. Such a calibration should not only be reliable within a determined spectrometer but also allow precise comparisons of independently obtained data.

It has been suggested[261] that a series of electron standard lines be established by determining, with great precision, consistent values of binding energies; thus the Au $4f$, Ag $3d$, Cu $2p$, Pd $3d$, and Fe $2p$ levels as well as the Cu $L_3M_{4,5}M_{4,5}$ Auger line, determined relative to the Fermi level of Pd, have been proposed. However, it has been found that many uncertainties still remain; they arise from the determination of the work function of the analyzer and systematic errors introduced by neglecting relativistic contributions to the electron energies; consequently, an attempt has been made to develop a calibration method which does not depend on an exact knowledge of the above-mentioned factors. The method, which has been developed for a spherical analyzer, has been based on the determination of the true electron energy as a function of the sphere voltage, and on a defined zero point for the scale of binding energies.[262]

The method of calibration based on accurate measurements of the work functions[263] is of interest as it makes possible precise determinations of ionization energies relative to the vacuum, thus allowing comparisons between gas-phase and solid-state results. On the other hand, the Fermi level of a conducting sample has sometimes been proposed as a reference level.[264] Observation of the slow-electron energy distribution (SEED) has also been suggested as a means of measuring the work function and the vacuum level of the specimen.[265]

In spite of some difficulties, the determination of the kinetic energies of conducting samples is not too troublesome; however, when nonconducting samples are studied, the removal of electrons can cause a positive charge to build up,[266,267] which reduces the kinetic energy of the emitted electrons. The obvious solution has been thought to reside in neutralization of the positive charge by bombardment of the sample with low-energy electrons ("electron floodgun")[268]; unfortunately such a procedure gives rise to a negative charge on the sample.

Since the kinetic energies of the photoelectrons are perturbed by the charging effects it is most important that the charge either must be removed or it must be accurately measured. Vapor deposition of gold on the surface of the nonconducting sample has been widely used for both alleviation and measurement of charging. Gold is assumed to form a conducting film over the surface of the sample; this film is in electrical contact with the base upon which the sample is mounted.[269] Such a technique has proved unsuccessful because of the difficulty in obtaining a continuous thin film of gold on the sample.

Various attempts have been made to measure charging effects by mounting the sample on a gold mesh and vapor-depositing gold onto the surface.[270,271] With this technique two pairs of $4f$ gold peaks are observed that correspond to the two different physical states of gold: the gold mesh which is in electrical contact with the spectrometer (hence free of charging) and the gold on the surface of the sample, which is insulated from the spectrometer.[272] It has been assumed that the difference in kinetic energy between the two pairs of peaks was a measure of the charging suffered by the sample; it seems, however, that this difference is a function of the amount of gold deposited on the surface.[273–275] Besides, chemical interactions may take place between the vapor-deposited gold standard and the specimen; such interactions have been reported for KCN, NaCl, $Na_2S_2O_3$, and copper phthalocyanine[276] and for phosphorus compounds.[277] These interactions are difficult to detect as the peaks usually retain their shapes and are simply shifted in energy. The alternative to the Au $4f$ line as a reference is the carbon $1s$ line, which has been extensively used.[278] In pioneer work the $1s$ line originating from hydrocarbon contamination on the surface has been used as a standard. In many cases this procedure has been rather unfortunate as it implicitly assumed the position of adlayer core level relative to the substrate Fermi level, to be independent of the substrate work function.

A graphite dilution method has been studied[279] and Sellotape has been suggested as a suitable source of carbon[280] but opinions diverge on the subject.[281] It has been shown that polymerized aliphatic hydrocarbons (such as "Scotch Tape," for example) present two distinct C $1s$ signals whose energy difference represents a fairly accurate estimate of the charging potential.[282,283]

An interesting study, liable to end the controversy, has recently been published.[284] In this work, the binding energies of the more intense photoelectron lines from 16 compounds have been measured with four spectrometers of different types; Al $K\alpha$ as well as Mg $K\alpha$ radiations have been used as sources and the investigations have extended over 18 months to test the stability. The results have been corrected by using the C $1s$ line of carbon contamination. In one of the spectrometers, gold coating has been performed as well and the energy of the $4f$ line has been measured. From these various experiments it is concluded that the agreement between them is excellent if the values C $1s$ = 285.0 eV and Au $4f$ = 83.8 eV are chosen for charge correction, provided that the amount of gold on the surface is minimized.

A method of calibration in which charging effects are compensated for has been suggested by Wagner,[285,286] who has developed the concept of the Auger parameter. Such a parameter is defined by the difference between the kinetic energies of a photoelectron line and an Auger line from the same element. Plots where the photoline position is the abscissa and the Auger line the ordinate have been drawn; the Auger parameter axis is then at 45°. Figure 12 shows such a plot for indium. Auger parameters have been calculated for 60 elements.[287]

The Auger parameter has the advantages that static charge corrections cancel; this parameter is unique for each compound

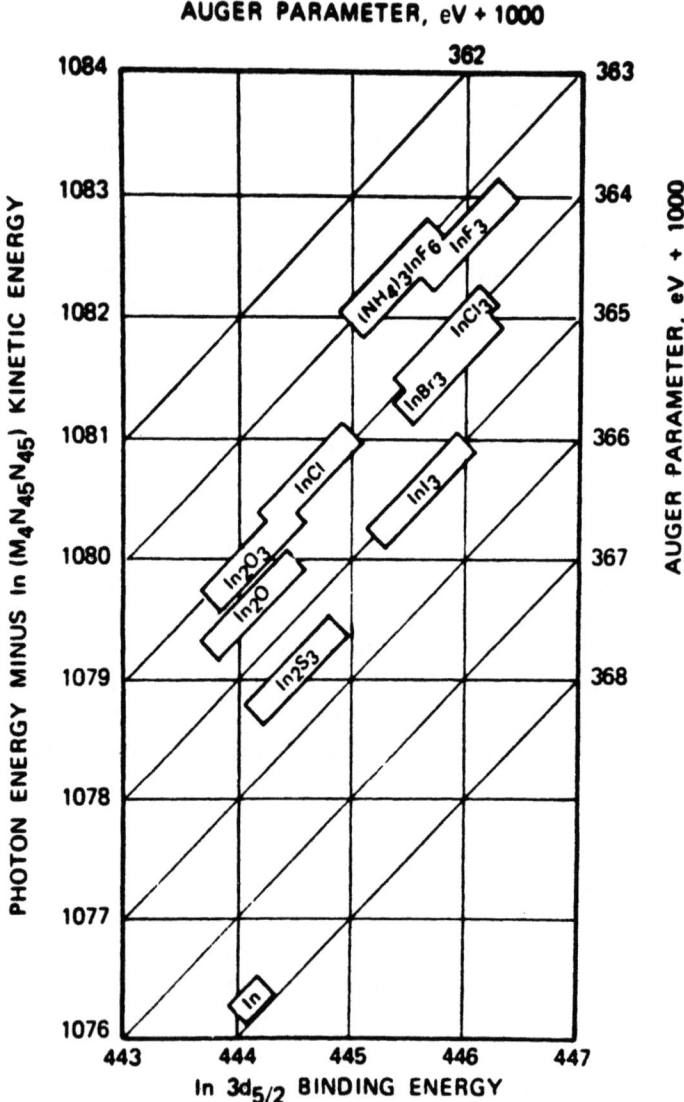

Figure 12. Chemical state plot for indium. Error bars indicate probable error limits. The number 1000 is added to make the parameter a positive number (from Ref. 287).

and can be measured more accurately than absolute line positions which must be corrected for static charge.

(ii) Chemical Shift

A change in the chemical environment of an atom induces variation in the electronic density of this atom; as a result, the corresponding photoelectron peaks are shifted on the binding energy scale (chemical shift). The chemical shift can be correlated with the oxidation state of the element or its fractional atomic charge, an increase in formal positive charge being generally associated with an increase in binding energy. However, in many cases the measured energy shift is not a straightforward representation of the variation in oxidation state (see Fig. 13). Chemical shifts may be interpreted either in terms of charge calculation (induction)[288,289] or, considering the final state, in terms of relaxation (polarization).[290-293]

In many solids the molecular potential experienced by core electrons is partly due to the influence of charged adjacent atoms; in these cases a correction in terms of the Madelung potential must be applied. The influence of the Madelung potential on inner-shell energies has been extensively studied; for example, the chemical shifts measured from a series of fluorine compounds have been correlated with this potential[294]; it has also been observed that the binding energy value for the metal atom is lower in PbO_2 than in any other Pb compound. Calculations have shown that the Madelung correction affects Pb(II) and Pb(IV) energies in opposite directions to such an extent that they coincide; as a result, a single lead peak is observed in the Pb_3O_4 spectrum.[295]

However, it soon became evident that the fractional charge model was not sufficient to explain the direction and magnitude of the chemical shifts. For example, we observed[294] that, for a large number of potassium(I) compounds, the energy of the potassium $2p$ line varied in a way corresponding to the chemical polarizability (Pearson's softness) of the anions[296] and in most cases in the direction opposite to that of the Madelung potential. Relaxation effects are responsible for such a variation. When an electron from a core level is excited, the outer-orbital electrons

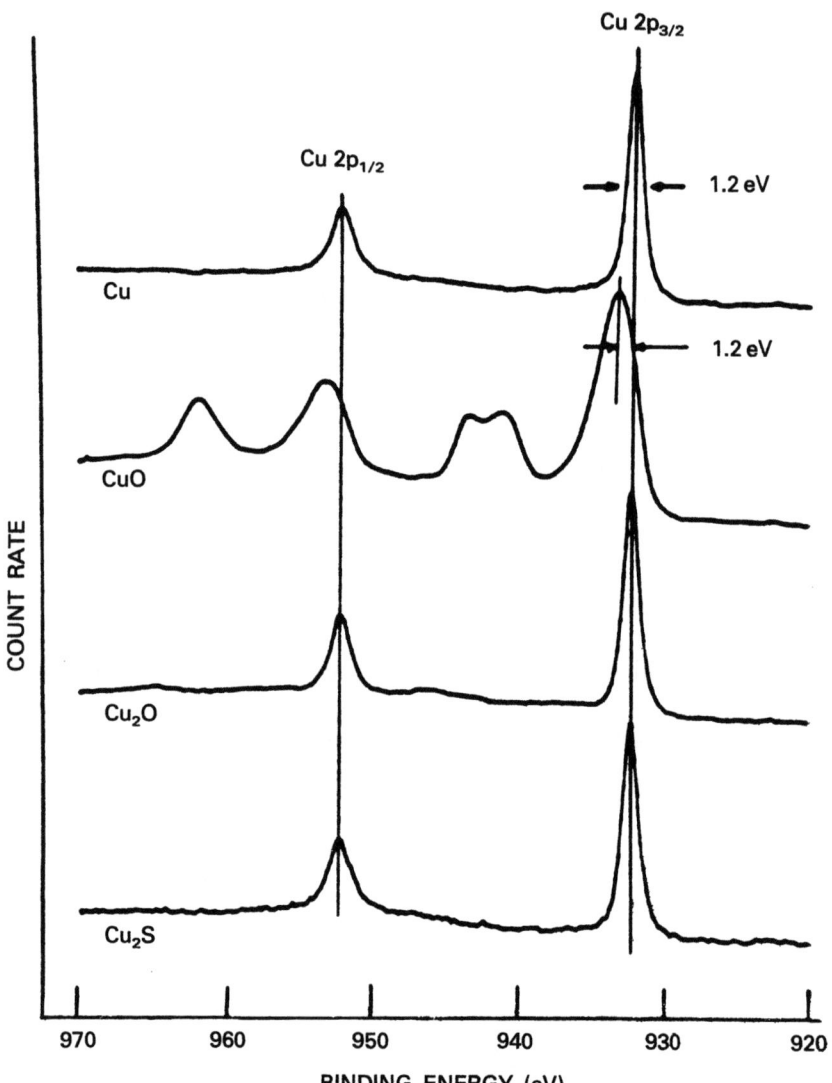

Figure 13. Cu 2p spectra from Cu, CuO, Cu_2O, and Cu_2S thin films excited by Mg $K\alpha$ radiation. Only CuO has a significant chemical shift, multiplet broadening, and shake-up satellites. (From Ref. 527.)

relax rapidly towards the created hole in order to screen the positive charge and the photoelectron acquires additional kinetic energy. Assuming that the relaxation is adiabatic and complete, an expression for binding energy, taking into account the polarization contribution, has been developed.[297,298]

In solids or condensed phases in general, not only the atomic but also the extra-atomic relaxation should be considered.[106,299-301] Extra-atomic relaxation in metals is manifested both as a contribution to the bonding energy and as a line-shape asymmetry[302,303] and it is observed to be large[304]: the magnitude, in eV, of the relaxation energy is approximatively 75–85% of the square root of the final energy of the ion.[305]

Core-level energies in metallic solids are always smaller than in free atoms. The difference can be described in terms of "screening" of the core-hole state charge by polarization of itinerant valence band electrons forming a "semilocalized exciton."[304] If we take the example of the $3d$ group, it is seen that binding energies are lowered most in metals for which d-wave screening yields large extra-atomic relaxation energies; binding energies drop dramatically from nickel to copper. Relaxation effects are also responsible for the discrepancies between XPS and AES values of chemical shift.[306]

Electron polarization during photoemission has a very important effect on multiplet splitting in small paramagnetic molecules: it seems that as electronic charge moves towards the core hole in bonding orbitals, the unpaired valence electron spin moves away, reducing the splitting.[307]

A very large number of compilations exists listing series of chemical shifts, measured for various compounds of almost every element. The study of Jørgensen and Berthou, which includes data for more than 600 nonmetallic compounds of 77 elements,[308] to which xenon has been added,[309] deserves mention.

(iii) Energy Shifts Induced by Chemisorption

All the values mentioned above have been measured from bulk compounds. One would expect that electronic states of the surface atoms would be severely affected by the surface discontinuity and therefore be quite different from their counterpart in the bulk solid.[310] Such a difference should be reflected in a binding energy shift in the XPS spectrum. As a matter of fact, the terms of the problem are not unequivocal and various controversial experimental results have been put forward. However, studies taking into account relaxation effects have allowed clarification of the situation and Shirley[311] has devised a series of useful rules and predictions about the XPS spectrum of a "metal-plus-adsorbate" system:

(1) The metal peaks should be essentially unshifted.

(2) Adsorbate core-level binding energies, referred to the vacuum level, should generally be lower than that of free molecules; for instance, a value of 537 eV has been found for the $1s$ line of atomic oxygen adsorbed on graphite[312] while the value of 546 eV is expected for nonadsorbed atomic oxygen.

(3) Adsorbate σ-orbital binding energies should also be lowered relative to gas-phase values, but the reduction should be less than in core levels.[313]

(4) Adsorbate π-orbital binding energies should be, in most cases, strongly affected by adsorption.[314]

(5) Molecular orientation should play an important part in the adsorbate spectrum; for example, a diatomic molecule standing up on a surface should show less reduction in core-level binding energy of the upper atom than of the lower atom, relative to the gas-phase value.

Clearly, exceptions could probably be found to these rules but numerous recent careful experiments seem to confirm these expectations.

7. Determination of Valence Band Density of States

The use of XPS for determining densities of states (DOS) has been recently reported for a large number of solids: metal, alloys, semiconductors, and insulators; a chapter of Hercules' exhaustive review on XPS has been concerned with such measurements[315] (see Fig. 14). From these various works it appears that XPS is very well suited for this kind of study since, at the X-ray energies used, the matrix-element modulation effects are very small and the density function of the final states is expected to

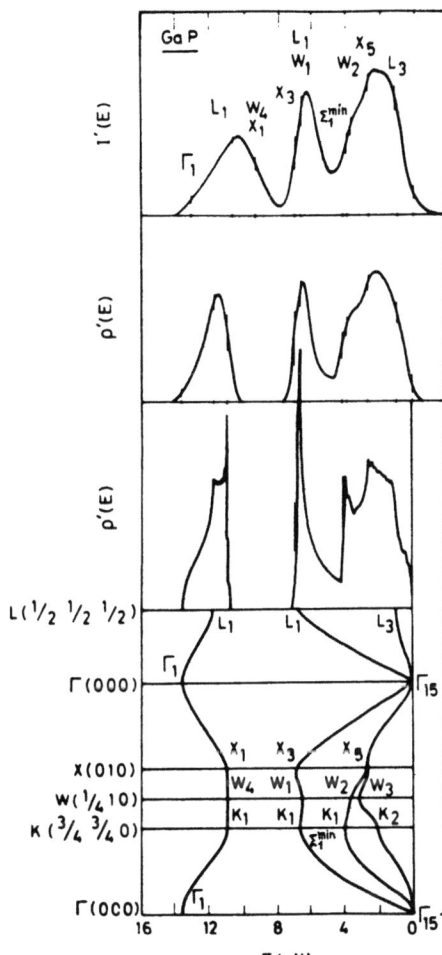

Figure 14. Band structure and density of states (E) of GaP, broadened density of states (E), and corrected XPS spectrum (after Ley et al.[301] and Chelikowsky et al.[319] (from Ref. 17).

become relatively flat and unstructured.[316] Furthermore, momentum conservation is easily obeyed and all transitions may be considered to be nondistorted. It does not follow, however, that XPS intensity will directly reflect the valence band DOS; cross-section modulation may emphasize some bands relative to others, as observed in diamond, silicon, and germanium,[317] or the probability of excitation from e_g states exceed that from t_{2g}, as in the case of copper, silver, and palladium.[318]

Two approaches have been used to interpret valence band spectra: a band picture using calculated energy bands (thus assuming delocalization of the valence electrons[301,319]) and a picture using crystal-field theory, assuming the occurrence of final hole-state splitting.[307,320] The general features of the measured DOS curves are comparable to theoretically calculated valence band DOS distributions.

8. Deconvolution Methods

Deconvolution studies appear to be less numerous in XPS than in AES. The reason for this is perhaps that photoelectron lines usually are narrower and better resolved than Auger peaks and that signal-to-noise ratio is more favorable in the former technique; nevertheless, deconvolution becomes essential whenever badly resolved lines appear.

Dynamic backgroud subtraction technique (Section II.7) can be applied to XPS and generally requires only a single integration. Convolution and deconvolution problems in XPS valence band spectra have been examined[321,322] and the applicability of Fourier transforms discussed. Resolution enhancement by transform techniques is suspect to many experimentalists; we must admit that results obtained in this way often contain spurious features and the solutions have been shown to be nonunique. However, criteria exist which can be used as valuable guides to data treatment; they have been critically examined in an excellent review by Wertheim.[323]

9. X-Ray-Excited AES

X-ray-induced Auger electrons can be observed in number of photoelectron spectra. In opposition to what is observed for electron-excited Auger transitions, core-level ionization cross sections are more important than valence band cross sections under photon irradiation (see Fig. 15).

Some requirements are imposed for the observation of Auger lines in XPS spectra.[324] First of all, the initial photoionization must be of high probability; for this reason, with elements

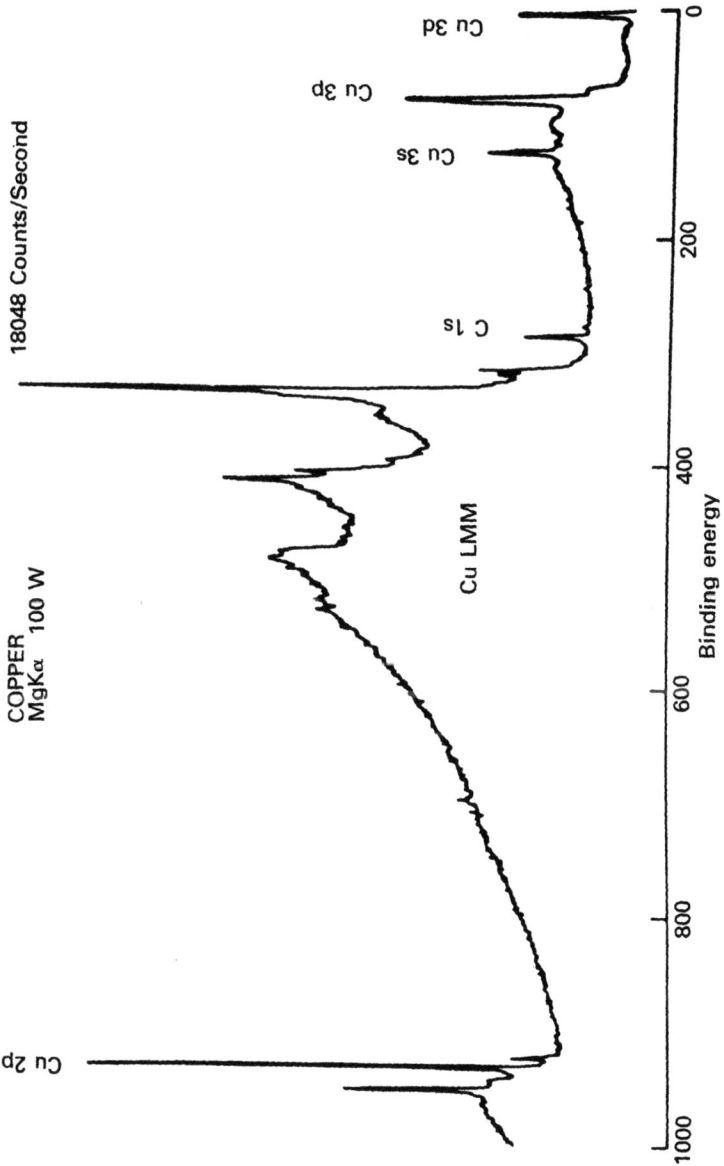

Figure 15. Full spectrum from copper showing Cu photoelectronic transitions and X-ray-excited LMM Auger transitions (from Ref. 527).

from lithium to magnesium, the Auger electron must result from initial ionization in the 1s shell; for elements from aluminum to selenium, the 2p shell would be more likely, etc. A convenient rule of thumb[325] is that the cross section for photoexcitation falls off rapidly with an approximate inverse cube dependence on the reduced energy. It should, however, be noted that the cross section is not only a direct function of the binding energy but also depends strongly on the angular momentum quantum number for photons in the energy range; thus the $3d_{5/2}$ line is more intense than the $3p_{1/2}$ line for all elements with $Z \geq 35$. It has also been shown that an Auger transition will not be easily observed if its kinetic energy is less than 100 eV.

X-ray excited Auger line intensities have been measured[326,327] and chemical shifts observed in a large number of compounds.[315,328–330] As a rule, Auger shifts are approximatively of the same order of magnitude and in the same direction as photoelectron line shifts but, because of large relaxation effects, many Auger peaks exhibit far larger chemical shifts than the photoelectron lines of the same element.[307] Such an effect is especially interesting for copper, zinc, silver, and cadmium analyses where the chemical shifts of the photoelectron lines are too small to be of use.

10. X-Ray-Induced Damage

In XPS and X–AES, X-ray interaction may be detrimental to the surface under investigation; surface reductions have been observed by several experimenters.[331–335] Specific investigations on the reduction of solid 3d transition metal compounds have been carried out by Wallbank et al.[336] who followed changes in the metal spectra of two cupric halides and potassium ferrocyanide. All the peaks showed on the low binding energy side of the parent 2p lines, additional lines whose relative intensities increased with elapsed irradiation time (Fig. 16). Similar examinations for K_2CrO_4 and CrO_3 have been performed and an important reduction of the photosensitive Cr(VI) cation observed[337]; a mechanism has also been proposed to explain the reduction process of Pt(IV) compounds during photoelectron measurements.[338]

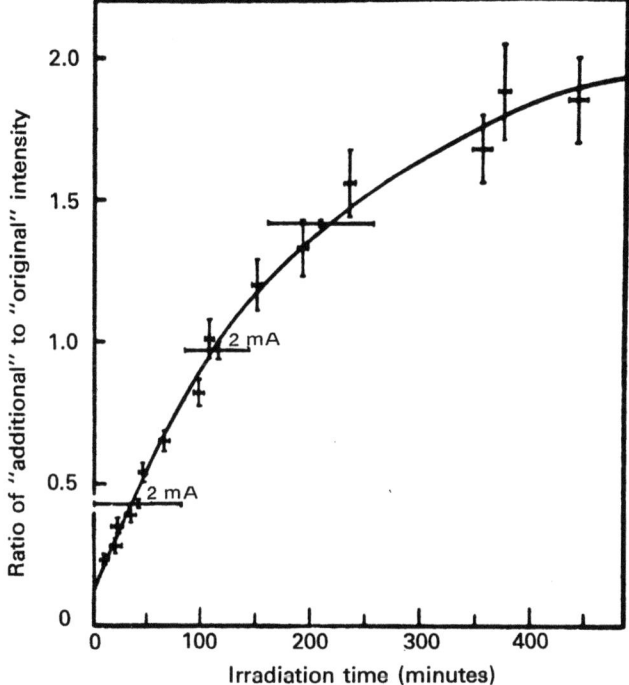

Figure 16. Reduction as a result of X-ray irradiation. Plot of the ratio of the additional intensity to "original" intensity against X-ray irradiation time for the Cu $2p_{3/2}$ line in cupric fluoride (from Ref. 336).

As various authors have pointed out, changes in the $3s$ and $3p$ regions of metals are likely to be much more insidious than those arising in the $2p$ region since the chemical shifts produced are then of the same order of magnitude as the multiplet and spin–orbit splitting; the additional peak may then be concealed in a broadening of the original peak. In conclusion, investigations in this region should always be accompanied by an examination of the core lines for possible reduction effects.

IV. AES AND XPS FOR SURFACE ANALYSIS

Although AES and XPS present many different features and for this reason are complementary, both are electron spectroscopic techniques and hence share the same general approach to various analytical problems. Among such problems, the question of the "third dimension" in chemical analysis is of particular importance: it involves the definition of the detected volume, hence the escape depth of the electrons, the distinction between surface and bulk contributions to the detected spectrum, the depth distribution of various elements, and the general topography of the surface, including the determination of the structure and the various reaction sites.

In all these determinations the position of the exciting source and that of the collector relative to the solid surface play an important part. Variations of either the angle of incidence or the collection angle and subsequent measurements of the new electronic spatial distribution may provide various types of information. Low-incidence angle or collection at grazing incidence enhance selectively the contribution from atoms situated nearer the surface; surface concentration profiles may be deduced from quantitative assessment of the relative importance of such contributions; quantitative analysis of the data obtained yields escape depth parameters; more qualitatively, surface roughness contours may be determined. Finally, by measurements on single-crystal specimens, fine structures and changes in peak intensities with angle may be related to surface structure and orbital symmetries.

1. Escape Depth

In electron spectroscopic measurements, the outermost layers of a solid contribute to the electron current in an exponentially decreasing manner,[339,340] for successively deeper layers[341] with a characteristic decay distance d.[5] Thus, the extent to which such measurements are specific to the surface region depends on the ejection depth d of the Auger electron or photoelectron. This escape depth, which is determined by the most probable inelastic electron–electron scattering, is a function of the kinetic energy of the escaping electron and, to a smaller extent, of the density of valence electrons[342]: for example, it seems that the trivalent metals, gadolinium and cerium, have larger scattering lengths than the divalent metals barium, strontium, and ytterbium.[343,344]

Several authors have compiled escape depth data as a function of the kinetic energy of the emitted electrons[345-348] (see Fig. 17). The curves initially show a sharp decrease with increasing electron energy up to 30 eV, then remain relatively flat from 30 to 500 eV, between which energies the mean escape depth d is estimated to be 5 ± 2 Å in densely packed inorganic solids; above 500 eV the escape depths increase rapidly with energy and are found to be proportional to $E_{kin}^{1/2}$; at 1000 eV the electron escape length is estimated to be between 15 and 20 Å. Since many

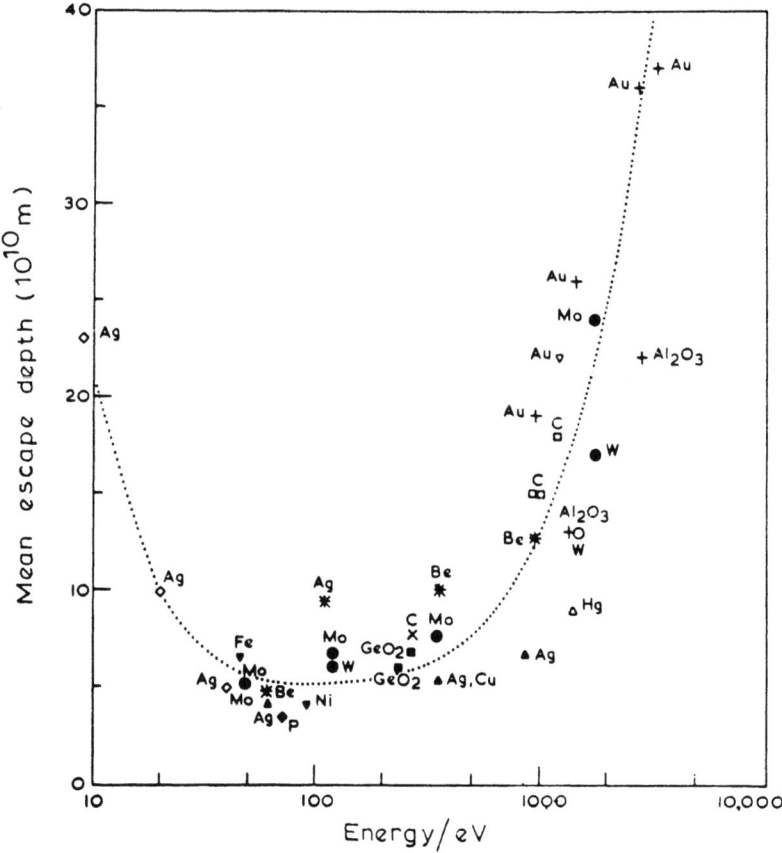

Figure 17. The variation of mean escape depth with kinetic energy of the escaping electron (from Ref. 346).

Auger transitions are in the energy region 50–500 eV, it is admissible, in the first approximation, to neglect the variation of the escape depth with energy when comparing amplitudes of the Auger peaks between those energies.[346,349]

In practice, escape depths are usually determined by depositing on the material of interest a uniform overlayer of another material[350–353] and monitoring the intensities of the substrate Auger or photoelectrons as a function of the thickness of the condensate. The now well-known method of Todd and Heckingbottom,[354] based on a comparison between the intensities of the photoelectron peaks and associated Auger peaks from the same atom and avoiding thus the problem of thickness measurements, is also worthy of mention. Quantitative analysis of electron angular distribution may yield electron attenuation lengths[355]; the increase of photoelectron yield with low X-ray incidence angle[356] and the variation of photoelectron current intensity with emission angle[357] were used to derive escape depth data. Information on electron attenuation lengths was also obtained by measuring the variation of surface-to-bulk photoemission ratio as a function of photon energies.[358]

In Auger work the most convenient unit to use for the electron mean free path, d, is the monolayer which is assumed to contain $\sim 1 \times 10^{15}$ atoms cm^{-2} and to be ~ 2.5 Å thick.[49] In the first approximation, for Auger electrons whose kinetic energy is between 100 and 1500 eV, the mean free path, in monolayer units, is given by [359]

$$d = 0.2(E_{kin})^{1/2} \qquad (6)$$

with E_{kin} expressed in eV. As most spectrometers collect electrons leaving the surface at various angles, the value $1.33d$ is thought to provide a fair approximation to the escape depth averaged over these angles.[49,360]

2. Distinction between Surface and Bulk Contributions

From the range of escape depths quoted above, it is obvious that the average depth of emission may be as high as ten atomic layers or more. Therefore two major aims of surface analytical techniques are, firstly, to enhance selectively the near-surface contribution to the spectrum and, secondly, to differentiate between surface and bulk spectral contributions; a further step in this direction is to establish a profile of elemental concentrations with depth. For the condition of low, or grazing, electron exit angle, significant increases are observed in the relative intensities of near-surface atoms[361]; hence decreasing the average angle between the electron escape direction and the surface provides a mean of enhancing the surface layer contribution by at least one order of magnitude. In XPS such a procedure was first introduced by Fadley and Bergström[362]; collection at low takeoff angles was subsequently developed by several authors who presented models for thick substrates with thin uniform surface layers[355,363,364]; in such a technique X-ray reflection and refraction[365] and the effect of a large solid angle[366] are important; the presence of nonuniform overlayers, due to islands ("patches") or interdiffusion of adsorbate and substrate should be carefully considered as they induce severe distortions in the angular dependence model.[357]

In AES, the same technique was introduced by Harris.[367] Rusch et al.[368] constructed a simple mathematical model to predict the angular dependence of the Auger signal intensity from a contaminant localized in the outermost atomic layer.

Excitation at grazing X-ray incidence provides another method for increasing significantly the surface-atom relative intensities in XPS spectra and for distinguishing between surface and bulk contributions.[365,369,370] In fact, at such angles, the concomitant effects of X-ray reflection and refraction on the surface lead, at the same time, to an increase of the total photoelectron intensity[356,371] and a reduction of the average X-ray penetration depth. Another way of distinguishing between surface and subsurface atoms used in AES is to study the effect, on Auger yield, of variation in the primary-electron energy.[372]

3. Profile

When adsorption is not confined to the first surface layer but is distributed through the surface region, a new variable, distribution depth, must be determined to achieve complete analytical information.[373] Such information can be established by nondestructive methods, as generally used in the presence of thin films; if one is dealing with thicker surface phases, destructive techniques such as ion sputtering have to be employed.

(i) Nondestructive Techniques

A method for obtaining relative element depth profiles is to utilize the variation in escape depths of different core-level electrons.[374,375] As electrons of different energies have different inelastic mean free paths in a given material,[376] the Auger electron spectrum of an element exhibiting more than one peak contains, *per se*, information about that element distribution with depth. Seah[377] proposed an algebraic expression which allows the film thickness and the contribution from the bulk to be calculated. This method involves measurements of the relative intensities of the low- and high-energy Auger peaks for the adsorbate and comparison with the standard spectrum of the pure bulk element. For adsorbates exhibiting only one set of Auger peaks, the same author proposed two other techniques useful for distinguishing between adsorbed monolayers and three-dimensional nuclei: the first method is based on observations of the plasmon losses of the adsorbate Auger electrons; in the second one, observations are made of the inelastic scattering features in the $N(E)$ Auger spectrum.

More quantitatively, calculations of intensities of Auger plasmon satellites as a function of depth of the emitter in metal substrate and as a function of the Auger line energy give the possibility of determining the adatom concentration versus depth distribution[378–380]; if data on escape depth are available, variation of signal intensities with angle provides quantitative information on depth distribution.[381,382] Varying the energy of the primary electrons and measuring at glancing incidence the decrease in the ratio of the Auger peaks at low and high primary energies[383] provides another nondestructive way of obtaining in-depth information. It is also possible, as has been done for platinum/tin alloys,[384] to utilize the difference in escape depths of an XPS signal and an Auger signal; then, if core-level ionization cross sections are available, absolute concentration versus depth measurement becomes feasible.[385]

(ii) Destructive Techniques

The most widespread method for obtaining depth profiles is the combination of noble-gas ion bombardment with AES or XPS techniques. With this method, successive atomic layers of the material under study are removed by sputtering with noble-gas ions and the chemical composition of the freshly exposed surface is directly determined from its Auger or photoelectron spectrum.[386,387] The final result expected from depth profile measurements is the concentration of an element as a function of its distance from the surface.[388] Actually, the experiments yield a relation between electron current intensity and sputtering time; if concentrations are proportional to the detected electron current and if the erosion rate remains constant with time, the desired concentration profile is obtained.[389]

Chemical depth profiles are commonly presented by plotting the measured electron peak heights for several elements as functions of the sputtering time.[390] (see Fig. 18). In AES the use of the deflection peak-to-peak as a measure of the detected Auger current is accurate as long as the Auger peak shapes remain unchanged during profiling, but situations do arise where Auger peak changes occur due to changes in bonding (for example, in going from an oxidized overlayer to the substrate[391]) or to different energy-loss mechanisms during the escape of Auger electrons.[392] Clearly, comparisons of measured Auger signals with different shapes produce quite erroneous results. These sorts of problems can virtually be eliminated by using integration techniques: suitable integrals can be obtained directly from the phase-sensitive detector by using the tailored modulation technique (TMT)[131] which was recently applied to correct automatically the effects of Auger line changes on depth profiles.[392] When the sputter-profiling technique is coupled with XPS measurements, another problem arises from the chemical shift resulting from changes in chemical environment of the element; recently, however, a method has been published which allows the true depth

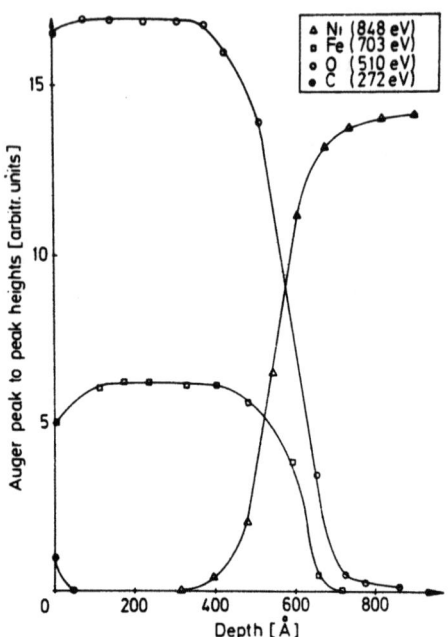

Figure 18. Depth profile of a sputter-deposited iron oxide layer on a nickel substrate (1 keV Ar^+ ion sputtering) (from Ref. 388).

distribution of an element to be calculated from raw data of XPS signal intensity as a function of sputtered thickness.[393]

Uniform erosion rate is the prerequisite for any reliable depth analysis by sputtering. Such a constant rate depends on the properties of the sample itself and on instrumental conditions; above all, no concentration gradient parallel to the sample surface must exist; the specimen must also be structurally homogenous[394] and free from subsurface defects[395]; because the sputtering yield is a function of the angle of incidence,[396,397] a rough surface will lead to different local erosion rates; besides, a constant erosion rate is possible only if all the constituents present in the sample have identical sputtering yields. When sputtering yields are different (as occurs in many alloys), the surface under bombardment becomes progressively depleted in the component with the highest sputtering yield (preferred or selective sputtering).[398,399] until a steady state is eventually reached[400–402] in which the product of sputter yield and surface coverage is constant for each component.[403–405] The number of layers that should be removed to establish steady-state conditions can be calculated.[406]

Changes in relative concentrations as a result of preferential sputtering affect a thin surface layer only. The altered layer thickness is an important parameter to measure; under ideal conditions it is estimated to be about a monolayer,[407] but measured values may range from some tens of angstroms[408] up to 7000 Å.[409] A way to minimize depletion effects is to use Auger or photoelectron lines which have kinetic energies sufficiently high that several atomic layer thicknesses[410] can be sampled.

For every analysis the sputter yields themselves have to be measured. If a steady state exists, sputtering yield ratios are relatively easy to deduce[38] from calculations of the fractional coverage as a function of time.[407,411] However, difficulties may arise from the fact that the sputter yields of the components in the same matrix are generally different from the sputter yields of the individual components in the pure materials. As a consequence, measurements have to be made for every individual material.[412] In such yield measurements one must take care that the value for the altered layer, which is observed to increase with sputter-ion energy,[413] is at least comparable to the sampling depth of the AES or XPS measurements in order to assure that the spectroscopic data provide a true measure of the sputtering effects.[414]

When the sputtering ratio is determined and the ion-current density is known, the eroded-layer thickness can be calculated. However, more reliable determinations can be made by comparison with standards.[388] Calibration of the sputter device can be made by sputtering away a layer of known thickness (calibrated metallic foils as supplied by many manufacturers are extremely convenient for this purpose) or by mechanical or interferometric determination of the step height at the edge of the erosion crater after prolonged sputtering.[415,416]

During spectroscopic measurements, one must take care that the area examined be smaller than the sputtered area in order to avoid edge effects which may lead to a large background signal. Such a condition is easily realized in AES where the primary electron beam diameter is less than 0.1 mm[417]; it is less easy to attain with XPS measurements because of the difficulties encountered in focusing the X-ray beam.

From the instrumental point of view, the appearance of commercially available systems permitting simultaneous removal of the outermost surface layers and the analysis of the freshly exposed surface was decisive for the development of the sputter-profiling technique. In the pioneer Auger work, profiling of elemental concentration with depth was achieved by performing consecutive argon-ion bombardment and Auger analysis in a stepwise fashion, but to such experiments two major objections may be raised: firstly, the difficulty of returning to the same area of analysis (following movement of the sample away from the analyzer for bombardment) and, secondly, the time gap between bombardment and analysis[346] which might result in contamination from the residual gas or surface segregation from the bulk. The objections were overcome with the use of a technique developed by Palmberg[418] where simultaneous ion bombardment and AES analysis are performed.

When using such a technique, it is essential that the ion gun be equipped with electrostatic shielding. The use of unshielded ion guns may cause large errors in beam-current measurements made without a Faraday cup[49]; decreases in Auger signal size as large as 50% may be observed under these conditions, the decrease being larger the smaller the Auger energy. It should also be noted that further errors may arise due to argon-ion-excited Auger transitions.[420]

As every technique does, sputter etching presents some disadvantages: it causes considerable damage in the surface region of the specimen. This damage consists in the creation of lattice defects, the implantation of ions, and the modification of surface composition. In multiple-component samples, selective sputtering may drastically alter surface topography and lead to the formation of cones,[421] rods, and hillocks[422]; the formation of a rough surface then leads to variations in erosion rate. Implantation of foreign ions, usually generated from the source, is often observed; such an inconvenience can be avoided by using a mass-separator device located between the ion gun and the sample.[423] Every surface, however, is contaminated in time by the residual gas present in the specimen chamber; the upper limit for the residual gas pressure can be calculated from the inequality

$$10^{-2} J_p > p \qquad (7)$$

where J_p is the primary ion-current density in $A\,cm^{-2}$ and p the gas pressure in Torr, the sputtering and sticking coefficients assumed to be unity.[424]

It should be emphasized that many time-consuming measurements and determinations in ion-sputtering work could be avoided if comparisons of chemical depth profiles reported in the literature were not made so difficult for want of a standardized presentation: as suggested by Windawi,[425] is is desirable in publications that in addition to data such as thickness, angle of incidence, ion species, and ion-beam energy, a new parameter, the ion dose, should be presented; its unit might be called the Wehner, defined as

$$1\text{ Wehner} = 1\ \mu A\,cm^{-2}\,min$$

which corresponds to an ionic beam bombardment of $1\,cm^2$ area with $1\,\mu A$ for 1 min duration. For an argon beam of 2 keV energy at normal incidence, a Wehner corresponds to sputtering depths of the order of magnitude of an atomic layer from the surface.

4. Surface Roughness

Most of the models used in electron spectroscopy to calculate electron current densities, escape depths or probe penetration have been devised by assuming atomically flat surfaces for the specimens under consideration. This assumption is no longer valid for surfaces such as those developed by machining, fracture, or ion sputtering during profile analyses.

Surface roughness affects the electron emission process mainly in two ways: Firstly, either by shading part of the surface from the exciting radiation or shielding the escaping electrons from detection and, secondly, by modifying both the incidence and the electron-escape angles with respect to a smooth surface.[426,427]

The effects of surface roughness on photoelectron angular

distribution were reviewed by Fadley et al.[428–430] Models for arbitrary and periodical contours were developed[428,431] and equations predicting photoelectron intensities were modified to take into account this new parameter.[432–434] It should be noted that significant shading occurs only if the characteristic dimensions of the roughness are of the order of, or greater than the electron mean free path.[428] Practically it should be possible to determine the average dimensional ratios characterizing the roughness, such as height-to-width, from a detailed study of angular distributions from certain types of rough surfaces.[428]

The influence of surface roughness in AES has been investigated both mathematically and experimentally by Holloway,[435] who proposed a model to calculate the intensities of Auger signals arising from a rough surface; it was assumed that the surface roughness was two-dimensionally isotropic and a mean surface level[436] was defined.

It has been shown that the average depth of emission from a rough surface is inferior to that originating in a flat surface. As surface enhancement is influenced by directional asymmetry in roughness distribution, it has been suggested that unidirectional polishing followed by angular distribution measurement parallel to the polishing direction might constitute a method for increasing the surface-atom contribution to the peak at low angles.[429]

5. Structure

Many surface properties are sensitive to surface structure; it is then essential to include it in surface characterization. Low-energy-electron diffraction (LEED) is the most widely used technique[437] for determination of the structure of the surface region and is often used in conjunction with AES.[438] LEED, however, has serious limitations: it is only sensitive to adsorbates having translational symmetry parallel to the surface and such a technique usually requires adsorbate overlayers with long-range order. Several investigations have shown that the Auger electron emission from single crystals is quite anisotropic,[439,440] and it has been suggested that the angular dependence of Auger emissions from adsorbates on single-crystal surfaces could serve as a useful means for identifying adsorption site positions.[441] Hence, an alternative approach to surface structure determination is to observe Auger emission angular distributions which originate from the competitive contributions of the diffraction effects[442,443] and the inherent angular dependence.[444,445] The predominance of one or the other effects depends on the electron energy: for electrons in the region 50–300 eV, the elastic scattering effect would be dominant[446,447]; the location of ordered adsorbates may be determined by analyzing the angular profiles of diffracted electrons (see Fig. 19).[448–450] In the very low energy region (less than or equal to 20 eV, hence accessible to UPS), where crystallographic effects generally appear to be negligibly small and the inherent angular dependence may be strong,[451] Auger emissions

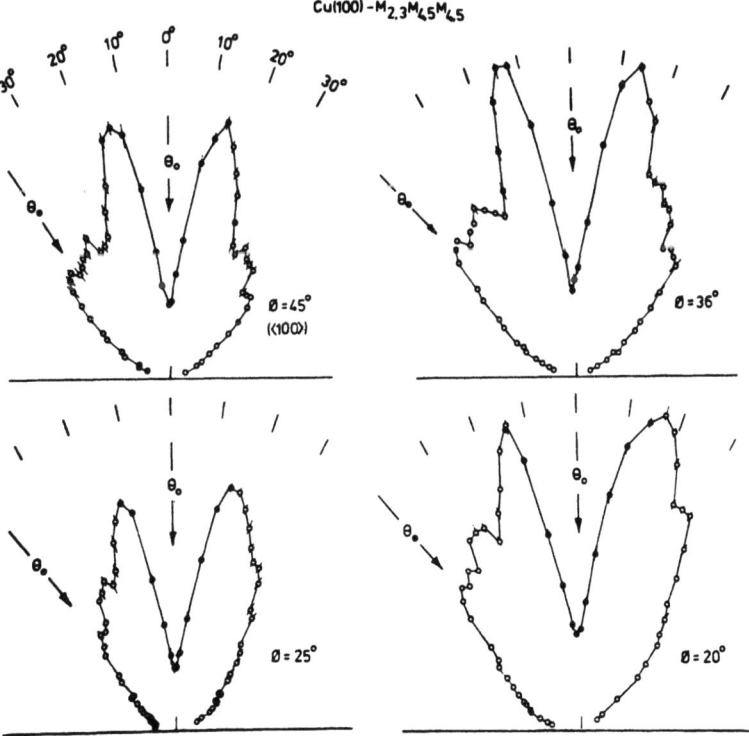

Figure 19. Experimental Auger electron emission profiles for several azimuths from a Cu(100) surface (from Ref. 448).

from the valence states of adsorbate species reflect the bond geometry of surface atoms.[452]

Since the various crystallographic planes not only differ in the number and geometrical arrangement of the surface atoms but also in the number and spatial orientation of the orbitals emerging from one surface atom,[453] different behavior is to be expected from different surface planes. For example, sulfur adsorption on gold takes place much less readily on faces of low density.[454] An explanation has been proposed[455] that considers only zone-averaged matrix elements for tight-binding d orbitals; it was predicted that only e_g character can be active along (100) and only t_{2g} character along (111) planes. The partial density of states for e_g and t_{2g} were in good agreement with gold valence photoelectron spectra obtained at various angles along the (100) and (111) directions.[456] In consequence, the dependence of signal intensity on the incidence direction of the exciting beam relative to the crystal axes may prove to be a useful source of additional data for determining surface structures. Another possibility of extracting structural information concerning the atom distribution near the surface or the position of the unit cell of various species detected by AES has been suggested.[457] The method is based on a correlation established between the variation of the backscattered electron signal intensity and of the Auger signal intensity as a function of the direction of the incident electron beam.[458,459]

V. SURFACE ANALYSIS APPLIED TO ELECTROCATALYSTS

Owing to the fact that XPS and AES analyses are normally performed outside an electrochemical cell, under high vacuum conditions, these techniques are, in a general way, not the best suited for the detection of adsorbed intermediates produced in the electrode reactions. As a matter of fact, only a very few studies of this kind (which will be discussed below) have been reported. The situation is more favorable in the case of reactions involving a stable surface phase as the electrode reactant or product, where reliable results concerning its chemical nature as well as amount of the surface species may be obtained by both

XPS and AES. This is illustrated, for example, by several XPS investigations on thin oxide (hydroxide) layers formed on noble metals under anodic polarization. Although the behavior of these layers, which influence in a large measure the electrocatalytic properties of the corresponding metals, has been extensively studied both by electrochemical and optical methods,[460-462] their exact nature is still far from being completely understood.

1. Oxide Layers on Noble Metals

(i) Platinum

The principal applications have been devoted to platinum, gold, and palladium. Kim et al.[463] have invesgitated, using XPS, the surface region of a platinum electrode oxidized anodically in 1 M aq. $HClO_4$ at 0.7, 1.2, and 2.2 V vs. SCE. Comparing the spectra obtained for the Pt electrodes with those of PtO and $PtO_2 \cdot H_2O$ prepared chemically, these authors have identified Pt metal and PtO (phase oxide)* on the electrode polarized at 1.2 V, and Pt metal, PtO, and PtO_2 following oxidation at 2.2 V. Since the peak corresponding to the metal is still distinctly observable after treatment (3 min) at 2.2 V, the thickness of the oxide layer (PtO and PtO_2) present on the platinum electrode at this potential is apparently less than the mean escape depth of photoelectrons estimated as about 20 Å. Kim et al. also mentioned the existence of a fourth Pt species (besides Pt metal, PtO, and PtO_2), appearing as a shoulder to the higher-binding-energy side of the main 4f peaks of the Pt metal. They assigned this shoulder (which has even been observed for Pt vapor-deposited within the sample chamber of the spectrometer at 10^{-5} Torr) to platinum atoms with chemisorbed oxygen. More recently, however, in contradiction to this suggestion, the asymmetric shape of the Pt metal photoelectron signals has been ascribed[464] to coupling of the core hole with the conduction electrons. This seems to be also supported by the observation of Dickinson et al.,[465] who have found that, for all the samples examined in their XPS study of Pt electrode, the additional-platinum-"satellite" peak represented a

---

* A trace amount of PtO was already observed after polarization at 0.7 V.

constant fraction of the area of the metal peak. A typical photoelectron spectrum of the Pt 4f region, obtained by these authors for the platinum electrode before anodic oxidation, is shown in Fig. 20a. This electrode was first cleaned in hot concentrated sulfuric acid, washed with purified water and then polarized in 0.5 M aq. H_2SO_4 at 0.15 V SCE, i.e., in the potential region where no Pt surface oxidation takes place. The asymmetric Pt 4f signals, $4f(\frac{7}{2})$ and $4f(\frac{5}{2})$, apparent in Fig. 20a, are deconvoluted into two peaks, a main peak and a smaller Pt "satellite" peak. Dickinson et al. have investigated systematically[465] the surface composition of a large number of Pt electrodes polarized in 0.5 M aq. H_2SO_4 solution over a wide range of potentials, from

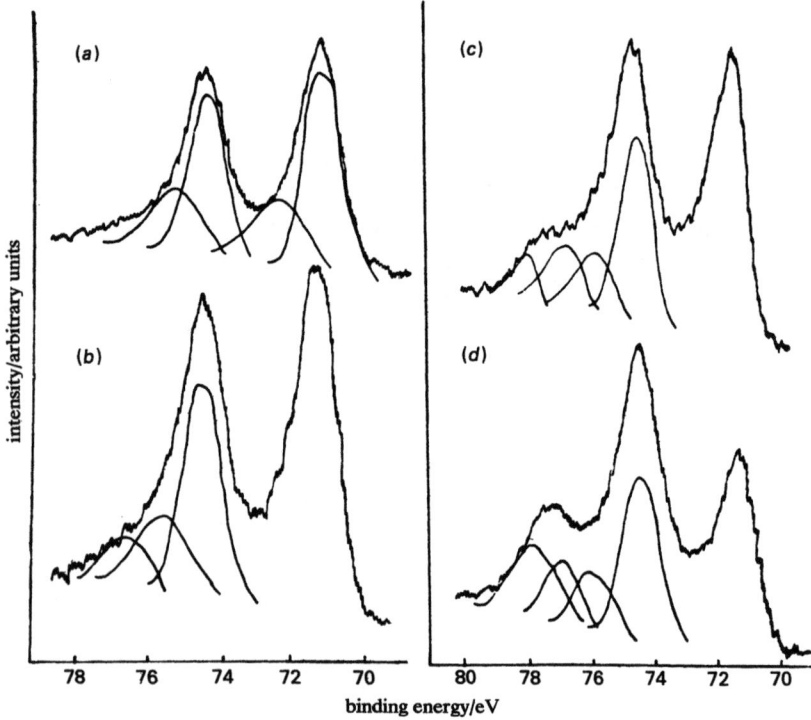

Figure 20. The platinum 4f electron peaks for the platinum electrode after polarization for 15 min: (a) at +0.15 V, (b) +1.00 V, (c) +1.50 V, and (d) +2.00 V in 0.5 mol dm^{-3} sulfuric acid. The deconvolutions are shown only for the $4f_{5/2}$ electron peaks in (b), (c), and (d) for the sake of clarity (from Ref. 465).

−0.3 to +2.6 V SCE. The formation of PtO became detectable on the Pt $4f$ electron spectrum (Fig. 20b) for the electrode pretreated at 0.8 V, although an additional O $1s$ electron peak due to oxide has already been observed for the Pt sample prepolarized at 0.7 V. These results are in reasonable agreement with those of Kim et al.,[463] who also detected traces of PtO on an electrode polarized at 0.7 V in 1 M aq. $HClO_4$. On the other hand, the electrochemical studies of Kozlowska, Conway, and Sharp using the potentiodynamic method[462,466] show that the platinum electrode in acid solution becomes progressively covered with OH species from 0.8 V SHE, the limit of monolayer coverage* (corresponding to 1 e per Pt atom) being apparently attained at 1.1 V. However, according to the results of ellipsometric measurements,[461,467] a true phase oxide should be present on the Pt surface only at potentials above about 1.0 V versus the hydrogen electrode in the same solution.† The latter value is consistent with the smaller anodic potential (i.e., 0.97 V vs. HE) for which the existence of platinum oxide was observed by photoelectron spectroscopy. This suggests that the presence at the electrode surface of adsorbed oxygen species (below 1.0 V) has no effect upon the Pt $4f$ electron spectrum and that it is only the oxide formation‡ that modifies sufficiently the chemical environment of Pt atoms to cause the observed chemical shift.

When evaluating the sensitivity of the XPS technique for such a case as the detection of the oxygen surface species at platinum, it must be taken into account that some reduction may occur in the sample chamber during recording of the spectra. The latter possibility was examined by Dickinson et al.[465] by comparing the quantities of electricity needed to reduce the oxide film immediately after anodic polarization and after the sample was used for recording the spectra. The results of these measurements,

---

* The monolayer is composed of some adatoms of O or OH together with two-dimensional phase oxide (see ‡ below).
† The hydrogen reference electrode in the same solution will be referred to as HE.
‡ According to widely accepted views,[461,466,468] at potentials above 1.0 V, Pt atoms leave their regular position in the metal lattice to form with O atoms some kind of two-dimensional nuclei of the oxide. This is based on the evidence, provided by ellipsometry,[467] and cyclic voltammetry at low temperature[466] for the appearance of a new phase at the Pt surface, above 1.0 V.

Table 1
Extent of Reduction of Platinum Oxide during Recording of XPS Spectra (from Ref.[465])

Potential of polarization(V)	Quantity of electricity needed to reduce oxide film($\mu C\,cm^{-2}$)	
	(1)[a]	(2)[b]
+0.90	46	0
+1.00	330	160
+1.30	810	280
+1.50	900	880
+2.00	1150	890

[a] Sample reduced immediately after polarization.
[b] Sample used for XPS spectra and then reduced.

shown in Table 1, indicate an appreciable extent of reduction of platinum oxide for the second group of samples.

This effect has been attributed chiefly to the X rays since much more reduction took place when the X-ray power was increased. In practice, the reduction occurring on the sample surface can be minimized by using moderate power of irradiation and short time of exposure to X rays. In spite of the partial reduction taking place during the recording of the spectra, Dickinson et al. were able to detect, for the Pt electrodes polarized at higher anodic potentials, the presence of the second platinum oxide—PtO_2. The binding energy corresponding to this additional Pt 4f peak (Fig. 20c, 20d) coincides, indeed, with the values measured for $PtO_2 \cdot H_2O$ powder and for PtO_2 chemically formed on Pt metal, as well as with the previously reported value[463] for PtO_2 produced electrochemically. The least anodic potential for which Dickinson et al. have observed PtO_2 was 1.3 V SCE, and at 1.75 V it became a predominant species within the oxide layer. However, even for the sample prepolarized at 2.6 V, the Pt 4f electron signal of PtO was still visible as well as that due to Pt metal. This implies that the thickness of the oxide layer formed on Pt under the experimental conditions used by Dickinson et al. did not exceed 15 Å, as may be verified from electrochemical charge measurements.[469]

In the connection it should be noted that no direct evidence existed, from previous electrochemical and optical studies, for the formation of PtO_2 at potentials above 1.3 V. The results of coulometric measurements[469] indicate that a limiting coverage, corresponding to approximately two oxygen atoms (four electrons) per Pt surface atom, is reached in 1 M H_2SO_4 at about 2.3 V HE. This limiting coverage, observed for rather short oxidation times, has been interpreted in terms of a chemisorbed oxygen film[469] or two layers of PtO.[470,471] According to Biegler et al.,[469] during longer oxidation at potentials above 2.3 V the chemisorbed oxygen film is converted into a phase oxide having different electrochemical characteristics. This suggestion appears to be consistent with the uniformity of optical properties of the Pt surface up to potentials of 2.4 V as reported by Parsons and Visscher.[470]

A more complex picture, including three distinct stages of the surface oxide growth on Pt, results from the ellipsometric study of Vinnikov et al.[472] The first stage, which starts at 0.8 V HE, is completed at about 1.5 V with the formation of an approximately 5-Å-thick film equivalent to monolayer of PtO ($2e$ per Pt atom in the surface). Oxidation at higher anodic potentials would produce on the Pt surface two other types of oxide:* the first of them found above 1.5 V attains, according to the ellipsometric data, a limiting thickness of about 13 Å; the second, formed at still higher potentials, seems to grow up to 60 Å. The change in parameters of the elliptically polarized light, observed by Vinnikov et al. above 1.5 V and corresponding to the second stage of oxidation, may be connected with the formation, at the Pt surface, of PtO_2 which, according to the XPS data, takes place above 1.5 V HE. The thickness of the oxide layer estimated from the photoelectron spectra of the Pt electrode is close to the value indicated by ellipsometry for the second stage of oxidation; however, the further growth of the oxide up to 60 Å, occurring during long time oxidation,[472] has not been observed under conditions of the XPS measurements.

If the principal conclusions of Dickinson et al. appear in a good agreement with the earlier results of Kim et al.,[463] they are

---

* This is also consistent with the oxidation scheme proposed earlier by Shibata and Sumino.[473]

in serious disagreement with the conclusions of another XPS investigation of the Pt electrode reported by Allen et al.[471] These authors have identified PtO_2 after prolonged oxidation at 4 V, but were unable to detect this species on the electrode prepolarized in 0.5 M aq. H_2SO_4 at 2.4 V HE. A possible reason for this discrepancy may lie in different experimental conditions existing during recording of the spectra, such as the intensity of X-ray flux, the pressure and temperature in the sample chamber and, the duration of recording, all of which are expected to affect the extent of PtO_2 reduction or (and) decomposition.

In spite of the controversies mentioned above, the XPS measurements provide a certain amount of new information on the surface constitution of the oxidized Pt electrode in acid solutions, as well as a confirmation of some earlier results obtained by electrochemical and optical methods:

(1) The formation of phase oxide at potentials above 1.0 V HE, suggested on the basis of the ellipsometric studies,[461,467] has been confirmed by photoelectron spectroscopy. This two-dimensional oxide seems to be chemically identical* (the same binding energies for the Pt 4f and O 1s electrons, respectively) with bulk oxide PtO.

(2) The change in the optical properties of the Pt surface above 1.5 V, believed[472] to be connected with the transformation of chemisorbed oxygen in the surface into a phase oxide, seems to correspond, in reality, according to the XPS data, to the formation of PtO_2. Also, in this case, the "surface PtO_2" appears to be chemically similar to the bulk compound $PtO_2 \cdot H_2O$.

(3) The chemisorbed OH species, known[460,461,466] to be present at the Pt surface below 1.0 V (from 0.8 V onward), could not be distinguished, on the basis of photoelectron spectra, from the adsorbed oxygen species found on the Pt samples in the absence of anodic oxidation.

(ii) Gold

Using the same experimental technique as they did in the case of the Pt electrode, Dickinson et al.[465] have also studied the

---

* The fact that Allen et al.[471] have determined some different binding energy for the Pt 4f electron, which they tentatively ascribed to $Pt(OH)_2$, is clearly connected with the method of deconvolution of the spectra.

changes in the surface composition of the Au electrode polarized in 0.5 M H_2SO_4 through a series of potentials ranging from -0.6 V to $+2.0$ V vs. SCE. In a general way, the picture of the gold surface, arising from XPS measurements, is significantly less complex than that presented by the platinum surface. The spectra of the Au electrodes polarized at low potentials—up to 1.0 V— showed in the Au $4f$ region (Fig. 21) two symmetrical peaks characteristic of gold metal (with the binding energy of the

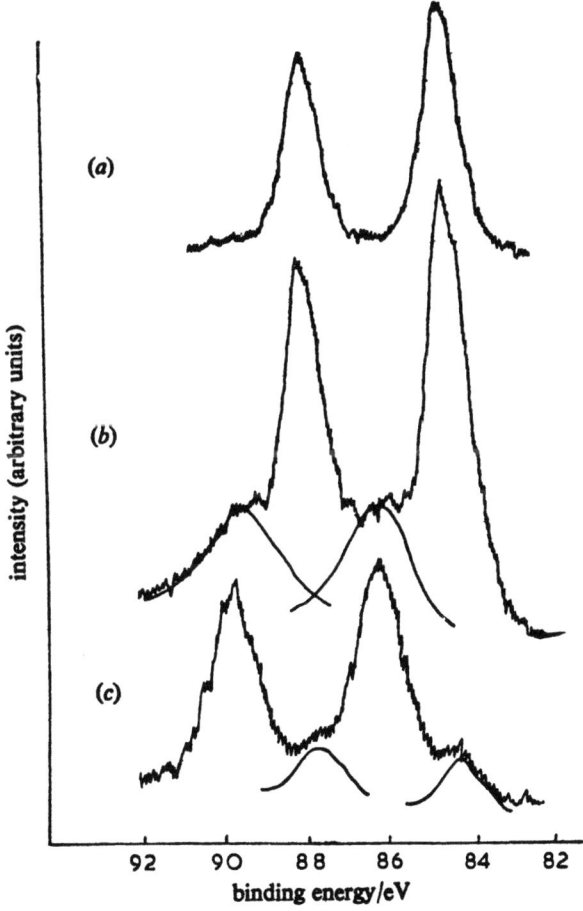

Figure 21. The gold $4f$ electron peaks of the gold electrode after 15 min polarization at a potential of (a) $+0.20$ V, (b) $+1.50$ V, and (c) $+1.70$ V in 0.5 mol dm^{-3} sulfuric acid (from Ref. 465).

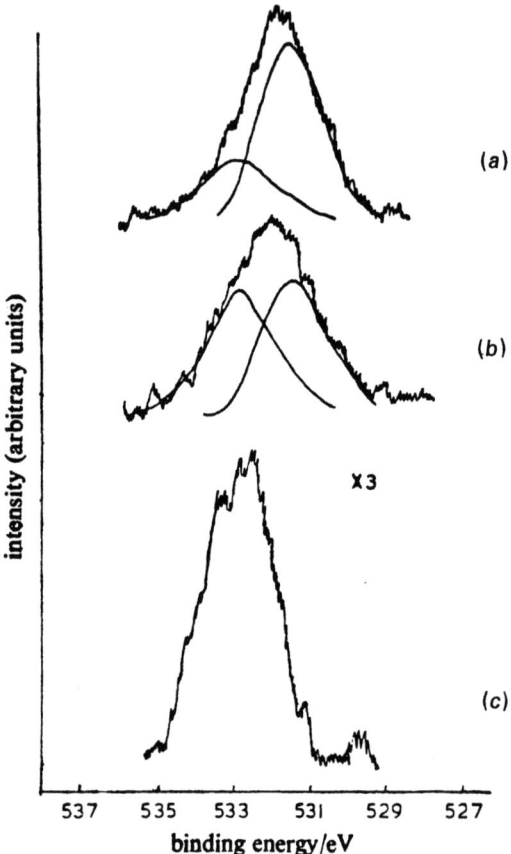

Figure 22. The oxygen 1s electron spectrum for the gold electrode after polarization for 15 min at a potential of (a) +1.70 V, (b) +1.50 V, and (c) −0.60 V in 0.5 mol dm^{-3} sulfuric acid (from Ref. 465).

Au $4f(\frac{7}{2})$ electron equal to 84.0 eV) and in the O 1s region (Fig. 22) a single peak normally assigned (see, for example, Ref. 474) to adsorbed oxygen and (or) water.*

The presence of gold oxide was detected on the O 1s electron spectrum with the appearance of an additional, lower bind-

---

* This peak being also observed for the "clean" gold, and in a general way— "clean" metal surfaces—it yields no specific information about metal–solution interactions.

ing energy, peak* for the electrode oxidized at 1.1 V and subsequently—on the Au $4f$ electron spectrum as a higher binding energy shoulder to the metal peaks—for the electrode polarized at 1.2 V. Comparison of the Au $4f$ binding energies (corresponding to the shoulders from Fig. 21) with the values characteristic of bulk Au_2O_3 and $Au(OH)_3$, indicates that the surface film on the gold electrode is chemically identical to Au_2O_3. The XPS data support, therefore, the view, based on the results of ellipsometric[475,476] and electrochemical[477,478] studies, according to which anodic oxidation of gold in 0.5 M H_2SO_4, above 1.35 V HE,† produces a phase oxide rather than chemisorbed oxygen. The fact that no modifications in the positions of the Au $4f$ photoelectron peaks have been observed for the gold electrodes oxidized up to 2.0 V SCE implies that the nature of the surface film remains essentially unchanged (i.e., only one kind of oxide is formed) in all of the potential range examined. This is also confirmed by the correlation of the coulometric data with the relative intensities of the photoelectron signals, which yields a linear relationship‡[465] between the amount of oxide measured electrochemically and the amount of oxide measured by XPS. These observations are consistent with the type of mechanism proposed by Laitinen and Chao,[480] involving a continuous growth of the oxide film with potential.

(iii) Palladium

The surface of the palladium electrode oxidized at various anodic potentials in 0.5 M aq. H_2SO_4 has been investigated by Kim et al.[481] Identification of the electrochemically formed palladium–oxygen species was based on the comparison with XPS data obtained for Pd metal (cleaned by argon ion bombardment)

---

* This peak is generally assigned, in the case of oxides, to lattice oxygen ions, O^{2-}.

† Some lower potential value (1.1 V), corresponding to the begining of oxide formation on Au, has been obtained in earlier ellipsometric work[479] using 1 M H_2SO_4 solution.

‡ This relationship has been demonstrated to be valid up to 1.6 V: above this potential, the Au $4f$ electron peak due to the metal is progressively attenuated and finally obscured for the sample oxidized at 1.8 V.

and PdO produced by heating Pd in air at high temperatures. Thus, for the Pd electrode prepolarized at 0.9 V SHE, the phase oxide PdO together with the Pd metal have been detected. Curiously enough, the Pd $3d$ electron region of the spectrum recorded for this electrode (Fig. 23) indicated, furthermore, the presence of a third Pd species, characterized by the small peak shifted to higher binding energy with respect to PdO. The assignment of this peak was complicated by the lack of data for bulk PdO_2, since the chemically prepared palladium dioxide decomposes under high vacuum conditions in the spectrometer.[481] However, since the chemical shift of the high binding energy Pd form with respect to PdO and Pd metal (Fig. 23) is quite similar to that observed[463] between PtO_2, PtO, and Pt metal (for Pt $4f$ electrons), a reasonable basis seems to exist for the presumption of a PdO_2-type species.

The apparent stability of this form of PdO_2, under conditions existing in the sample chamber of the spectrometer, may be related to the significant extent of hydration of the electrode surface. The exact form in which water was present therein

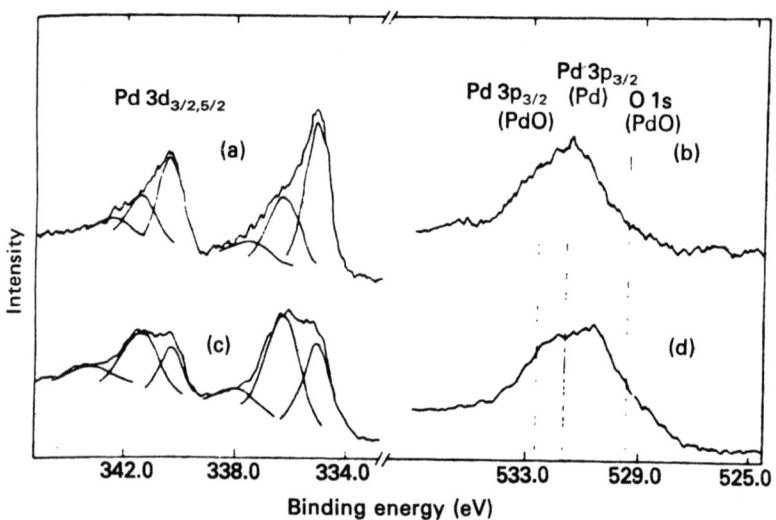

Figure 23. X-ray photoelectron spectra of (a) Pd $3d_{5/2,3/2}$ and (b) O $1s$ and Pd $3p_{3/2}$ electrons for a Pd foil oxidized at +0.90 V for 1000 sec in 1 N H_2SO_4. In (c) and (d), a foil was oxidized at +1.28 V for 60 sec (from Ref. 481).

[H_2O_{ads}, hydrated PdO and (or) PdO_2, or the corresponding hydroxide $Pd(OH)_2$ or $Pd(OH)_4$] has not been determined because of the difficulties encountered in deconvoluting the O 1s electron spectrum, due to the interference from the Pd $3p^{3/2}$ line. Assuming the approximate escape depth of 1000 eV photoelectrons through PdO to be about 15 Å,[482] the surface concentration of PdO species on the electrode oxidized at 0.9 V has been estimated,[481] from the area ratio of the $3d^{5/2}$ peaks metal/oxide, as equivalent to about two monolayers. This is consistent with the results of previous galvanostatic studies[483,484] indicating that PdO begins to form at about 0.8 V (first plateau in the charging curves). The presence, besides PdO, of trace amount of PdO_2 at a potential of 0.9 V may be explained on the basis of the reaction sequence suggested by Hoare[485] to account for the rest potential of the Pd electrode in O_2-saturated 1 M aq. H_2SO_4 (0.87 V). This reaction sequence includes the formation of PdO_2, via oxidation of Pd^{2+} ions produced by anodic dissolution of the metal.

The Pd $3d$ electron spectrum (Fig. 24) of the Pd electrode prepolarized at 1.71 V, i.e., above the steady-state potential of the Pd/PdO_2 couple, 1.47 V,[485] shows the peaks due to two palladium oxides—PdO and PdO_2. It is to be noted that Pd metal is here completely obscured, which indicates that the thickness of the oxide layer is at least 40 Å,[481] i.e., much larger than that found in the case of Pt electrodes.[465] A relatively small PdO_2/PdO ratio, characterizing the oxide layer formed on Pd at this high anodic potential, seems to confirm a significant extent of decomposition of PdO_2 (to form PdO and O_2) consistent with an oxide mechanism[486] of oxygen evolution. However, some PdO_2 could also decompose outside the electrochemical cell, during the transfer of the sample, and/or inside the spectrometer.

The examples of XPS studies discussed above concerning oxide films on noble-metal electrodes demonstrate that photoelectron spectroscopy may yield in some cases unique information about the true chemical nature of the surface species. XPS measurements have shown, first of all, that the constituents of the surface films on Pt, Au, and Pd are practically characterized by the same binding energies as the *bulk* oxides of the corresponding metals. This observation, applying even to the coverages less than a monolayer, supports the view[461,467,468] according to which

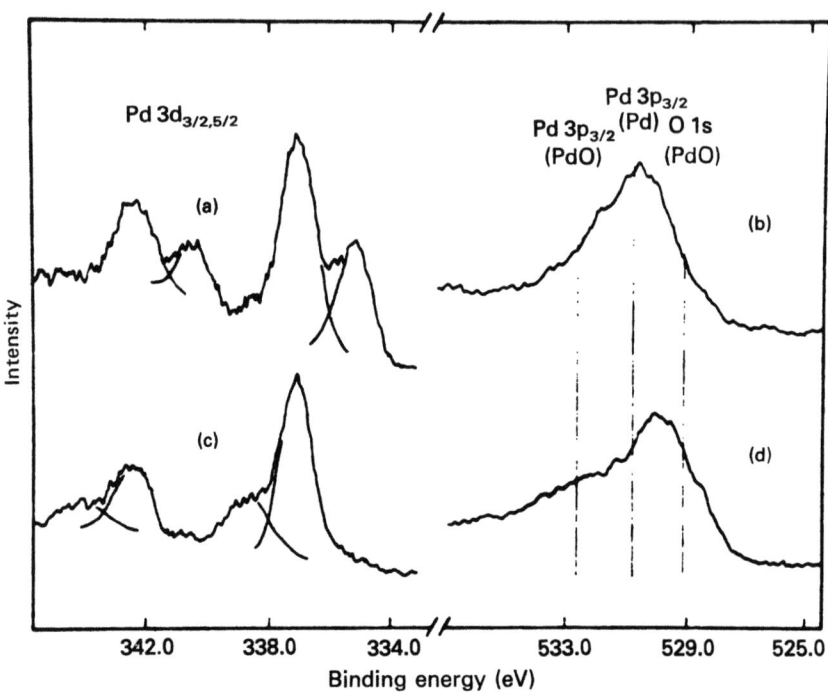

Figure 24. X-ray photoelectron spectra of (a) Pd $3d_{5/2,3/2}$ and (b) O $1s$ and Pd $3p_{3/2}$ electrons for a Pd foil exposed to a potential increasing from +0.0 V to +1.71 V at 40 mV/sec. In (c) and (d), a foil was exposed to a constant potential of 1.71 V for 30 sec (from Ref. 481).

the noble metals are covered by phase oxides or rearranged oxide lattices[466] already during initial stages of surface oxidation. Photoelectron spectroscopy has also allowed the detection of higher oxides, PtO_2 on Pt and PdO_2 on Pd, which are apparently formed at much lower anodic potentials than generally considered.

In spite of the fact that the analyses by electron spectroscopy are carried out *ex situ* under high-vacuum conditions, good agreement between the XPS data and the previous results of *in situ* optical and electrochemical studies has generally been found. In the cases where anodic oxidation produces stable oxide films on the electrode surface, electron spectroscopy (and especially

XPS) appears as a particularly suitable complementary technique facilitating, for example, the interpretation of ellipsometric data. Changes in composition of the electrode surface, induced by the procedure of XPS analysis, have been shown[465] to depend largely on the choice of experimental conditions and, especially, on the X-ray intensity. In certain cases (e.g., for Pt and Au) the extent of X-ray (and high-vacuum) reduction of the surface species can be evaluated by means of coulometric measurements and consequently taken into account in calculating the actual concentrations.

Contrary to what has been observed for the surface films composed of phase oxides, no meaningful results can be expected from the XPS or AES analysis of electrochemically formed chemisorbed-oxygen species. Thus, in the initial-reversible stages[465,466,487] of surface oxidation, the interruption of the potential control, caused by removal of the electrode from solution, is supposed to modify substantially the original state of the surface, as demonstrated, for example, in the case of Au.[488] Furthermore, even in the absence of electrochemical treatment, the surfaces of the solid samples are usually contaminated with adsorbed oxygen and water. The resolution obtained at the present time with XPS and AES measurements is insufficient to allow distinction to be made between contaminations of this type and the chemisorbed oxygen species originating from anodic oxidation.

2. RuO_2-Based Film Electrodes

The most significant achievement of electrocatalysis in recent years is probably connected with the successful large-scale application of RuO_2-based film electrodes in chlor–alkali cells. Paradoxically enough, little definite information exists on the mechanism of chlorine evolution reaction at this kind of electrode. This situation is partially caused by the lack of data on the surface structure of such electrodes; because of the badly defined state of their surface and the existence of parallel faradic processes, other than purely electrochemical techniques are required in order to characterize the adsorbed intermediates of the reaction.

The chemical constitution of the 20 –30 Å outermost layer of

RuO$_2$ and RuO$_2$–TiO$_2$ film electrodes, prepared by thermal decomposition of the corresponding chlorides on titanium supports, has been investigated by Augustynski *et al.*[489] using photoelectron spectroscopy. The analyses were carried out with the electrodes (i) as prepared and (ii) after prolonged electrolysis in 4 M NaCl solution. Figure 25 shows typical photoelectron spectra of the Ru $3p$ region for commercial RuO$_2$ powder and RuO$_2$ deposited on titanium. The spectra of both samples are quite similar, each of them being composed of double signals indicating

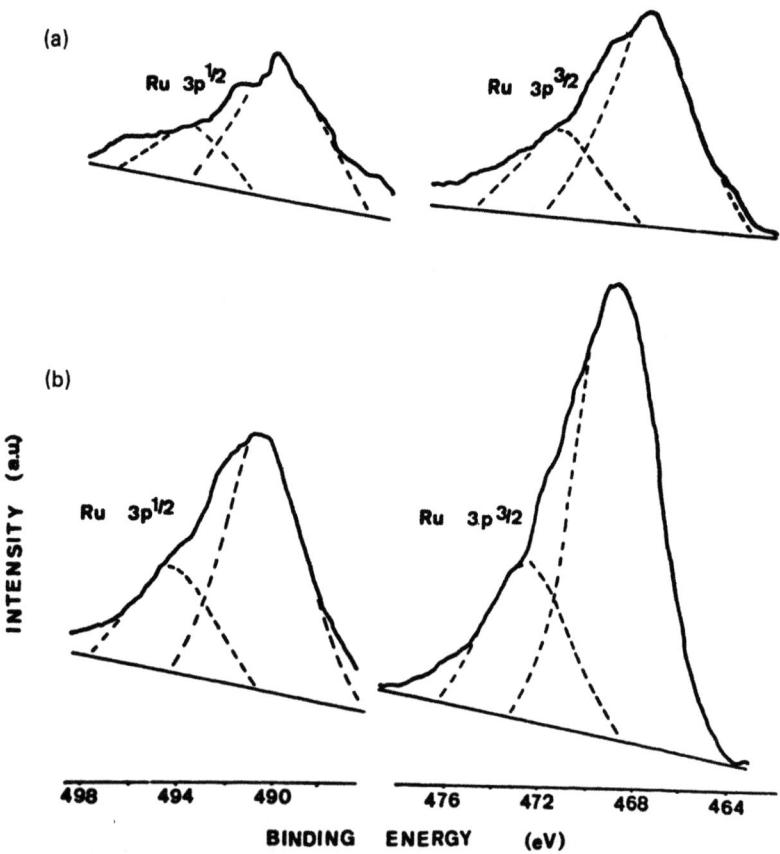

Figure 25. X-ray photoelectron spectra of Ru $3p$ region of (a) RuO$_2$ powder, and (b) RuO$_2$ deposit on titanium (from Ref. 489).

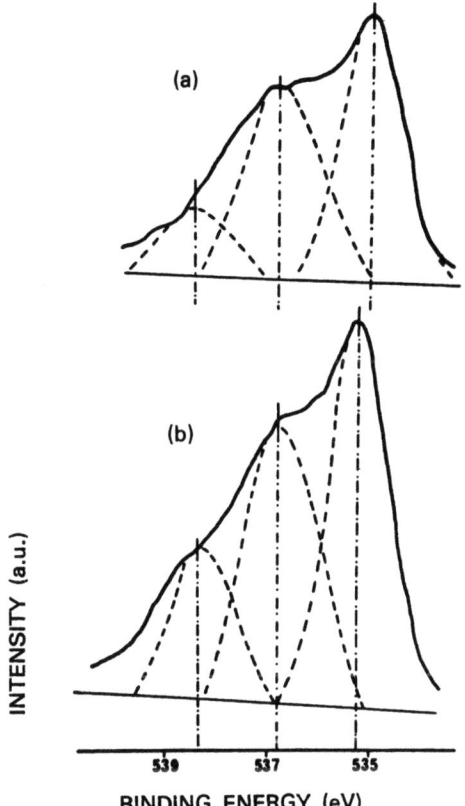

Figure 26. Oxygen 1s photoelectron signals for (a) RuO_2 powder, and (b) RuO_2 deposit on titanium (from Ref. 489).

the presence of two different ruthenium species. Also the corresponding O 1s electron spectra (Fig. 26) show two large peaks associated with lattice oxygen, independently of the usual smaller signal at higher binding energies due to adsorbed gases and water. The existence on the surface of RuO_2 (powder and single crystal) of an additional, higher valency ruthenium compound has been first reported by Kim and Winograd[490] on the basis of XPS measurements. This compound has been assigned to ruthenium trioxide, RuO_3. The hypothesis concerning the possible presence of RuO_3 on the surface of RuO_2-film electrodes receives confirmation in the fact that the Ru/O ratios, obtained from relative intensities of the photoelectron peaks, are close to $\frac{1}{2}$ and $\frac{1}{3}$[489] for the largest and the smaller Ru and O signals, respectively. This is also supported by the similarity of binding energies for various Ru electron levels, between the "surface" RuO_3 and a Ru(VI) compound $BaRuO_4 \cdot H_2O$. The double Ru signals, characteristic of the presence of RuO_2 and RuO_3, have been also observed in the spectra of RuO_2-TiO_2 film electrodes.

These spectra have revealed an interesting fact, namely, that the Ru/Ti ratio calculated from the intensities of the corresponding photoelectron peaks (about 0.15) was appreciably lower than the mean film composition (Ru/Ti equal to 0.28). The enrichment in TiO_2 in the surface region of the RuO_2-TiO_2 electrodes must be connected with some preferential diffusion process taking place during the final heat treatment.

No detectable changes in the Ru binding energies and the Ru(VI)/Ru(IV) and Ru/Ti ratios have been observed for the RuO_2 and RuO_2-TiO_2 deposits on titanium employed as anodes in the electrolysis of 4 M NaCl. On the other hand, as expected, the O 1s photoelectron signals assigned to water and adsorbed oxygen were clearly enlarged for both these electrodes. Furthermore, as a result of electrolysis (and of immersion in NaCl solution), a large amount of chlorine species was detected on the surface of the anodes. In this connection, it should be mentioned that the freshly prepared RuO_2 deposit already contains a certain amount of chlorides[491,492] arising from incomplete decomposition of $RuCl_3$ during heating at 450°C. After electrolysis, the relative concentration of chlorides in the surface region of the RuO_2 electrode increases [from about 4 at.% vs. Ru(IV) before electrolysis] to about 6 at.% and, in addition, on the higher-binding-energy side of the Cl 2p spectrum* (Fig. 27) appears another chemical form of chlorine.

Taking into account that (i) the binding energy for a given electron level generally increases together with the oxidation state of the element and that (ii) the higher-binding-energy Cl

---

* Since the Cl 2p electron spectrum consists normally of two signals, due to spin–orbit splitting, the larger $2p^{3/2}$ and the smaller $2p^{1/2}$ placed at higher binding energies, the middle (largest) peak at Fig. 27 must result from the superposition of the $2p^{1/2}$ signal arising from the first Cl species and the $2p^{3/2}$ signal characteristic of the second Cl species.

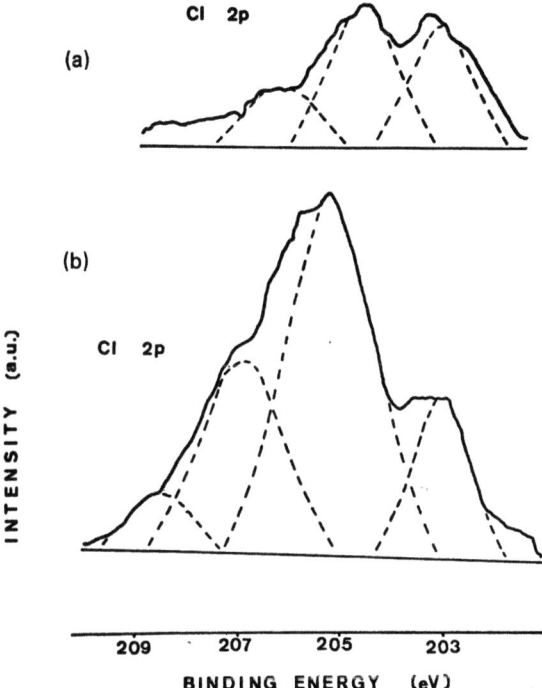

Figure 27. X-ray photoelectron spectra of Cl 2p region for RuO$_2$ deposit on titanium: (a) as prepared, (b) used as an anode for chlorine evolution (from Ref. 489).

form arises from the process of chlorine evolution, the latter species have been assigned[489] to adsorbed atomic chlorine Cl$_{ads}$ or ClO$_{ads}^-$ ions. The values of relative concentrations of Cl$_{ads}$ or ClO$_{ads}^-$, calculated from photoelectron spectra [about 15 at.% vs. Ru(IV) for a RuO$_2$ electrode, and near 32 at.% vs. the sum of Ti(IV) and Ru(IV) for a RuO$_2$–TiO$_2$ electrode], indicate a high degree of surface coverage by these species for both types of anodes examined here.

The latter observation is not consistent with the mechanistic considerations,[493,494] concerning the chlorine evolution reaction on RuO$_2$-based electrodes, founded on the results of electrochemical investigations, which imply rather low coverages by Cl$_{ads}$ on the electrode. It seems also that further kinetic studies should involve, as suggested by Augustynski *et al.*, the examination of a possible role which Ru(VI) species could play in the chlorine evolution process, e.g., through reaction sequences of the kind

$$2Cl^- + RuO_3 + 2H_3O^+ \to (RuO_2)2Cl_{ads} + 3H_2O \quad (8)$$

and

$$RuO_2 + 3H_2O \to RuO_3 + 2H_3O^+ + 2e \quad (9)$$

or

$$Cl^- + RuO_3 \to (RuO_2)ClO_{ads}^- \quad (10)$$

followed by (9) and

$$ClO_{ads}^- + Cl_{ads}^- (\text{or } Cl_{aq}^-) + 2H_3O^+ \to Cl_2 + 3H_2O \quad (11)$$

As a conclusion to this section, it should be mentioned that the coexistence, in the surface region, of different oxidation states of the metal does not appear to be an exclusive feature of the RuO_2 electrode; as a matter of fact, two chemical forms of iridium have been detected by photoelectron spectroscopy[495] on the surface of IrO_2 film electrodes.

3. Characterization of Electrode Surfaces in Connection with Adsorption Studies

This kind of application of electron spectroscopy is well exemplified in the approach taken by O'Grady *et al.*[496] for the study of electrochemical adsorption of hydrogen on single-crystal platinum. These authors have used LEED (low-energy electron diffraction) and Auger spectroscopy combined with a thin-layer cell*; a special system has been developed for direct transfer of the sample between the high-vacuum chamber of the spectrometer and the electrochemical cell, in order to avoid any external contamination.

According to the experimental procedure employed,[496] the

---

* The use of the thin-layer cell in conjunction with LEED–AES was originally suggested by Hubbard.[497]

single-crystal surface is initially cleaned by argon-ion sputtering in the sample preparation chamber and then characterized by means of the Auger spectrum and the LEED pattern.* Afterwards the sample is directly transferred to the thin-layer cell, where electrochemical measurements (e.g., cyclic voltammetry) are carried out. Finally, the sample is returned to the spectrometer chamber to be analyzed again by LEED and Auger. The purpose of this sequence of operations is twofold: it allows, firstly, a check on the cleanliness of the metal surface and, subsequently, the changes in its constitution and electronic structure introduced by electrochemical reaction to be studied. Furthermore, the use of argon-ion sputtering allows an initial cycling of the electrode potential to high anodic values in order to oxidize and desorb impurities from the surface to be avoided. As a matter of fact, this kind of treatment is likely to induce restructuring upon return to the potential region of hydrogen adsorption.[496]

In the preliminary experiments with a clean argon-ion sputtered Pt(111) surface O'Grady et al. have, however, encountered difficulties in forming a stable electrolyte film between the two electrodes in the thin-layer cell. Further measurements, carried out using the Pt(111) electrode deliberately covered with uniform carbon layer (about 0.3 of a monolayer), have shown distinct modifications in the behavior of hydrogen adsorption.

VI. APPLICATION OF AES AND XPS TO PASSIVITY AND CORROSION STUDIES

It is not surprising to see that among the applications of AES and XPS to electrochemical problems, the major part is connected with studies of corrosion and passivity. As is well known, these processes involve, in general, the formation of relatively stable solid layers, well suited for analysis by electron spectroscopy; knowledge of chemical composition of these layers is an important condition for better understanding of the mechanism of passivity and the formation and breakdown of passivating films.

---

* Before recording the LEED pattern, the platinum sample is annealed at 900°C.

1. AES Studies

Auger electron spectroscopy was adopted relatively early as an analytical tool for the study of passive films on metals and alloys; in particular, the application of AES combined with the ion-sputtering technique (see Section IV.3) has been of great utility in the determination of the in-depth composition profiles of surface layers. More recently, evidence of a number of electron-beam-induced secondary effects, established by different authors (see Section II.8 and references therein) has, however, demonstrated the necessity for cautious interpretation of the results obtained with this technique. In order to illustrate this kind of application, a few examples of recent AES studies will be discussed.

Seo et al.[498] have investigated, using AES and argon-ion-etching, the elemental composition of oxide films formed on iron polarized at various anodic potentials in a boric acid-sodium borate solution. The inner layer of the films produced at potentials lower than the Flade potential (0 V vs. SCE) has been found to contain a notable amount of ferrous species, whereas the same region of the films formed at potentials above 0.5 V consisted essentially of ferric oxide.* The identification of Fe(II) and Fe(III) species was based on the analysis of (i) the chemical shift of low-energy Auger peaks of iron and (ii) the O/Fe peak-to-peak height ratios. It should be noted, in this connection, that the assignment of the low-energy Auger peaks due to Fe species and the interpretation of the corresponding chemical shifts have been subject to controversy[500,498] and that some uncertainty may exist about the possible reducing effects of the electron beam and argon-ion bombardment.

da Cunha Belo et al.[501] have used Auger spectroscopy for studying the chemical constitution of passive films formed on ferritic stainless steels (26% Cr with or without 1% of Mo) in 3.5% NaCl solution at various potentials. On the basis of the concentration profiles of different elements, obtained by combination of AES with argon-ion sputtering, these authors con-

---

* These results are consistent with the model proposed by Wagner,[499] attributing iron passivity to a thin film of Fe-deficient magnetite with a composition between that of Fe_3O_4 and γ-Fe_2O_3.

cluded that the composition of the passive film, having a thickness of approximately 10 atomic layers, changes continuously with depth.* With regard to the chromium concentration, (i) a chromium enrichment, especially in the inner layer of the film, and (ii) a chromium depletion inside the substrate alloy near the metal-oxide interface have been observed. No molybdenum enrichment, with respect to the bulk composition, has been found: small amounts of this element have been detected in the inner part of the film only. At the potentials examined, chlorine (chloride ions originating from the solution) has been shown to be localized in the outer layers of the film. Finally, indirect evidence of the porous structure of the passive film has been provided by measurements of the absolute concentrations[501] of the elements present in the films, indicating lower numbers of atoms per layer than in the metallic matrix.

The composition of passive films on a series of Fe–Cr–Mo ferritic stainless steels, with varying chromium and molybdenum contents (up to 25% and 5%, respectively), has been extensively studied by Yaniv et al.[503] as a function of pH of the solution, presence of chloride, and potential; the concentration depth profiles have been determined through simultaneous etching of the alloy surface by argon-ion bombardment and AES analysis. In the case of passive films formed in acid sulfate solutions, with or without chloride addition, the general trends have been similar to these observed by da Cunha Belo et al., i.e., (i) absence of molybdenum in the outermost layers and no enrichment in the inner layers of the film; (ii) chromium enrichment and iron depletion inside the film. The enrichment of chromium in the films produced on different alloys of the series was proportional to the content of this element in the matrix. Both sulfur and chlorine have been detected in the outer region of the film. On the other hand, for the samples polarized anodically in 1 M NaOH, a depletion of chromium and an absence of molybdenum in the film has been ascertained.

Separate experiments carried out by Yaniv et al. have shown that the observed small concentration level of molybdenum was

---

* This picture is similar to that arising from the analyses of air-formed films on iron–chromium alloys, performed[502] with ion-scattering spectrometry.

attributable neither to the electron beam effect nor to the occurrence of preferential sputtering, but was an inherent characteristic of the passive films on Fe–Cr–Mo alloys.[504] Thus, the results of AES analyses clearly refute previous assertions[505,506] correlating the improved corrosion resistance of molybdenum-containing steels in chloride and acid solutions with Mo enrichment of the surface films.

2. XPS Studies

If, in the applications of AES mentioned above, the emphasis was on elemental in-depth analysis of surface layers, the principal interest of XPS studies, discussed below, consists rather in determination and interpretation of chemical shifts.

Castle and Clayton[507] have investigated by the XPS method the chemical composition of passive layers formed on 18/8 austenitic stainless steel as a result of immersion in low-conductivity deoxygenated water. An aim of their study was to test the ability of photoelectron spectroscopy to detect changes in the thin film composition due to variable environmental factors. Prior to the exposure to water, the samples were etched in the spectrometer preparation chamber by argon-ion bombardment. No evidence of surface segregation of the alloying components, as a result of the ion bombardment procedure (5 μA, 3.5 keV, for 1 hr), could be detected. The concentrations of Cr and Ni relative to Fe, calculated from the photoelectron spectra of the etched samples, were, indeed, quite close to those for the bulk alloy.

The passive films produced during immersion of stainless steel in water have been shown to be composed of Fe^{2+}, Fe^{3+}, Cr^{3+}, O^{2-}, and OH^- species, and to be less than 30 Å thick.[507] It has been deduced, from the spectra recorded as a function of collection angle and from the concentration profiles obtained by argon-ion etching, that the outermost layer differed from the rest of the passive film by being particularly rich in OH^- species. This outer layer, it has been suggested, consists of bound water* which could be stabilized by adsorbed organic species, whose presence

---

* The presence of large amounts of bound water in the oxide film on stainless steel had earlier been reported by Okamoto.[508]

was deduced from the observed carbon 1s photoelectron peak. Some particular experiments carried out by Castle and Clayton[507] have also demonstrated that the composition of these thin oxide films, formed on stainless steel during immersion in water, was not influenced to a significant extent by the unavoidable exposure of the sample to air during its transfer from the cell to the spectrometer. These results seem to indicate that, in spite of their *ex situ* character, the XPS analyses can yield, in a number of cases, meaningful information about chemical constitution and mean thickness of passive films.

The composition of surface films formed on binary iron–chromium alloys during anodic polarization in 1 M H_2SO_4, has been investigated, using the XPS technique, by Asami et al.[509] In the case of Fe–Cr alloys containing at least 13 at.% of Cr and polarized in the passive region, the protective film has been shown to consist mainly of hydrated chromium(III) oxy-hydroxide. On the other hand, the passive films on the low (less than 13 at.%) chromium alloys, having considerably lower corrosion resistance, were essentially composed of iron oxy-hydroxide. It has been also found that the Cr/Fe ratio in the surface films formed on mechanically polished iron–chromium alloys, exposed briefly to air[510] or polarized in 1 M H_2SO_4 below the passive region,[509] was practically unchanged with respect to the bulk alloys. The results of Asami et al. and those, analogously, of other authors,[511–513] indicate that Cr enrichment in passive films is a common feature of corrosion-resistant alloys and steels containing chromium; they confirm also the suitability of photoelectron spectroscopy as a semiquantitative method for analysis of thin passive films.

Dickinson et al.[514] have studied by XPS the nature of surface species formed on nickel electrodes polarized in 0.5 M H_2SO_4, in the active, passive, and transpassive regions. Different Ni, O, and S signals have been resolved by comparing them with the spectra of a series of bulk nickel compounds. The presence of an $Ni(OH)_2$ film in the prepassive region, suggested earlier by Bockris et al.[515] on the basis of ellipsometric studies, has been confirmed. Furthermore, two different sulfur species, incorporated in the film and assigned to sulfate and sulfide, have been detected. The presence of the latter species implies that SO_4^{2-} ions undergo reduction on the nickel electrode in the active (prepassive) potential region. Both NiO and Ni(OH)$_2$ have been observed on the surface of the nickel electrode polarized at the potentials in the passive region. This suggests that the occurrence of passivation is connected with the appearance of NiO. The thickness of the passive film has been estimated* as 11–17 Å, assuming a homogeneous mixture of species on the surface. This range of values is in reasonable agreement with those proviously suggested by other authors.[516,517] The information concerning transpassive and second passive regions is less clear[514]; the presence of Ni(III) oxide species with or without NiO could not be ascertained. The reason for this is the great complexity of the Ni $2p$ electron spectra caused by multiplet splitting.[514]

A similar investigation has been reported[518] for tin electrodes polarized anodically in 0.5 M NaOH solution. Stannous oxide and hydroxide have been detected in the prepassive region. For electrodes polarized at higher potentials, where onset of passivation is observed, stannic oxide and hydroxide have been shown to be formed. At still higher anodic potentials, it has been shown that the oxide layer is composed of Sn(IV) species only; an increase in the thickness and, simultaneously, in the oxide/hydroxide ratio with increasing potential have been also observed.

Augustynski et al.[519,520] and Koudelkova et al.[521] have investigated in some detail the interactions between various aggressive and inhibiting anions and oxide-covered aluminum metal. A number of XPS analyses were carried out for aluminum samples exposed to solutions containing chlorides, perchlorates, sulfates, nitrates, and chromates in the absence of external polarization (i.e., at open-circuit potential, E_{st}) or polarized anodically at potentials situated in the passive region, in proximity of the critical pitting potential, E_{cr}.

In order to characterize the initial state of the aluminum surface prior to immersion in the solution, the XPS spectra of freshly prepared, mechanically polished samples were recorded (Fig. 28). Two separate peaks were observed in the Al $2s$ and

---

* These calculations were based on the comparison of the relative intensities of Ni(II) and Ni(metal) photoelectron signals.

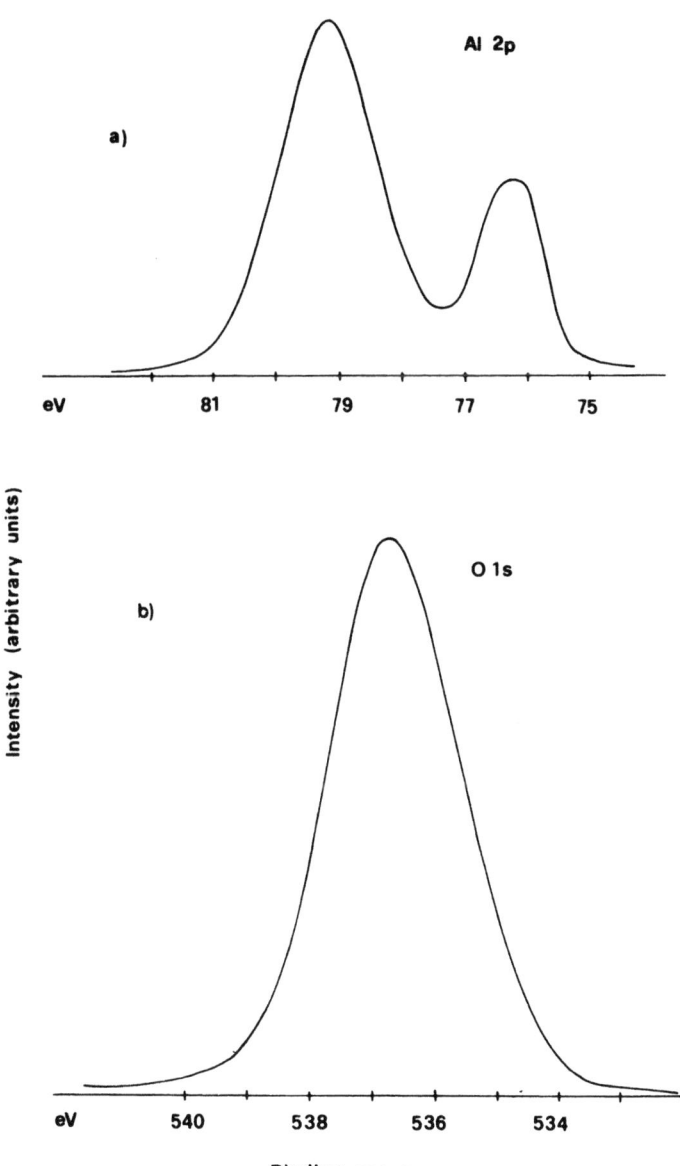

Figure 28. The Al 2p and O 1s regions of the XPS spectra obtained for mechanically polished aluminum (from Ref. 519).

Al $2p$ regions, indicating the presence of an oxide on the metal surface. The shape and position of the O $1s$ peak, too, were quite characteristic of an hydrated aluminum oxide. The photoelectron spectra recorded for aluminum samples exposed to chloride solutions showed, besides the Al and O peaks similar to those from Fig. 28, a Cl $2p$ signal.

The concentration of chloride found in the oxide film and calculated relative to Al(III) was about 3 at.% at E_{st} (after 18 hr immersion in 0.1 M NaCl) and about 12 at.% near the critical pitting potential ($E_{cr} - 0.07$ V).[520] Analogous measurements have also been performed[520] with the specimens exposed to sulfate/chloride solutions. In the first series of experiments, the aluminum samples (supporting an air-formed film) were exposed directly to 1 M SO_4^{2-}/0.1 M Cl^- solution during 18 hr; both SO_4^{2-} and Cl^- ions were incorporated, under these conditions, within the oxide film, the relative SO_4^{2-} concentration being twice that of Cl^-. The second set of immersion tests was intended to decide whether the observed adsorption of Cl^- ions at the oxide–solution interface was really significant from the point of view of the passivity breakdown mechanism or whether Cl^- ions from the electrolyte were incorporated within the aluminum oxide lattice as a result of the film thickening process. Thus, the Al samples were first exposed, for 18 hr, to 1 M SO_4^{2-} solution and then, for another 18 hr, to the solution containing both sulfate and chloride ions (1 M SO_4^{2-}/0.1 M Cl^-). Though the relative amount of SO_4^{2-} ions found in the surface film had, in this case, increased up to 20–30 at.%, that of Cl^- remained apparently unchanged at about 3 at.%. These results indicate that, in spite of their large extent of incorporation within the oxide film on aluminum, the SO_4^{2-} ions do not impede the simultaneous adsorption of Cl^- ions on the oxide–solution interface and their entry into the film.

XPS analyses of the aluminum specimens exposed to perchlorate solutions produced rather unexpected results[522,520]; one single signal, characterized by the same binding energy as the signal previously assigned to $Cl^{(-I)}$ species, has been detected in the Cl $2p$ region. Since the chemical shift between the experimentally observed Cl $2p^{3/2}$ binding energy and that expected for adsorbed perchlorate ions was close to 10 eV (Table 2), it was inferred that ClO_4^- ions were quantitatively reduced inside the

Table 2
Binding Energies of the Cl $2p^{3/2}$ Electron of Some Chlorine Compounds

Species	Binding energy (eV) ±0.2 eV
$Cl^{(-I)}$ detected in the oxide film on Al	203.4
$AlCl_3$	204.1
KCl	205.5
$KClO_4$	215.0

passive film, possibly according to an electrochemical mechanism

$$3ClO_4^- + 8Al^{3+} + 24e \rightarrow 3Cl^{(-I)} + 4Al_2O_3$$

and

$$Al \rightarrow Al^{3+} + 3e \quad \text{(at the metal–oxide interface)}$$

An interesting aspect of the above reduction of ClO_4^- ions resides in the fact that, besides producing aggressive Cl^- species, it is simultaneously a film-forming, i.e., passivating reaction. It should be remembered, in this connection, that ClO_4^- ions are able to induce, in a way analogous to that of Cl^- ions, the passivity breakdown and pitting corrosion of aluminum. However, the E_{cr} of Al observed in perchlorate solutions is distinctly less negative than that in chloride solutions.

This difference between the anodic behavior of aluminum in chloride and perchlorate solutions is yet more marked under transient galvanostatic conditions where, in the ClO_4^- solution, the potential of Al rises to very high anodic values before reaching the stationary dissolution potential.[523]

The photoelectron spectra performed on aluminum samples exposed to nitrate and nitrate/chloride solutions[519] have demonstrated that the reduction of oxyanions by aluminum was, in fact, quite a usual phenomenon. The N $1s$ peak recorded for these samples was shifted about 7–8 eV to lower binding energies in comparison with the nitrate region. Since, on the other hand, the binding energy of these N species was near that of NH_4Cl (Table 3), it was postulated that NO_3^- ions were reduced to NH_4^+ or NH_3, the latter species being able to replace bound water in

Table 3
Binding Energies of the N 1s Electron of Certain Nitrogen Compounds

Compound	Binding energy (eV) ±0.2 eV
KNO_3	413.1
$Cr(NO_3)_3 \cdot 9 H_2O$	412.6
NH_4Cl	406.85
AlN	402.85
N species detected in the oxide films on Al samples exposed to the following solutions: 1 M $NaNO_3$, 0.33 M $Al(NO_3)_3$, 1 M NH_4Cl, and 1 M NH_4NO_3	405.25

the aluminum oxide lattice. This is consistent with the fact that, apparently, the same N species were also present in the oxide film on Al samples exposed to NH_4Cl solution (Fig. 29).

XPS analyses of the passive films formed on aluminum in solutions containing another inhibiting anion, CrO_4^{2-}, have shown that these films typically include a mixture of hydrated chromium(III) and aluminum(III) oxides, as well as a significant amount of adsorbed (incorporated) chromium(VI) species (see Table 4).[521]

Although a number of authors have postulated (or agreed)[524,525] that chromate reduction must play an important role in the mechanism of the inhibiting action, no experimental evidence for the presence of Cr(III) species in the surface oxide film on Al has been previously reported. In concludion, all the results of XPS analyses of Al samples show clearly that the interactions between aluminum and various aggressive (chloride) or inhibiting ions from the solution already take place at potentials distinctly more negative than the critical potential. On the other hand, it has been confirmed that Cl⁻ ions penetrate (at E_{st} in deaerated solutions) not only the air-formed oxide films on aluminum but also the films preformed in the sulfate or chromate solutions.

As indicated by the composition profiles obtained from XPS analyses combined with argon-ion bombardment,[526] chloride ions incorporated at E_{st} are mainly located in the first 10–15 Å of the

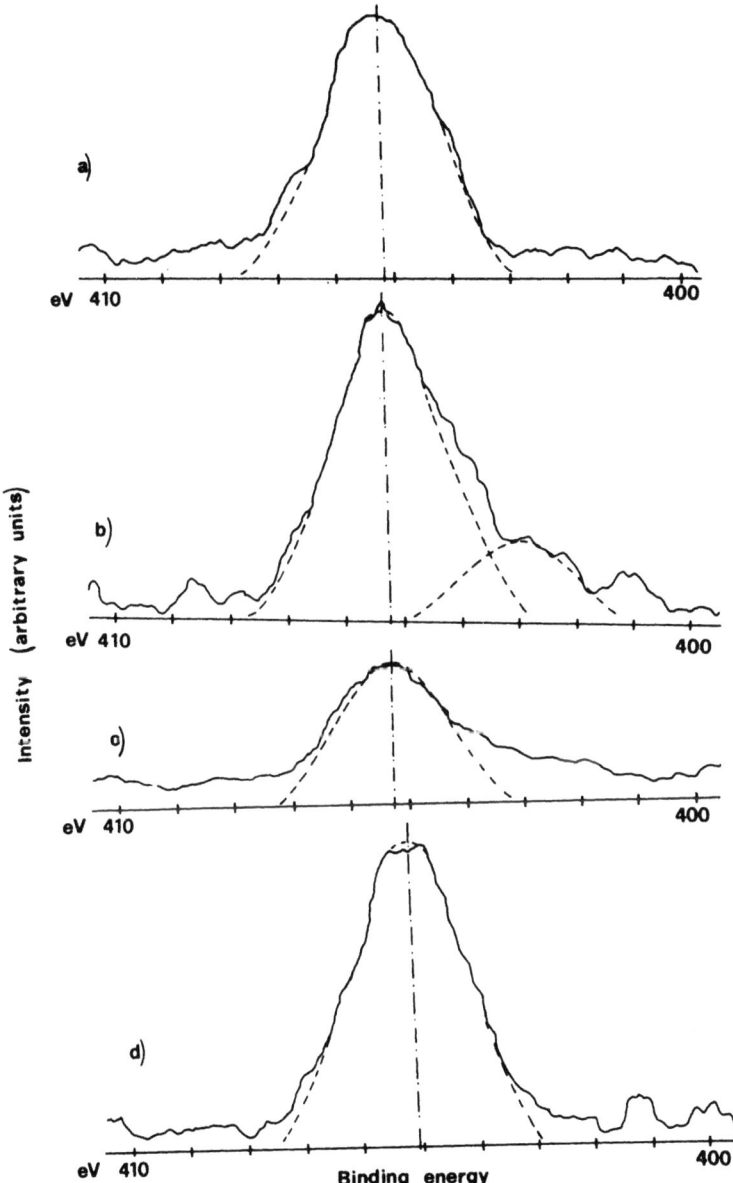

Figure 29. Nitrogen 1s photoelectron signals for aluminum samples exposed during 15 hr to the following solutions: (a) 1 M $NaNO_3$; (b) 1 M $NaNO_3$ + 0.1 M NaCl; (c) 1 M NH_4Cl; (d) 1 M NH_4NO_3 (from Ref. 519).

Table 4
Relative Amounts of Different Species Detected in the Passive Films Formed on Al Samples in Chromate Solutions

Solutions	Relative concentrations (at.%)[a]				
	Al^0	Al^{III}	Cr^{III}	Cr^{VI}	O
1 M Na_2CrO_4	—	16	66	18	330
0.2 M Na_2CrO_4	5	24	56	15	325
0.2 M Na_2CrO_4; pH = 7	8	44	38	10	310

[a] The sum of concentrations of the metallic species (Al^0, Al^{III}, Cr^{III}, and Cr^{VI}) was taken as 100 at.%. The concentration of O species was referenced to the sum of Al^{III}, Cr^{III}, and Cr^{VI}.

oxide film, at the proximity of the oxide–solution interface. At more anodic potentials, significant amounts of chloride seem also to be present inside the oxide film, suggesting that the events that determine the critical pitting potential take place rather at the metal–oxide than at the oxide–solution interface. This conclusion is also supported by the fact that sulfate ions, in spite of their large extent of incorporation in the outer region of the film, cause practically no shift of the critical pitting potential of aluminum in chloride solutions.

On the other hand, the behavior of such anions as nitrates, chromates (and also molybdates and acetates) is characterized by the fact that they all undergo reduction inside the protective oxide film. It seems, therefore, that the inhibiting effect exerted by these anions on the pitting corrosion of aluminum is to be related to the film-forming and proton-consuming character of their reduction.[520]

VII. CONCLUDING REMARKS

Auger electron spectroscopy and photoelectron spectroscopy view the surface from a somewhat different perspective; each provides somewhat different information and each is hampered by its own particular uncertainties.

AES has the advantage that, under identical experimental

conditions, it is usually more sensitive than XPS over a wide range of elements.[528] However, the ionization cross sections for Auger excitation decrease with increasing binding energy, hence with atomic number Z, while those for XPS increase. So, for high atomic numbers ($Z > 45$), XPS will be more sensitive than AES. In addition, the number of deexcitation channels increases faster with Z for Auger than for photoelectronic processes, so that the primary excitation is dissipated into more numerous but less intense transitions.

AES remains quite unique whenever spatial resolution is of the essence; the incident electron beam can be focused, if necessary, to a very fine spot. For example, AES is routinely used for analyzing extremely small areas such as grain boundaries or fracture surfaces.

Chemical effects are not easy to interpret in AES as Auger emission involves three electronic levels, whereas one only is concerned in XPS. Also, Auger peaks are usually broader than the corresponding photoelectron lines which makes isolation of chemical effects difficult. However, the ability to vary at will the primary energy enables separation of Auger and loss peaks where there is ambiguity. The use of synchrotron radiation[529,530] will indubitably give the same ability to photoelectron spectroscopy, though the access to such a radiation source will remain, for a long time, the privilege of a very restricted number of laboratories.

Another advantage of AES is the speed of the analysis; the CMA possesses the ability to examine the complete Auger spectrum in less than 1 sec; rapid data acquisition may be critically important in kinetic studies. This very high-speed analysis is even more important in the presence of electron beam effects. Electron-induced desorption and decomposition are indeed the major drawbacks of AES (Section II.8). AES can only provide reliable analysis of a virgin surface at primary beam current densities of about 3 μA cm^{-2}, thus allowing 600 sec for analysis; quite evidently, at such reduced beam conditions, the advantage of speed of AES over XPS largely disappears. It is, however, fair to remember that XPS itself is not entirely devoid of such inconveniences (cf. Section III.9).

In conclusion XPS offers major advantages whenever chemical information is wanted: the photoelectron lines offer better resolution and are easier to interpret; besides, the specimen is much less liable to be damaged during analysis. However, AES offers greater sensitivity and possesses decisive advantages whenever speed and spatial resolution are of necessity.

The application of electron spectroscopy to electrochemical problems can hardly be considered in terms of advantages and drawbacks with respect to electrochemical methods used in surface studies. As a matter of fact, from the point of view of an electrochemist, the major interest of Auger spectroscopy and photoelectron spectroscopy seems to reside in their unique ability for direct identification of the chemical state of surface species. AES and XPS appear, therefore, as typically *complementary* techniques with respect to electrochemical and coupled (*in situ*) optical methods. Several electrochemical methods allow accurate measurement of the surface coverage θ by adsorbed electroactive species down to $\theta = 0.02$ and determination of relative free energies of adsorption with an accuracy of 0.2 kcal.[4] If the electron spectroscopic techniques can be expected, in general, to provide a comparable sensitivity for concentration measurements, they seem unable to distinguish, with the same accuracy, between various states of adsorption at the electrode surface. For example, Hammond and Winograd,[531] studying the underpotential deposition of monolayer quantities of silver and copper on platinum electrodes by means of XPS and AES, have found, in the core photoelectron specra, no evidence of multiple adatom surface states which are observable in the cyclic voltammograms. The recorded photoelectron spectra present, for both underpotential deposited metals, single peaks shifted by less than 1 eV to lower binding energies with respect to the bulk metals. These chemical shifts were apparently constant for coverages of Ag and Cu atoms up to a monolayer. A detection limit of approximately 0.01 of a monolayer of the metal has been indicated.

However, there are also a number of cases, such as certain nonfaradaic adsorption processses or building-up, via electrochemical reactions, of solid layers on the electrode surfaces, where electrochemical methods are unable to provide meaningful information about the existence and relative amount of surface species, and consequently, are not at all competitive with respect to AES and XPS. One can mention, as an example, the incorporation of certain ions into passive films or conducting (or semiconducting) oxide electrodes.

It must be emphasized that no technique is complete in itself for the requirements of surface analysis; each provides information about a particular characteristic but does not present the total picture. For this reason it is becoming increasingly popular for two or even three instruments to be included in an analytical system.[532] The most popular among such instruments combine AES, XPS, and ultraviolet photoelectron spectroscopy (UPS). UPS differs from XPS only by the energy of the X-ray exciting radiation (usually He I (22.1 eV) or He II (40.8 eV)); though, since the availability of synchrotron continuous radiation as an exciting source, the distinction between them has become somewhat blurred. Although UPS can be almost ruled out as a tool for general elemental analysis or as a quantitative analytical technique, it possesses many desirable features for chemisorption and surface reaction studies. UPS is an extremely sensitive technique for examining unlocalized states or states that possess low angular momenta.

It has been extensively used to map density of states in the valence band, though, in such studies, XPS spectra are usually easier to interpret, as in UPS final-state densities and momentum conservation have to be taken into account, complicating the analysis. One of the major advantages of UPS over the two other techniques is that photodissociation and photodesorption cross sections are very small indeed for adsorbed species, including sensitive adsorbates such as organics.

Other techniques, which have arisen from AES, may also provide useful complementary information when used in conjunction with AES or XPS. For example, ionization spectrocopy (IS), sometimes called characteristic loss spectroscopy (CLS),[533] which is an empty band probe, can be used to provide information similar to XPS if no source of soft X-rays is available, but it is no real substitute for XPS because of the difficulties in interpretation. It is also a precious auxiliary to AES in situations where Auger peaks overlap undesirably.

Soft X-ray appearance potential spectroscopy (SXAPS)[534] is also an empty band probe with the advantage over IS that there is no restriction on the positions of the empty states that can be studied. It has been extremely useful in mapping the density of states in the unoccupied portions of the $3d$ bands in transition metals. The range of application of SXAPS is, however, restricted because too many elements have been shown to be observable either with only very low sensitivity or not at all.

CLS and SXAPS have some valuable applications but it seems doubtful, because of their reduced versatility, that they will attain the same popularity as XPS or AES.

We would not close this brief survey without mentioning two very different techniques, which have gained a reputation of their own, and which are very often associated with electron spectroscopies; secondary ion mass spectroscopy (SIMS) and low-energy electron spectroscopy (LEED).

In SIMS, the sample is bombarded with a beam of primary ions; particles that are sputtered away from the sample as a result of this bombardment are analyzed by a mass spectrometer; instruments exist in which SIMS and AES can be used in turn without moving the specimen. SIMS provides semiquantitative analysis in the monolayer range and will certainly become even more widely used when the rather complicated processes, which are responsible for the emission of positive or negative ions, are fully elucidated.[535,536]

LEED is the oldest and most versatile tool for the investigation of surface structures. In a LEED experiment, a monoenergetic beam of electrons is directed towards the crystal surface and the elastic component of the backscattered electrons is observed.[437] If the surface atoms are arranged in a periodic lattice they act as a grating for the electron waves and diffract them. Though in early years AES was considered a mere auxiliary technique to be performed with a LEED instrument, it has rapidly taken precedence over the latter, and for some time, especially after the appearance of the CMA, LEED has seemed to lose favor with spectroscopists. However, it was soon recognized that no basic surface studies in the monolayer range, nor any proper understanding of the complex interactions taking place in the surface region, could be attempted without a thorough knowledge of the surface structure; it is essentially for this reason that the RFA instrument, with its dual use as a combined AES/LEED instrument, has continued popularity. The better clue to a revival of interest in LEED is that several papers showing how to adapt a LEED system on a CMA have recently been published. We have no doubt that in the near future AES and XPS, complemented with other surface techniques such as LEED, will help to make some electrochemical concepts, still based on circumstantial evidence, amenable to more direct investigation.

REFERENCES

[1] B. G. Baker, in *Modern Aspects of Electrochemistry*, No. 10, Ed. by J. O'M. Bockris and B. E. Conway, Plenum Press, New York, 1975.
[2] L. A. Harris, *J. Appl. Phys.* **39** (1968) 1419.
[3] R. E. Weber and W. T. Peria, *J. Appl. Phys.* **38** (1967) 4355.
[4] B. E. Conway, H. Angerstein-Kozlowska, and F. C. Ho, *J. Vac. Sci. Technol.* **14** (1977) 351.
[5] D. E. Eastman and M. I. Nathan, *Phys. Today* **28** (1975) 44.
[6] J. C. Rivière, AERE Report AERE-R 7368 (1973).
[7] R. L. Park, *Phys. Today* **28** (1975) 52.
[8] H. E. Bishop and J. C. Rivière, *J. Appl. Phys.* **40** (1969) 1740.
[9] A. P. Janssen, R. C. Schoonmaker, J. A. D. Matthew, and A. Chambers, *Solid State Commun.* **14** (1974) 1263.
[10] W. Schröder, E. Peters, and J. Hölzl, *Appl. Phys.* **3** (1974) 135.
[11] F. P. Netzer and J. A. D. Matthew, *Surf. Sci.* **51** (1975) 352.
[12] H. Papp and J. Pritchard, *Surf. Sci.* **53** (1975) 371.
[13] A. E. Morgan and W. J. M. van Velzen, *Surf. Sci.* **40** (1973) 360.
[14] J. A. D. Matthew and C. M. K. Watts, *Phys. Lett.* **37A** (1971) 239.
[15] L. H. Jenkins, D. M. Zehner, and M. F. Chung, *Surf. Sci.* **38** (1973) 327.
[16] T. A. Carlson, *Radiat. Res.* **64** (1975) 53.
[17] L. Fiermans, R. Hoogewijs, and J. Vennik, *Surf. Sci.* **47** (1975) 2.
[18] D. A. Shirley, *Phys. Rev. A* **9** (1974) 1549.
[19] O. Werme, T. Bergmark, and K. Siegbahn, *Phys. Scr.* **6** (1972) 141.
[20] N. V. Smith, G. K. Wertheim, S. Hüfner, and M. M. Traum *Phys. Rev. B* **10** (1974) 3197.
[21] T. E. Gallon and J. D. Nuttall, *Surf. Sci.* **53** (1975) 698.
[22] H. Raether, *Springer Tracts Mod. Phys.* **38** (1965) 84.
[23] C. J. Todd, *Vacuum* **23** (1973) 195, and references therein.
[24] J. Koch, *Rev. Sci. Instrum.* **45** (1974) 1212.
[25] H. E. Bishop, J. P. Coad, and J. C. Rivière, *J. Electron Spectrosc.* **1** (1972/3) 389.
[26] D. J. Poker, *Rev. Sci. Instrum.* **46** (1975) 105.
[27] S. Aksela, *Rev. Sci. Instrum.* **43** (1972) 1350.
[28] E. B. Bas, U. Banninger, and P. Keller, *J. Vac. Sci. Technol.* **9** (1972) 306.
[29] E. N. Sickafus and D. M. Holloway, *Surf. Sci.* **51** (1975) 131.

[30] H. Z. Sar-El, *Rev. Sci. Instrum.* **38** (1967) 1210.
[31] J. S. Risley, *Rev. Sci. Instrum.* **43** (1972) 95.
[32] P. W. Palmberg, G. E. Riach, R. E. Weber, and N. C. McDonald, *Handbook of Auger Electron Spectroscopy*, Physical Electronic Inc., 1972.
[33] W. A. Coghlan and R. E. Clausing, *Surf. Sci.* **33** (1972) 411; and W. A. Coghlan and R. E. Clausing, "Catalog of Calculated Auger Transitions for the Elements," USAEC Report ORNL-TM-3576.
[34] C. C. Chang, *Surf. Sci.* **25** (1971) 53.
[35] G. Schön, *J. Electron Spectrosc.* **1** (1972/3) 377.
[36] D. A. Shirley, *Chem. Phys. Lett.* **17** (1972) 312.
[37] D. A. Shirley, *Phys. Rev.* A **7** (1973) 1520.
[38] L. A. West, *J. Vac. Sci. Technol.* **13** (1976) 198.
[39] C. Burggraf, B. Carrière, and S. Goldsztaub, *Ref. Phys. Appl.* **11** (1976) 13.
[40] N. J. Taylor, *Rev. Sci. Instrum.* **40** (1969) 792.
[41] R. E. Weber and A. L. Johnson, *J. Appl. Phys.* **40** (1969) 314.
[42] K. Ueda and R. Shimizu, *Surf. Sci.* **36** (1973) 789.
[43] C. A. Shell and J. C. Rivière, *Surf. Sci.* **40** (1973) 149.
[44] R. Bouwman, L. H. Toneman, and A. A. Holscher, *Vacuum* **23** (1973) 163.
[45] R. Bouwman, L. H. Toneman, and A. A. Holscher, *Surf. Sci.* **35** (1973) 8.
[46] R. Bouwman, J. B. van Mechelen, and A. A. Holscher, *Surf. Sci.* **57** (1976) 441.
[47] G. A. Somorjai and S. H. Overbury, *Discuss. Faraday Soc.* **60** (1975) 279.
[48] J. M. Morabito, *Surf. Sci.* **49**, (1975) 318.
[49] C. C. Chang, *Surf. Sci.* **48** (1975) 9.
[50] J. J. Vrakking and F. Meyer, *Appl. Phys. Lett.* **18** (1971) 226.
[51] F. Meyer and J. J. Vrakking, *Surf. Sci.* **33** (1972) 271.
[52] M. Perdereau, *Surf. Sci.* **24** (1971) 239.
[53] J. W. T. Ridgway and D. Haneman, *Surf. Sci.* **24** (1971) 251.
[54] L. L. Levenson, L. E. Davis, C. E. Bryson III, J. J. Melles, and V. H. Kou, *J. Vac. Sci. Technol.* **9** (1972) 608.
[55] R. W. Joyner, C. S. McKee, and M. W. Roberts, *Surf. Sci.* **27** (1971) 279.
[56] H. Pollard, *Surf. Sci.* **20** (1970) 269.
[57] P. W. Palmberg and T. N. Rhodin, *J. Appl. Phys.* **30** (1968) 2425.
[58] J. V. Florio and W. D. Roberston, *Surf. Sci.* **18** (1969) 398.
[59] S. Thomas and T. W. Haas, *J. Vac. Sci. Technol.* **9** (1971) 840.
[60] R. W. Joyner, J. Rickman and M. W. Roberts, *J.C.S. Faraday I*, **70** (1974) 1825.
[61] E. Taglauer and W. Heiland, *Appl. Phys. Lett.* **24** (1974) 437.
[62] R. Anton and M. Harsdorff, *Thin Solid Films* **22** (1974) 523.
[63] J. M. Morabito, *Anal. Chem.* **46** (1974) 189.
[64] D. T. Quinto, V. S. Sundaram, and W. D. Roberston, *Surf. Sci.* **28** (1971) 504.
[65] C. Leygraf, G. Hultquist, and S. Ekelund, *Surf. Sci.* **46** (1974) 152.
[66] P. Braun and W. Färber, *Surf. Sci.* **47** (1975) 57.
[67] A. van Santen, L. H. Toneman, and R. Bouwman, *Surf. Sci.* **47** (1975) 64.
[68] C. Argile and G. E. Rhead, *Surf. Sci.* **53** (1975) 659.
[69] C. Argile and G. E. Rhead, *J. Phys.* C **7** (1974) L261.
[70] G. E. Rhead, *J. Vac. Sci. Technol.* **13** (1976) 603.
[71] D. H. Holloway, *Surf. Sci.* **66** (1977) 479.
[72] R. L. Gerlach and A. R. DuCharme, *Jap. J. Appl. Phys. Suppl.* 2, Pt. 2 (1974) 675.
[73] J. J. Vrakking and F. Meyer, *Phys. Rev.* A **9** (1974) 1932.

[74] C. J. Powell, *J. Vac. Sci. Technol.* **13** (1976) 219.
[75] J. C. Tracy, *Surf. Sci.* **38** (1973) 265.
[76] A. R. DuCharme and R. L. Gerlach, *J. Vac. Sci. Technol.* **11** (1974) 281.
[77] P. H. Holloway, *Solid State Commun.* **19** (1976) 729.
[78] T. E. Gallon, *J. Phys. D* **5** (1972) 822.
[79] D. M. Smith and T. E. Gallon, *J. Phys. D* **7** (1974) 151.
[80] J. J. Vrakking and F. Meyer, *Surf. Sci.* **47** (1975) 50.
[81] K. Goto, K. Ishikawa, T. Koshikawa, and R. Shimizu, *Surf. Sci.* **47** (1975) 477.
[82] P. W. Palmberg, *J. Vac. Sci. Technol.* **13** (1976) 214.
[83] M. L. Tarng and G. K. Wehner, *J. Appl. Phys.* **44** (1973) 1534.
[84] F. J. Szalkowski and G. A. Somorjai, *J. Chem. Phys.* **56** (1972) 6097.
[85] F. J. Szalkowski and G. A. Somorjai, *Surf. Sci.* **52** (1975) 431.
[86] F. J. Szalkowski and G. A. Somorjai, *J. Chem. Phys.* **61** (1974) 2064.
[87] J. C. Philips, *Rev. Mod. Phys.* **42** (1970) 317; J. A. van Vechten, *Phys. Rev.* **182** (1969) 891.
[88] H. H. Farrell, *Surf. Sci.* **34** (1973) 465.
[89] R. Manne, *J. Chem. Phys.* **46** (1967) 4645.
[90] B. Carrière, J. P. Deville, and S. Goldsztaub, *Silicates Industriels* **11** (1974) 313.
[91] J. S. Solomon and W. L. Baun, *Surf. Sci.* **51** (1975) 228.
[92] A. P. Janssen, R. C. Schoonmaker, and A. Chambers, *Surf. Sci.* **47** (1975) 41.
[93] J. W. May and C. E. Carroll, *Surf. Sci.* **29** (1972) 85.
[94] A. P. Janssen, R. C. Schoonmaker, A. Chambers, and M. Prutton, *Surf. Sci.* **45** (1974) 45.
[95] G. F. Amelio, *Surf. Sci.* **22** (1970) 301.
[96] E. N. Sickafus and F. Steinrisser, *J. Vac. Sci. Technol.* **10** (1973) 43.
[97] D. R. Arnott and D. Haneman, *Surf. Sci.* **45** (1974) 128.
[98] S. Ferrer, A. M. Baro, and M. Salmeron, *Solid State Commun.* **16** (1975) 65.
[99] D. W. Fischer, *Phys. Rev. B* **5** (1972) 4219.
[100] J. P. Coad and J. C. Rivière, *Proc. Roy. Soc. London Ser. A* **331** (1972) 403.
[101] G. E. Becker and H. D. Hagstrum, *Surf. Sci.* **30** (1972) 505.
[102] F. Steinrisser and E. N. Sickafus, *Phys. Rev. Lett.* **27** (1971) 992.
[103] M. Salmerón and A. M. Baró, *Surf. Sci.* **49** (1975) 356.
[104] C. J. Powell, *Phys. Rev. Lett.* **30** (1973) 1179.
[105] P. J. Bassett, T. E. Gallon, J. A. D. Matthew, and M. Prutton, *Surf. Sci.* **35** (1973) 63.
[106] S. P. Kowalczyk, R. A. Pollak, F. R. McFeely, L. Leyland, and D. A. Shirley, *Phys. Rev. B* **8** (1973) 2387.
[107] J. J. Lander, *Phys. Rev.* **91** (1953) 1382.
[108] N. R. Avery, *Surf. Sci.* **61** (1976) 391.
[109] N. R. Avery, *Phys. Rev. Lett.* **32** (1974) 1248.
[110] M. Salmerón, A. M. Baró, and J. M. Rojo, *Surf. Sci.* **53** (1975) 689.
[111] A. M. Baró, M. Salmerón, and J. M. Rojo, *J. Phys. F* **5** (1975) 826.
[112] P. H. Citrin, J. E. Rowe, and S. B. Christman, *Phys. Rev. B* **14** (1976) 2642.
[113] H. H. Madden and J. E. Houston, *J. Vac. Sci. Technol.* **14** (1977) 412.
[114] T. W. Haas, J. T. Grant, and G. J. Dooley, *Proceedings of the Second International Symposium on Adsorption/Desorption Phenomena*, Ed. by F. Ricca, Academic Press, London, 1973.
[115] B. J. Hopkins, A. R. Jones, and R. I. Winton, *Surf. Sci.* **57** (1976) 266.
[116] J. P. Coad and J. C. Rivière, *Surf. Sci.* **24** (1971) 1.
[117] K. Ueda and R. Shimizu, *Surf. Sci.* **43** (1974) 77.

[118] H. D. Shih, K. O. Legg, and F. Jona, *Surf. Sci.* **54** (1976) 355.
[119] P. J. Bassett and T. E. Gallon, *J. Electron Spectrosc.* **2** (1973) 101.
[120] E. N. Sickafus, *Rev. Sci. Instrum.* **42** (1971) 933.
[121] L. Fiermans and G. Vennik, *Surf. Sci.* **38** (1973) 357.
[122] P. Staib and J. Kirschner, *Appl. Phys.* **3** (1974) 421.
[123] P. Staib, *Surf. Sci.* **53** (1975) 582.
[124] J. E. Houston, *Surf. Sci.* **38** (1973) 283.
[125] J. T. Grant, T. W. Haas, and J. E. Houston, *J. Vac. Sci. Technol.* **11** (1974) 227.
[126] J. E. Houston, *Rev. Sci. Instrum.* **45** (1974) 897.
[127] J. E. Houston, *Appl. Phys.* **6** (1975) 281.
[128] J. T. Grant, T. W. Haas, and J. E. Houston, *Surf. Sci.* **42** (1974) 1.
[129] J. T. Grant, T. W. Haas, and J. E. Houston, *Phys. Lett.* **45A** (1973) 309.
[130] J. T. Grant and T. W. Haas, *Surf. Sci.* **44** (1974) 617.
[131] J. E. Houston and R. L. Park, *Rev. Sci. Instrum.* **43** (1972) 1437.
[132] J. E. Houston, *Appl. Phys. Lett.* **24** (1974) 42.
[133] J. E. Houston, *J. Vac. Sci. Technol.* **12** (1975) 255.
[134] K. Goto, K. Ishikawa, T. Koshikawa, and R. Shimizu, *Surf. Sci.* **47** (1975) 477.
[135] J. T. Grant, M. P. Hooker, and T. W. Haas, *Surf. Sci.* **46** (1974) 672.
[136] L. C. Isett and J. M. Blakely, *J. Vac. Sci. Technol.* **11** (1974) 1282.
[137] R. W. Springer, D. J. Poker, and T. W. Haas, *Appl. Phys. Lett.* **27** (1975) 368.
[138] R. W. Springer and D. J. Poker, *Rev. Sci. Instrum.* **48** (1977) 74.
[139] J. T. Grant, M. P. Hooker, R. W. Springer, and T. W. Haas, *Surf. Sci.* **60** (1976) 1.
[140] D. J. Poker, R. W. Springer, F. E. Ruttenberg, and T. W. Haas, *J. Vac. Sci. Technol.* **13** (1976) 507.
[141] J. T. Grant, M. P. Hooker, and T. W. Haas, *Surf. Sci.* **51** (1975) 318.
[142] M. N. Varma, A. Joshi, M. Strongin, and V. Radeka, *Rev. Sci. Instrum.* **44** (1973) 1643.
[143] P. H. van Cittert, *Z. Phys.* **69** (1931) 298; H. C. Burger and P. H. van Cittert, *Z. Phys.* **79** (1932) 722; H. C. Burger and P. H. van Cittert, *Z. Phys.* **81** (1933) 428.
[144] H. H. Madden and J. E. Houston, *J. Appl. Phys.* **47** (1976) 3071.
[145] T. N. Taylor and P. J. Estrup, *J. Vac. Sci. Technol.* **10** (1973) 261.
[146] K. Christmann, O. Schober and G. Ertl, *J. Chem. Phys.* **60** (1974) 4719.
[147] R. M. Lambert and C. M. Comrie, *Surf. Sci.* **38** (1973) 197.
[148] J. Tracy and P. W. Palmberg, *J. Chem. Phys.* **51** (1969) 4852.
[149] J. M. Martinez and J. B. Hudson, *J. Vac. Sci. Technol.* **10** (1973) 35.
[150] H. H. Madden and G. Ertl, *Surf. Sci.* **35** (1973) 211.
[151] H. Tokutaka, M. Prutton, I. G. Higginbotham, and T. E. Gallon, *Surf. Sci.* **21** (1970) 23.
[152] L. S. Cota Araiza and B. D. Powell, *Surf. Sci.* **51** (1975) 504.
[153] S. Thomas, *J. Appl. Phys.* **45** (1974) 161.
[154] R. E. Kirby and D. Lichtman, *Surf. Sci.* **41** (1974) 447.
[155] R. E. Kirby and J. W. Dieball, *Surf. Sci.* **41** (1974) 467.
[156] M. Salmerón and A. M. Baró, *Surf. Sci.* **29** (1972) 300.
[157] A. P. Janssen, R. C. Schoonmaker, and A. Chambers, *Surf. Sci.* **49** (1975) 143.
[158] P. H. Holloway, *Surf. Sci.* **54** (1976) 506.
[159] D. Lichtman and J. Campuzano, *Jap. J. Appl. Phys. Suppl.* **2** Pt. 2 (1974) 24.
[160] A. P. Janssen, *Surf. Sci.* **52** (1975) 230.
[161] C. T. H. Stoddart and E. D. Hondros, *Trans. Brit. Ceram. Soc.* **73** (1973) 61.

[162] Y. P. Zingerman, *Sov. Phys. Solid State* **16** (1974) 1168.
[163] M. Nishijima, K. Fujiwara, and T. Murotani, *J. Appl. Phys.* **46** (1975) 3089.
[164] W. Ranke and K. Jacobi, *Surf. Sci.* **47** (1975) 525.
[165] P. A. Redhead, *Can. J. Phys.* **42** (1964) 886.
[166] Y. Margoninski, D. Segal, and R. E. Kirby, *Surf. Sci.* **53** (1975) 488.
[167] D. Menzel, P. Kronauer, and W. Jelend, *Ber. Bunsenges. Phys. Chem.* **75** (1971) 1074.
[168] E. Bauer, *Vacuum* **22** (1972) 539.
[169] W. Jelend and D. Menzel, *Vide* **28** (1973) 86.
[170] D. Menzel, *Surf. Sci.* **47** (1975) 370.
[171] W. Brenig, *Z. Phys. B* **23** (1976) 361.
[172] M. Gettings and J. P. Coad, A.E.R.E. Harwell Report No. R 8288 (1976)
[173] J. P. Coad, M. Gettings, and J. C. Rivière, *Faraday Discuss. Chem. Soc.* **60** (1975) 269.
[174] J. P. Rynd and A. K. Rastogi, *Surf. Sci.* **48** (1975) 22.
[175] A. E. Clark Jr, C. G. Pantano, and L. L. Hench, *J. Am. Ceram. Soc.* **59** (1976) 37.
[176] S. T. Mason, *J. Electron Spectrosc.* **9** (1976) 21.
[177] T. A. Carlson, *Faraday Discuss. Chem. Soc.* **60** (1975) 30.
[178] D. P. Spears, H. J. Fischbeck, and T. A. Carlson, *Phys. Rev. A* **9** (1974) 1603.
[179] T. A. Carlson, J. C. Carver, L. J. Saethre, F. G. Santibanez, and G. A. Vernon, *J. Electron Spectrosc.* **5** (1974) 247.
[180] L. Yin, T. Tsang, and I. Adler, *J. Electron Spectrosc.* **9** (1976) 67.
[181] J. Escard, G. Mavel, J. E. Guerchais, and R. Kergoat, *Inorg. Chem.* **13** (1974) 695.
[182] B. Wallbank, I. G. Main, and C. E. Johnson, *J. Electron Spectrosc.* **5** (1974) 259.
[183] K. S. Kim, *J. Electron Spectrosc.* **3** (1974) 217.
[184] M. Brisk and A. D. Baker, *J. Electron Spectrosc.* **7** (1975) 81.
[185] K. S. Kim and N. Winograd, *Chem. Phys. Lett.* **31** (1975) 312.
[186] S. K. Sen, J. Riga, and J. Verbist, *Chem. Phys. Lett.* **39** (1976) 560.
[187] D. C. Frost, C. A. McDowell, and B. Wallbank, *Chem. Phys. Lett.* **40** (1976) 189.
[188] C. K. Jørgensen and H. Berthou, *Chem. Phys. Lett.* **13** (1972) 186.
[189] H. Berthou, C. K. Jørgensen, and C. Bonnelle, *Chem. Phys. Lett.* **38** (1976) 199.
[190] C. K. Jørgensen, *Struct. Bonding (Berlin)* **30** (1976) 141.
[191] M. A. Brisk and A. D. Baker, *J. Electron Spectrosc.* **7** (1975) 197.
[192] A. J. Signorelli and R. G. Hayes, *Phys. Rev. B* **8** (1973) 81.
[193] L. Yin, I. Adler, T. Tsang, L. J. Matienzo, and S. O. Grim, *Chem. Phys. Lett.* **24** (1974) 81.
[194] G. A. Vernon, G. Stucky, and T. A. Carlson, *Inorg. Chem.* **15** (1976) 278.
[195] T. Robert and G. Offergeld, *Chem. Phys. Lett.* **29** (1974) 606.
[196] S. Larsson, *Chem. Phys. Lett.* **32** (1975) 401.
[197] K. S. Kim, *Phys. Rev. B* **11** (1975) 2177.
[198] J. C. Helmer, *J. Electron Spectrosc.* **1** (1972/3) 259.
[199] G. C. Allen and P. M. Tucker, *Inorg. Chim. Acta* **16** (1976) 41.
[200] C. S. Fadley, in *Electron Emission Spectroscopy*, Ed. by W. Dekeyser, L. Fiermans, G. Vanderkelen, and J. Vennik, Reidel, Dordrecht, 1973, p. 151.
[201] C. K. Jørgensen, *Struct. Bonding (Berlin)* **24** (1975) 1.
[202] P. A. Cox, *Struct. Bonding (Berlin)* **24** (1975) 59.
[203] P. A. Cox, Y. Baer, and C. K. Jørgensen, *Chem. Phys. Lett.* **22** (1973) 433.

[204] C. K. Jørgensen, *Fresenius, Z. Anal. Chem.* **288** (1977) 161.
[205] R. L. Cohen, G. K. Wertheim, A. Rosenwaig, and H. J. Guggenheim, *Phys. Rev. B* **5** (1972) 1037.
[206] F. R. McFeely, S. P. Kowalczyk, L. Ley, and D. A. Shirley, *Phys. Lett.* **49A** (1974) 301.
[207] S. P. Kowalczyk, L. Ley, R. L. Martin, F. R. McFeely, and D. A. Shirley, *Faraday Discuss. Chem. Soc.* **60** (1975) 7.
[208] W. C. Lang, B. D. Padalia, L. M. Watson, D. M. Fabian, and P. R. Norris, *Faraday Discuss. Chem. Soc.* **60** (1975) 37.
[209] T. A. Carlson, J. C. Carver, and G. A. Vernon, *J. Chem. Phys.* **62** (1975) 932.
[210] D. R. Gupta and S. K. Sen, *Phys. Rev. B* **10** (1974) 71.
[211] D. C. Frost, C. A. McDowell, and I. S. Woolsey, *Chem. Phys. Lett.* **17** (1972) 320.
[212] D. C. Frost, C. A. McDowell, and I. S. Woolsey, *Mol. Phys.* **27** (1974) 1473.
[213] L. Kowalczyk, L. Ley, F. R. McFeely, and D. A. Shirley, *Phys. Rev. B* **11** (1975) 1721.
[214] J. A. Tossell, *J. Electron Spectrosc.* **10** (1977) 169.
[215] E. K. Viinikka and S. Larsson, *J. Electron Spectrosc.* **7** (1975) 163.
[216] J. C. Carver, G. K. Schweitzer, and T. A. Carlson. *J. Chem. Phys.* **57** (1972) 973.
[217] G. K. Wertheim, S. Hüfner, and H. J. Guggenheim, *Phys. Rev. B* **7** (1973) 556.
[218] I. Ikemoto, K. Ishii, S. Kinoshita, T. Fujikawa, and H. Kuroda, *Chem. Phys. Lett.* **38** (1976) 467.
[219] R. A. Pollak, L. Ley, F. R. McFeely, S. P. Kowalczyk, and D. A. Shirley, *J. Electron Spectrosc.* **3** (1974) 381.
[220] C. A. Lucchesi and J. E. Lester, *J. Chem. Educ.* **50** (1973) A 205 and A 269.
[221] B. Wannberg, U. Gelius, and K. Siegbahn, *J. Phys. E* **7** (1974) 149.
[222] P. C. Kemeny, A. D. McLachlan, F. L. Battye, R. T. Poole, R. C. G. Leckey, J. Liesegang, and J. G. Jenkin, *Rev. Sci. Instrum.* **44** (1973) 1197.
[223] M. E. Gellender and A. D. Baker, *J. Electron Spectrosc.* **4** (1974) 249.
[224] D. Betteridge and M. A. Williams, *Anal. Chem.* **46** (1974) 125 R.
[225] H. Fellner-Feldegg, U. Gelius, B. Wannberg, A. G. Nilsson, E. Basilier, and K. Siegbahn, *J. Electron Spectrosc.* **5** (1974) 643.
[226] M. J. Weiss, *J. Electron Spectrosc.* **1** (1972/3) 179.
[227] H. G. Nöller, H. D. Polaschegg, and H. Schillalies, *J. Electron Spectrosc.* **5** (1974) 705.
[228] J. F. McGilp and I. G. Main, *J. Electron Spectrosc.* **6** (1975) 397.
[229] S. Y. Yavor, I. A. Petrov, and E. P. Denisov, *Zh. Tekh. Fiz.* **41** (1971) 1839.
[230] J. L. Gardner and J. A. R. Samson, *J. Electron Spectrosc.* **2** (1973) 267.
[231] A. D. McLachlan, R. C. Leckey, J. G. Jenkin, and J. Liesegang, *Rev. Sci. Instrum.* **44** (1973) 873.
[232] P. W. Palmberg, *J. Electron Spectrosc.* **5** (1974) 691.
[233] C. S. Fadley, R. N. Healey, J. M. Hollander, and C. E. Miner *J. Appl. Phys.* **43** (1972) 1085.
[234] W. Denk and T. von Egidy, *Nucl. Instrum. Methods* **102** (1972) 281.
[235] P. Staib, *Vacuum* **22** (1972) 481.
[236] I. Lindau, J. C. Helmer, and J. Uebbing, *Rev. Sci. Instrum.* **44** (1973) 265.
[237] J. D. Lee, *Rev. Sci. Instrum.* **44** (1973) 893.
[238] K. Maeda and Y. Tada, *Jpn. J. Appl. Phys.* **11** (1972) 1059.
[239] K. Siegbahn, D. Hammond, H. Fellner-Feldegg, and E. F. Barnett, *Science* **176** (1972) 245.

[240] Y. Baer, G. Busch, and P. Cohn, *Rev. Sci. Instrum.* **47** (1975) 466.
[241] R. T. Poole, R. C. G. Leckey, J. Liesegang, and J. G. Jenkin, *J. Phys. E* **6** (1973) 226.
[242] J. L. Gardner and J. A. R. Samson, *J. Electron Spectrosc.* **6** (1975) 53.
[243] R. Prins, *Chem. Phys. Lett.* **19** (1973) 355.
[244] A. Calabrese and R. G. Hayes, *Chem. Phys. Lett.* **27** (1974) 376.
[245] A. Calabrese and R. G. Hayes, *J. Electron Spectrosc.* **6** (1975) 1.
[246] J. H. Scofield, *J. Electron Spectrosc.* **8** (1976) 129.
[247] J. T. J. Huang and F. O. Ellison, *J. Electron Spectrosc.* **4** (1974) 233.
[248] P. C. Kemeny, J. G. Jenkin, J. Liesegang, and R. C. G. Leckey, *Phys. Rev. B* **9** (1974) 5307.
[249] R. C. G. Leckey, *Phys. Rev. A* **13** (1976) 1043.
[250] R. S. Swingle III, *Anal. Chem.* **47** (1975) 21.
[251] S. T. Manson, *J. Electron Spectrosc.* **1** (1972/3) 413.
[252] T. K. Ng and D. M. Hercules, *J. Electron Spectrosc.* **7** (1975) 257.
[253] D. M. Wyatt, J. C. Carver, and D. M. Hercules, *Anal. Chem.* **47** (1975) 1297.
[254] L. E. Cox and D. M. Hercules, *J. Electron Spectrosc.* **1** (1972/3) 193.
[255] C. D. Wagner, *Anal. Chem.* **44** (1972) 1050.
[256] C. K. Jørgensen and H. Berthou, *Faraday Discuss. Chem. Soc.* **54** (1972) 269.
[257] H. Berthou and C. K. Jørgensen, *Anal. Chem.* **47** (1975) 482.
[258] V. I. Nefedov, N. P. Sergushin, I. M. Band, and M. B. Trzhaskovskaya, *J. Electron Spectrosc.* **2** (1973) 383.
[259] V. I. Nefedov, N. P. Sergushin, Y. V. Salyn, I. M. Band, and M. B. Trzhaskovskaya, *J. Electron Spectrosc.* **7** (1975) 175.
[260] L. J. Brillson and G. P. Ceasar, *Surf. Sci.* **58** (1976) 457.
[261] K. Asami, *J. Electron Spectrosc.* **9** (1976) 469.
[262] M. F. Ebel, *J. Electron Spectrosc.* **8** (1976) 213.
[263] L. Evans, *Chem. Phys. Lett.* **23** (1973) 134.
[264] A. F. Carley, R. W. Joyner, and M. W. Roberts, *Chem. Phys. Lett.* **27** (1974) 580.
[265] F. Ascarelli and G. Missoni, *J. Electron Spectrosc.* **5** (1974) 417.
[266] M. F. Ebel and H. Ebel, *J. Electron Spectrosc.* **3** (1974) 169.
[267] M. F. Ebel, *Vak.-Tech.* **23** (1974) 33.
[268] D. A. Huchital and R. T. McKeon, *Appl. Phys. Lett.* **20** (1972) 158.
[269] J. R. Lindsay, H. J. Rose, W. E. Swartz, P. H. Watts, and K. A. Rayburn, *Appl. Spectrosc.* **27** (1973) 1.
[270] D. J. Hnatowich, J. Hudis, M. L. Perlman, and R. C. Ragaini, *J. Appl. Phys.* **42** (1971) 4883.
[271] J. M. Thomas, E. L. Evans, M. Barber, and P. Swift, *Trans. Faraday Soc.* **67** (1971) 1875.
[272] D. Chadwick, *Chem. Phys. Lett.* **21** (1973) 291.
[273] G. Johansson, J. Hedman, A. Berndtsson, M. Klasson, and R. Nilsson, *J. Electron Spectrosc.* **2** (1973) 295.
[274] D. S. Urch and M. Webber, *J. Electron Spectrosc.* **5** (1974) 791.
[275] C. R. Ginnard and W. M. Riggs, *Anal. Chem.* **46** (1974) 1306.
[276] D. Betteridge, J. C. Carver, and D. M. Hercules, *J. Electron Spectrosc.* **2** (1973) 327.
[277] L. J. Matienzo and S. O. Grim, *Anal. Chem.* **46** (1974) 2052.
[278] J. P. Contour and G. Mouvier, *J. Electron Spectrosc.* **7** (1975) 85.
[279] A. D. Hamer, D. G. Tilsey, and R. A. Walton, *J.C.S. Dalton Trans. I* (1973) 116.
[280] C. K. Jørgensen, *Chimia* **25** (1971) 213.

[281] V. I. Nefedov, Y. A. Buslaev, and G. Kokunov, *J. Anorg. Chem.* **19** (1974) 116.
[282] C. K. Jørgensen and H. Berthou, *J. Fluorine Chem.* **2** (1975) 425.
[283] C. K. Jørgensen and H. Berthou, *Chem. Phys. Lett.* **31** (1975) 416.
[284] V. I. Nefedov, Y. V. Salyn, G. Leonhardt, and R. Scheibe, *J. Electron Spectrosc.* **10** (1977) 121.
[285] C. D. Wagner, *Anal. Chem.* **47** (1975) 1201.
[286] C. D. Wagner, *Faraday Discuss. Chem. Soc.* **60** (1975) 291.
[287] C. D. Wagner, *J. Electron Spectrosc.* **10** (1977) 305.
[288] W. L. Jolly, *Faraday Discuss. Chem. Soc.* **54** (1972) 13.
[289] K. L. Cheng and J. W. Prater, *Crit. Rev. Anal. Chem.* **5** (1975) 37.
[290] D. T. Clark, I. W. Scanian, and J. Muller, *Theor. Chim. Acta* **35** (1974) 341.
[291] G. Howat and O. Goscinski, *Chem. Phys. Lett.* **30** (1975) 87.
[292] R. E. Watson, M. L. Perlman, and J. F. Herbst, *Phys. Rev. B* **13** (1976) 2358.
[293] S. Svensson, N. Martensson, E. Basilier, P. A. Malmqvist, U. Gelius, and K. Siegbahn, *J. Electron Spectrosc.* **9** (1976) 51.
[294] C. K. Jørgensen, H. Berthou, and L. R. Balsenc, *J. Fluorine Chem.* **1** (1972) 327.
[295] J. M. Thomas, reported by D. Betteridge, *Analyst* **99** (1974) 994.
[296] C. K. Jørgensen, *Top. Curr. Chem.* **56** (1975) 1.
[297] L. Hedin and A. Johansson, *J. Phys. B* **2** (1969) 1336.
[298] C. H. Hodges, *J. Phys. C* **8** (1975) 1849.
[299] S. P. Kowalczyk, L. Ley, F. R. McFeely, R. A. Pollak, and D. A. Shirley, *Phys. Rev. B* **8** (1973) 3583.
[300] D. W. Davis and D. A. Shirley, *Chem. Phys. Lett.* **15** (1972) 185.
[301] L. Ley, R. A. Pollak, F. R. McFeely, S. P. Kowalczyk, and D. A. Shirley, *Phys. Rev. B* **9** (1974) 600.
[302] L. Ley, F. R. McFeely, S. P. Kowalczyk, J. G. Jenkin, and D. A. Shirley, *Phys. Rev. B* **11** (1975) 600.
[303] D. A. Shirley, *J. Electron Spectrosc.* **5** (1974) 135.
[304] L. Ley, S. P. Kowalczyk, F. R. McFeely, R. A. Pollak, and D. A. Shirley, *Phys. Rev. B* **8** (1973) 2392.
[305] U. Gelius, *Phys. Scr.* **9** (1974) 133.
[306] D. W. Davis, R. L. Martin, M. S. Banna, and D. A. Shirley, *J. Chem. Phys.* **59** (1974) 4235.
[307] S. P. Kowalczyk, L. Ley, F. R. McFeely, R. A. Pollak, and D. A. Shirley, *Phys. Rev. B* **9** (1974) 381.
[308] C. K. Jørgensen and H. Berthou, *Dan. Vidensk. Selk. Mat. Fys. Medd. K.* **38** No. 15 (1972).
[309] C. K. Jørgensen and H. Berthou, *Chem. Phys. Lett.* **36** (1975) 42.
[310] P. Biloen, *Surf. Sci.* **47** (1975) 48.
[311] D. A. Shirley, *J. Vac. Sci. Technol.* **12** (1975) 280.
[312] M. Barber, E. L. Evans, and J. M. Thomas, *Chem. Phys. Lett.* **18** (1973) 423.
[313] J. E. Demuth and D. E. Eastman, *Phys. Rev. Lett.* **32** (1974) 1123.
[314] J. W. Gadzuk, *J. Vac. Sci. Technol.* **13** (1976) 343.
[315] D. M. Hercules, *Anal. Chem.* **48** (1976) 294 R.
[316] C. J. Vesely and D. L. Kingston, *Phys. Rev. B* **8** (1973) 2685.
[317] R. G. Cavell, S. P. Kowalczyk, L. Ley, R. A. Pollak, B. Mills, D. A. Shirley, and W. Perry, *Phys. Rev. B* **7** (1973) 5313.
[318] V. V. Nemoshkalenko, V. G. Aleshin, Y. N. Kucherenko, and L. M. Sheludshenko, *J. Electron Spectrosc.* **6** (1975) 145.

[319] J. Chelikowsky, D. J. Chadi, and M. L. Cohen, *Phys. Rev. B* **8** (1973) 2786.
[320] G. K. Wertheim, H. J. Guggenheim, and S. Hüfner, *Phys. Rev. Lett.* **30** (1973) 1050.
[321] H. Ebel and N. Gurker, *J. Electron Spectrosc.* **5** (1974) 799.
[322] H. Ebel and N. Gurker, *Phys. Lett.* **50A** (1975) 449.
[323] G. K. Wertheim, *J. Electron Spectrosc.* **6** (1975) 239.
[324] C. D. Wagner, *Anal. Chem.* **44** (1972) 967.
[325] A. Barrie, *J. Electron Spectrosc.* **7** (1975) 75.
[326] C. D. Wagner, *Anal. Chem.* **44** (1972) 1050.
[327] H. Berthou and C. K. Jørgensen, *J. Electron Spectrosc.* **5** (1974) 935.
[328] J. C. Fuggle, L. M. Watson, D. J. Fabian, and S. J. Affrossman, *J. Phys. F* **5** (1975) 375.
[329] K. L. Wang and A. Joshi, *J. Vac. Sci. Technol.* **12** (1975) 927.
[330] J. Haber and L. Ungier, *J. Electron Spectrosc.* **12** (1977) 305.
[331] D. C. Frost, A. Ishitani, and C. A. McDowell, *Mol. Phys.* **24** (1972) 861.
[322] G. Kumar, J. R. Blackburn, R. G. Albrige, W. E. Moddeman, and M. M. Jones, *Inorg. Chem.* **11** (1972) 296.
[333] D. Cahen and J. E. Lester, *Chem. Phys. Lett.* **18** (1973) 108.
[334] P. Burroughs, A. Hamnett, and A. F. Orchard, *J.C.S. Dalton Trans.* (1974) 565.
[335] K. Hirokawa, F. Honda, and M. Oku, *J. Electron Spectrosc.* **6** (1975) 333.
[336] B. Wallbank, C. E. Johnson, and I.G. Main, *J. Electron Spectrosc.* **4** (1974) 263.
[337] B. A. De Angelis, *J. Electron Spectrosc.* **9** (1976) 81.
[338] P. Burroughs, A. Hamnett, J. F. McGilp, and A. F. Orchard, *J.C.S. Faraday Trans.* II, **71** (1975) 177.
[339] T. E. Gallon, *Surf. Sci.* **17** (1969) 486.
[340] M. P. Seah, *Surf. Sci.* **40** (1973) 381.
[341] D. C. Jackson, T. E. Gallon, and A. Chambers, *Surf. Sci.* **36** (1973) 381.
[342] J. C. Shelton, *Surf. Sci.* **44** (1974) 305.
[343] C. R. Helms and W. E. Spicer, *Phys. Lett. A* **57A** (1976) 369.
[344] G. Brodén, *Phys. Kondens. Mater.* **15** (1972) 171.
[345] J. C. Tracy, NATO Summer School Lectures, Ghent, 1972.
[346] J. C. Rivière, *Contemp. Phys.* **14** (1973) 513.
[347] C. J. Powell, *Surf. Sci.* **44** (1974) 29.
[348] I. Lindau and W. E. Spicer, *J. Electron Spectrosc.* **3** (1974) 409.
[349] D. R. Penn, *J. Vac. Sci. Technol.* **13** (1976) 221.
[350] M. P. Seah, *Vacuum* **22** (1972) 475.
[351] M. Klasson, J. Hedman, A. Berndtsson, R. Nilsson, C. Nordling, and P. Melnik, *Phys. Scr.* **5** (1972) 93.
[352] T. A. Carlson and G. E. McGuire, *J. Electron Spectrosc.* **1** (1972/73) 161.
[353] M. L. Tarng and G. K. Wehner, *J. Appl. Phys.* **44** (1973) 1534.
[354] C. J. Todd and R. Heckingbottom, *Phys. Lett.* **42A** (1973) 455.
[355] W. A. Fraser, J. V. Florio, W. N. Delgass, and W. D. Robertson, *Surf. Sci.* **36** (1973) 661.
[356] B. L. Henke, *Phys. Rev. A* **6** (1972) 94.
[357] J. Brunner and H. Zogg, *J. Electron Spectrosc.* **5** (1974) 911.
[358] H. Iwasaki and S. Nakamura, *Surf. Sci.* **65** (1977) 345.
[359] C. C. Chang, *Characterization of Solid Surfaces*, Ed. by P. F. Kane and G. B. Larrabee, Plenum, New York, 1974, Chap. 20.
[360] J. C. Shelton, *J. Electron Spectrosc.* **3** (1974) 417.

[361] R. J. Baird, C. S. Fadley, S. K. Kawamoto, M. Mehta, R. Alvarez, and J. A. Silva, *Anal. Chem.* **48** (1976) 843.
[362] C. S. Fadley and S. Å. L. Bergström, *Phys. Lett.* **35A** (1971) 375.
[363] C. S. Fadley, *J. Electron Spectrosc.* **5** (1974) 725.
[364] W. A. Fraser, J. V. Florio, W. N. Delgass, and W. D. Robertson, *Rev. Sci. Instrum.* **44** (1973) 1490.
[365] C. S. Fadley, R. J. Baird, W. Siekhaus, T. Novakov, and S. Å. L. Bergström, *J. Electron Spectrosc.* **4** (1974) 93.
[366] C. S. Fadley, *Progr. Solid State Chem.* **11** (1976) 265.
[367] L. A. Harris, *Surf. Sci.* **15** (1969) 77.
[368] T. W. Rusch, J. P. Bertino, and W. P. Ellis, *Appl. Phys. Lett.* **23** (1973) 359.
[369] M. Mehta and C. S. Fadley, *Phys. Lett.* **55A** (1975) 59.
[370] G. Allié, E. Blanc, D. Dufayard, and R. M. Stern, *Surf. Sci.* **46** (1974) 188.
[371] B. L. Henke, *J. Phys. (Paris) Colloq.* **4** (1971) 115.
[372] J. J. Vrakking and F. Meyer, *Surf. Sci.* **35** (1973) 34.
[373] C. R. Brundle, *Surf. Sci.* **48** (1975) 99.
[374] C. R. Brundle, *J. Vac. Sci. Technol.* **11** (1974) 212.
[375] G. A. Somorjai and S. H. Overbury, *Discuss Faraday Soc.* **60** (1975) 279.
[376] M. P. Seah, *Surf. Sci.* **32** (1972) 703.
[377] M. P. Seah, *J. Phys. F* **3** (1973) 1538.
[378] P. J. Feibelman, *Phys. Rev. B* **7** (1973) 2305.
[379] M. Šunjić and D. Šokčević, *J. Electron Spectrosc.* **5** (1974) 963.
[380] M. Šunjić and D. Šokčević, *Solid State Commun.* **18** (1976) 373.
[381] P. W. Palmberg, *Anal. Chem.* **45** (1973) 549 A.
[382] K. Watanabe, M. Hasiba, and T. Yamashina, *Surf. Sci.* **61** (1976) 483.
[383] F. Meyer and J. J. Vrakking, *Surf. Sci.* **45** (1974) 409.
[384] R. Bouwman and P. Biloen, *Surf. Sci.* **41** (1974) 348.
[385] C. R. Brundle, *J. Electron Spectrosc.* **5** (1974) 291.
[386] J. P. Coad and J. G. Cunningham, *J. Electron Spectrosc.* **3** (1974) 435.
[387] F. Pons, J. LeHericy, and J. P. Langeron, *J. Microsc. Spectrosc. El.* **2** (1977) 49.
[388] S. Hofman, *Appl. Phys.* **9** (1976) 59.
[389] H. W. Werner, *Vacuum* **24** (1974) 493.
[390] J. Johannessen, W. E. Spicer, and Y. E. Strausser, *J. Vac. Sci. Technol.* **13** (1976) 849.
[391] J. S. Solomon, *Appl. Spectrosc.* **30** (1976) 46.
[392] J. T. Grant, R. G. Wolfe, M. P. Hooker, R. W. Springer, and T. W. Haas, *J. Vac. Sci. Technol.* **14** (1977) 232.
[393] H. Iwasaki and S. Nakamura, *Surf. Sci.* **57** (1976) 779.
[394] G. K. Wehner and D. J. Hajicek, *J. Appl. Phys.* **42** (1971) 1145.
[395] R. S. Nelson and D. J. Mazey, *Radiat. Eff.* **18** (1973) 127.
[396] P. Sigmund, *J. Mater. Sci.* **8** (1973) 1545.
[397] M. J. Witcomb, *J. Mater. Sci.* **9** (1974) 1227.
[398] J. W. Coburn and E. Kay, *Crit. Rev. Solid. State Sci.* **4** (1974) 561.
[399] G. C. Nelson, *J. Vac. Sci. Technol.* **13** (1976) 974.
[400] J. M. Morabito, *Thin Solid Films* **29** (1973) 21.
[401] M. L. Tarng and G. K. Wehner, *J. Appl. Phys.* **42** (1971) 2449.
[402] A. Turos, W. F. van der Weg, S. Sigurd, and J. W. Mayer, *J. Appl. Phys.* **45** (1974) 2777.
[403] J. W. Coburn and E. Kay, *J. Vac. Sci. Technol.* **12** (1975) 403.
[404] V. E. Heinrich and J. C. C. Fan, *Surf. Sci.* **42** (1974) 139.

[405] H. C. Feng and J. M. Chen, *J. Phys. C* **7** (1974) L.75.
[406] A. van Oostrom, *J. Vac. Sci. Technol.* **13** (1976) 224.
[407] H. W. Werner and N. Warmoltz, *Surf. Sci.* **57** (1976) 706.
[408] D. T. Quinto, V. S. Sundaram, and W. D. Robertson, *Surf. Sci.* **28** (1971) 504.
[409] G. S. Anderson, *J. Appl. Phys.* **40** (1968) 2884.
[410] N. S. McIntyre and D. G. Zetaruk, *J. Vac. Sci. Technol.* **14** (1977) 181.
[411] H. Shimizu, M. Ono and K. Nakayama, *Surf. Sci.* **36** (1973) 817.
[412] J. R. Arthur and J. J. Le Pore, *J. Vac. Sci. Technol.* **13** (1976) 979.
[413] P. S. Ho, J. E. Lewis, H. S. Wildman and J. K. Howard, *Surf. Sci.* **57** (1976) 393.
[414] P. S. Ho, J. E. Lewis and J. K. Howard, *J. Vac. Sci. Technol.* **13** (1976) 322.
[415] F. Schulz, K. Wittmark, and J. Maul, *Radiat. Eff.* **18** (1973) 211.
[416] V. Leroy, J. P. Servais, and L. Habraken, *C. Res. Met.* **35** (1973) 69.
[417] R. Weber, *Res. Dev.* **23** (1972) 22.
[418] P. W. Palmberg, *J. Vac. Sci. Technol.* **9** (1972) 160.
[419] M. P. Hooker and J. T. Grant, *Surf. Sci.* **51** (1975) 328.
[420] J. T. Grant, M. P. Hooker, R. W. Springer, and T. W. Haas, *J. Vac. Sci. Technol.* **12** (1975) 481.
[421] R. Shimizu, *Jap. J. Appl. Phys.* **13** (1974) 228.
[422] R. S. Berg and G. J. Kominiak, *J. Vac. Sci. Technol.* **13** (1976) 403.
[423] H. Liebl, *Messtechnik (Halle)* **12** (1972) 358.
[424] A. Benninghoven, *Z. Naturforsch.* **22a** (1967) 841.
[425] H. M. Windawi, *J. Vac. Sci. Technol.* **13** (1976) 1195.
[426] R. J. Baird, C. S. Fadley, S. K. Kawamoto, and M. Mehta, *Chem. Phys. Lett.* **34** (1975) 49.
[427] R. J. Baird, C. S. Fadley, S. K. Kawamoto, M. Mehta, R. Alvarez, and J. A. Silva, *Anal. Chem.* **48** (1976) 843.
[428] C. S. Fadley, R. J. Baird, W. Siekhaus, T. Novakov, and S. Å. L. Bergström, *J. Electron Spectrosc.* **4** (1974) 93.
[429] C. S. Fadley, *Discuss. Faraday Soc.* **60** (1975) 18.
[430] C. S. Fadley, *Prog. Solid State Chem.* **11** (1976) 265.
[431] H. Ebel, M. F. Ebel, and E. Hillbrand, *J. Electron Spectrosc.* **2** (1973) 277.
[432] C. S. Fadley, *J. Electron Spectrosc.* **5** (1974) 725.
[433] H. Ebel and M. F. Ebel, *Mikrochim. Acta Suppl.* **5** (1973) 333.
[434] M. F. Ebel, *J. Electron Spectrosc.* **5** (1974) 837.
[435] P. H. Holloway, *J. Electron Spectrosc.* **7** (1975) 215.
[436] J. O. Porteus, *J. Opt. Soc. Am.* **53** (1963) 1394.
[437] P. J. Estrup, *Phys. Today* **28** (1975) 33.
[438] J. C. Tracy and J. M. Burksand, *Crit. Rev. Solid State Sci.* **4** (1974) 381, and references therein.
[439] T. W. Rusch and W. P. Ellis, *Appl. Phys. Lett.* **26** (1975) 44.
[440] T. Matsudaira, M. Watanabe. and M. Onchi, *Jap. J. Appl. Phys. Suppl.* **2**, Pt. 2 (1974) 181.
[441] J. R. Noonan, D. M. Zehner, and L. H. Jenkins, *J. Vac. Sci. Technol.* **13** (1976) 183.
[442] J. B. Pendry, *J. Phys. C* **8** (1975) 2413.
[443] B. W. Holland, *J. Phys. C* **8** (1975) 2679.
[444] G. Allié, E. Blanc, and D. Dufayard, *Surf. Sci.* **57** (1976) 293.
[445] D. Aberdam, R. Baudoing, E. Blanc, and G. Gaubert, *Surf. Sci.* **57** (1976) 306.
[446] L. McDonnell and D. P. Woodruff, *Vacuum* **22** (1972) 477.

[447] B. W. Holland, L. McDonnell, and D. P. Woodruff, *Solid State Commun.* **11** (1972) 991.
[448] L. McDonnell, D. P. Woodruff, and B. W. Holland, *Surf. Sci.* **51** (1975) 249.
[449] D. P. Woodruff, *Surf. Sci.* **53** (1975) 538.
[450] T. Matsudaira, N. Nishijima, and M. Onchi, *Surf. Sci.* **61** (1976) 651.
[451] J. W. Gadzuk, *Solid State Commun.* **15** (1974) 1011.
[452] M. M. Traum, N. V. Smith, and F. J. Di Salvo, *Phys. Rev. Lett.* **32** (1974) 1241.
[453] Z. Knor, *J. Vac. Sci. Technol.* **8** (1971) 57.
[454] M. Kostelitz and J. Oudar, *Surf. Sci.* **27** (1971) 176.
[455] F. R. McFeely, J. Stöhr, G. Apai, P. S. Wehner, and D. A. Shirley, *Phys. Rev. B* **14** (1976) 3273.
[456] R. J. Baird, L. F. Wagner, and C. S. Fadley, *Phys. Rev. Lett.* **37** (1976) 111.
[457] S. K. Andersen and A. Howie, *Surf. Sci.* **50** (1975) 197.
[458] M. Baines, A. Howie and S. K. Andersen, *Surf. Sci.* **53** (1975) 546.
[459] D. Aberdam, R. Baudoing, E. Blanc, and C. Gaubert, *Surf. Sci.* **65** (1977) 77.
[460] J. P. Hoare, *The Electrochemistry of Oxygen*, Wiley-Interscience, New York, 1968.
[461] A. Damjanovic, in *Modern Aspects of Electrochemistry*, No. 5, Ed. by J. O'M. Bockris and B. E. Conway, Plenum Press, New York, 1969, p. 369.
[462] B. E. Conway, in *MTP International Reviews of Science, Physical Chemistry Series One*, Vol. 6, Ed. by J. O'M Bockris, Butterworths, London, 1973, p. 41.
[463] K. S. Kim, N. Winograd and R. E. Davis, *J. Am. Chem. Soc.* **93** (1971) 6296.
[464] S. Hüefner, G. K. Wertheim, D. N. E. Buchanan, and K. W. West, *Phys. Lett.* **A46** (1974) 420.
[465] Th. Dickinson, A. F. Povey, and P. M. A. Sherwood, *J. Chem. Soc. Faraday Trans. I* **71** (1975) 298.
[466] H. A. Kozlowska, B. E. Conway, and W. B. A. Sharp, *J. Electroanal. Chem.* **43** (1973) 9.
[467] A. K. N. Reddy, M. A. Genshaw, and J. O'M. Bockris, *J. Chem. Phys.* **48** (1968) 671.
[468] A. J. Appleby, in *Modern Aspects of Electrochemistry*, No. 9, Ed. by J. O'M. Bockris and B. E. Conway, Plenum Press, New York, 1974, p. 369.
[469] T. Biegler, D. A. J. Rand, and R. Woods, *J. Electroanal. Chem.* **29** (1971) 269.
[470] R. Parsons and W. H. M. Visscher, *J. Electroanal. Chem.* **36** (1972) 329.
[471] G. C. Allen, P. M. Tucker, A. Capon, and R. Parsons, *J. Electroanal. Chem.* **50** (1974) 335.
[472] Y. Y. Vinnikov, V. A. Shepelin, and V. I. Veselovskii, *Elektrokhimiya* **9** (1973) 552 and 649.
[473] S. Shibata and M. P. Sumino, *Electrochim. Acta* **16** (1971) 1089.
[474] J. Haber, J. Stoch, and L. Ungier, *J. Electron Spectrosc.* **9** (1976) 459.
[475] R. S. Sirohi and M. A. Genshaw, *J. Electrochem. Soc.* **116** (1969) 910.
[476] Y. Y. Vinnikov, V. A. Shepelin, and V. I. Veselovskii, *Elektrokhimiya* **8** (1972) 1229 and 1384.
[477] G. Grueneberg, *Electrochim. Acta* **10** (1965) 339.
[478] J. W. Schultze and K. J. Vetter, *Ber. Bunsenges. Phys. Chem.* **75** (1971) 470.
[479] W. E. Reid and J. Kruger, *Nature* **203** (1964) 402.
[480] H. Laitinen and M. S. Chao, *J. Electrochem. Soc.* **108** (1961) 726.
[481] K. S. Kim, A. F. Gossmann, and N. Winograd, *Anal. Chem.* **46** (1974) 197.

[482] T. A. Carlson and G. E. McGuire, *J. Electron Spectrosc.* **1** (1972/73) 161.
[483] S. E. S. El Wakkad and A. M. Shams El Din, *J. Chem. Soc.* (1954) 3094.
[484] K. J. Vetter and D. Berndt, *Z. Elektrochem.* **62** (1958) 378.
[485] J. P. Hoare, *J. Electrochem. Soc.* **111** (1964) 610.
[486] W. G. Berl, *Trans. Faraday Soc.* **83** (1943) 253.
[487] K. J. Vetter and J. W. Schultze, *J. Electroanal. Chem.* **34** (1972) 131, 141.
[488] J. W. Schultze, *Electrochim. Acta* **17** (1972) 451.
[489] J. Augustynski, L. Balsenc, and J. Hinden, Presented at the Symposium on Novel Electrode Materials (Chemical Society), Brighton, September, 1975; *J. Electrochem. Soc.*, **125** (1978) 1093.
[490] K. S. Kim and N. Winograd, *J. Catalysis* **35** (1974) 66.
[491] S. Pizzini, G. Buzzanca, C. Mari, L. Rossi, and S. Torchio, *Mat. Res. Bull.* **7** (1972) 449.
[492] C. Iwakura, H. Tada, and H. Tamura, *Electrochim. Acta* **22** (1977) 217.
[493] G. Faita and G. Fiori, *J. Appl. Electrochem.* **2** (1972) 31.
[494] R. G. Erenburg, L. I. Krishtalik, and I. P. Yaroshevskaya, *Elektrokhimiya* **11** (1975) 1068, 1072.
[495] J. Augustynski and J. Hinden, unpublished results.
[496] W. E. O'Grady, M. Y. C. Woo, P. L. Hagans, and E. Yeager, *J. Vac. Sci. Technol.* **14** (1977) 365.
[497] A. T. Hubbard, *Crit. Rev. Anal. Chem.* **3** (1973) 201.
[498] M. Seo, M. Sato, J. B. Lumsden, and R. W. Staehle, *Corros. Sci.* **17** (1977) 209.
[499] C. Wagner, *Ber. Bunsenges. Phys. Chem.* **77** (1973) 1080.
[500] R. W. Revie, B. G. Baker, and J. O'M. Bockris, *J. Electrochem. Soc.* **122** (1975) 1460.
[501] M. da Cunha Belo, B. Rondot, F. Pons, J. Le Héricy, and J. P. Langeron, *J. Electrochem. Soc.* **124** (1977) 1317.
[502] R. P. Frankenthal and D. L. Malm, *J. Electrochem. Soc.* **123** (1976) 186.
[503] A. E. Yaniv, J. B. Lumsden, and R. W. Staehle, *J. Electrochem. Soc.* **124** (1977) 490.
[504] J. B. Lumsden and R. W. Staehle, *Scr. Metall.* **6** (1972) 1205.
[505] T. N. Rhodin, *Corrosion* **12** (1956) 465 t.
[506] N. A. Nielsen and T. N. Rhodin, *Z. Elektrochem.* **62** (1958) 707.
[507] J. E. Castle and C. R. Clayton, *Corros. Sci.* **17** (1977) 7.
[508] G. Okamoto, *Corros. Sci.* **13** (1973) 471.
[509] K. Asami, K. Hashimoto, and S. Shimodaira, *Corros. Sci.*, **18** (1978) 151.
[510] K. Asami, K. Hashimoto, and S. Shimodaira, *Corros. Sci.* **17** (1977) 713.
[511] (a) K. Hashimoto, T. Masumoto, and S. Shimodaira; (b) S. Shimodaira; (c) A. E. Yaniv, J. B. Lumsden, and R. W. Staehle; (d) H. Okada, H. Ogawa, I. Itoh, and H. Omata; Proc. USA–Japan Seminar on "Passivity and Its Breakdown on Iron and Iron Base Alloys," Honolulu, March, 1975, Ed. by R. W. Staehle and H. Okada, N.A.C.E., Houston, Texas, 1976, pp. 34, 38, 72, 82.
[512] I. Olefjord and H. Fishmeister, *Corros. Sci.* **15** (1975) 697.
[513] K. Asami, K. Hashimoto, T. Masumoto, and S. Shimodaira, *Corros. Sci* **16** (1976) 909.
[514] T. Dickinson, A. F. Povey, and P. M. A. Sherwood, *J. Chem. Soc. Faraday I*, **73** (1977) 327.
[515] J. O'M. Bockris, A. K. N. Reddy, and B. Rao, *J. Electrochem. Soc.* **113** (1966) 1133.

[516] J. Siejka, C. Cherki, and J. Yahalom, *J. Electrochem. Soc.* **119** (1972) 991.
[517] B. Macdougall and M. Cohen, *J. Electrochem. Soc.* **122** (1975) 383.
[518] R. O. Ansell, T. Dickinson, A. F. Povey, and P. M. A. Sherwood, *J. Electrochem. Soc.* **124** (1977) 1360.
[519] J. Augustynski, H. Berthou, and J. Painot, *Chem. Phys. Lett.* **44** (1976) 221.
[520] J. Augustynski, *Passivity of Metals*, Ed. by R. P. Frankenthal and J. Kruger, The Electrochemical Society, Inc., Princeton, 1978, p. 989.
[521] M. Koudelkova, J. Augustynski, and H. Berthou, *J. Electrochem. Soc.* **124** (1977) 1165.
[522] J. Painot and J. Augustynski, *Electrochim. Acta* **20** (1975) 747.
[523] J. Painot, Thesis, University of Geneva, (1974).
[524] C. Edeleanu and U. R. Evans, *Trans. Faraday Soc.* **65** (1949) 683.
[525] M. F. Abd Rabbo, J. A. Richardson, G. C. Wood, and C. K. Jackson, *Corros. Sci.* **16** (1976) 677.
[526] J. Augustynski, M. Koudelkova, and J. P. Coad, unpublished results.
[527] P. E. Larsson, *J. Electron Spectrosc.* **4** (1974) 213.
[528] L. J. Brillson and G. P. Ceasar, *J. Appl. Phys.* **47** (1976) 4195.
[529] D. E. Eastman and J. L. Freeouf, *Phys. Rev. Lett.* **34** (1975) 395.
[530] S. Flodstrom, R. Z. Bachrach, R. S. Bauer, J. C. McMenamin, and S. B. M. Hagström, *J. Vac. Sci. Technol.* **14** (1977) 303.
[531] J. S. Hammond and N. Winograd, *J. Electrochem. Soc.* **124** (1977) 826.
[532] J. H. Leck, *Vacuum* **26** (1976) 419.
[533] R. L. Gerlach, *J. Vac. Sci. Technol.* **8** (1971) 599.
[534] S. Andersson and C. Nyberg, *Surf. Sci.* **51** (1975) 489.
[535] H. W. Werner, *Surf. Sci.* **47** (1975) 301.
[536] R. L. Gerlach and L. E. Davis, *J. Vac. Sci. Technol.* **14** (1977) 339.
[537] J. T. Grant and T. W. Haas, *Surf. Sci.* **24** (1971) 332.

5

An Introduction to the Electrochemistry of Charge Transfer Complexes II*

Electrochemical Reactions Involving Charge Transfer Complexes

F. Gutmann and J.-P. Farges†

School of Chemistry, Macquarie University, North Ryde, New South Wales 2113, Australia

I. INTRODUCTION

Electrochemistry has been defined by Bockris[1] as "the study of phenomena at electrified interfaces," and we shall concern ourselves here with some of those aspects of charge transfer complex (CTC) physical chemistry that are relevant within this definition. Thus, the CTC may dissociate in solution giving rise to ions; their ionics as well as electrodics will be discussed. The CTC may form in the course of an electrode reaction, which may or may not involve an excited state of the complex; it may arise at a colloidal interface or on a surface, via a solvent interaction, or via a proton interaction. These matters will be dealt with in the various sections of this chapter which follow.

---

* Part I published in *Modern Aspects of Electrochemistry*, No. 12, Ed. by J. O'M. Bockris and B. E. Conway, Plenum Press, New York, 1977.

† Permanent address: Laboratoire de Biophysique, Université de Nice, Nice-Valrose, 06034 France.

II. IONICS

1. Donicity

Strong charge transfer complexes, with ground states that are mainly dative (ionic), dissociate in solvents of sufficiently high permittivity to give rise to electrolyte solutions. Weaker complexes may exhibit a like behavior if sufficient energy is externally applied to produce an excited state that is mainly ionic though its lifetime usually confines electrolyte behavior to the duration of the excitation.

However, solutions of even quite weak CTCs still exhibit a certain residual conductivity, in common with other organic liquids: they behave as liquid organic semiconductors. Their conductivity is ohmic at least for values of the applied field not so high as to cause space charge effects; the carriers are electrons and/or holes. Their behavior is best discussed in terms of (amorphous) semiconductor theory[2] and is thus beyond the scope of this review.

One would be tempted to equate the "donor strength" of a donor with its ionization potential. However, several donors of very nearly equal ionization potential appear to have quite differing "donor strengths"; this is so because the ionization potential measures the energy required to remove the electron from the gaseous molecule, in its ground state, to infinity while in the formation of a CTC the electron only has to be moved to the acceptor molecule, the process being carried out not *in vacuo* but in a material medium such as a solvent.

Especially for the study of solutions, it has thus been found convenient[3] to introduce the "donicity" DN of donor D, defined arbitrarily as the negative of the total enthalpy change measured in complexing with antimony pentachloride[3]:

$$DN \equiv -\Delta H_{D.SbCl_5}$$

As a reference, the solvent is chosen as 1,2-dichloroethane and the concentrations are assumed to be low. Some donicities are listed in Table 1. The donicity may be used only as an approximate expression of the donor strength of a molecule toward a given substrate, though it has been found to serve as a most

Table 1
Donicities of Electron Donors[a]

Solvent	DN_{SbCl_5}
Benzoylchloride	2.3
Nitromethane (NM)	2.7
Dichloroethylene carbonate (DEC)	3.2
Nitrobenzene (NB)	4.4
Acetic anhydride	10.5
Phosphorus oxychloride	11.7
Benzonitrile (BN)	11.9
Selenium oxychloride	12.2
Acetonitrile (AN)	14.1
Sulpholane	14.8
Propanediol-1,2-carbonate (PDC)	15.1
Benzylcyanide	15.1
Ethylenesulphite (ES)	15.3
iso-Butyronitrile	15.4
Propionitrile	16.1
Ethylenecarbonate (EC)	16.4
Phenylphosphonic difluoride	16.4
Methylacetate	16.5
n-Butyronitrile	16.6
Acetone	17.0
Ethylacetate	17.1
Water	18.0
Phenylphosphonic dichloride	18.5
Diethylether	19.2
Tetrahydrofurane (THF)	20.2
Diphenylphosphonic chloride	22.4
Trimethylphosphate (TMP)	23.0
Tributylphosphate (TBP)	23.7
Dimethylformamide (DMF)	26.6
N,N-Dimethylacetamide (DMA)	27.8
Dimethylsulphoxide (DMSO)	29.8
N,N-Diethylformamide	30.9
N,N-Diethylacetamide	32.2
Pyridine (py)	33.1
Hexamethylphosphoricamide (HMPA)	38.8
1,2-Dichlorethane	—
Sulfurylchloride	0.1
Thionylchloride	0.4
Acetylchloride	0.7
Tetrachloroethylene carbonate	0.8

[a] The acceptor is always antimony pentachloride and the solvent is 1,2 dichloroethane. After V. Gutmann, Ref. 3.

useful guide for the interpretation or prediction of a number of interactions,[3] such as:

(a) formation of donor-acceptor complexes,
(b) ionization of covalent compounds by means of donor molecules,
(c) autocomplex formation,
(d) rate of solvent substitution,
(e) solvation of metal ions, and
(f) redox-equilibria in nonaqueous media.

The electron affinity, likewise, should be expected to determine in part, the ease, or otherwise, of the complexation reaction in a more subtle way than one would expect; in fact it has been shown[4] to be related to the effective dipole moment of the acceptor. The electron affinities of π-CTC's have been critically discussed.[5]

2. Conductivities

We shall first discuss a solution of a strong CTC in a polar solvent, and, as an example, consider solutions of the phenothiazine–iodine CTC.

These are, in many respects, archetypical cases of CTCs between a strong donor and a strong acceptor.[6] The ground state of the solid complexes is dative; infrared spectroscopy supports[6] this for at least the 2:3 complex. Stoichiometries of 1:2 as well as of 2:3 have been reported.[6]

Dilution of the complex in CH_3CN solution with CH_3CN* results[8] in the equivalent conductivity remaining constant to $\pm 3\%$ at a value of $130\ \Omega^{-1}\ cm^2/eq$ within a concentration range of 3×10^{-3} to $10^{-6}\ M$ extrapolating to the same value for Λ_∞, viz., $130\ \Omega^{-1}\ cm^2$. The complex in acetonitrile thus appears to be completely dissociated.

The conductivity of the pre-prepared $1.6\times 10^{-4}\ M$ complex in CH_3CN increases but slightly with temperature. Its activation energy is about $0.0029\ eV$ while that of the conductivity of the

---

* Conductance in acetonitrile has been extensively studied.[7] It has a moderate permittivity that favors ion-pair formation and provides a good example for some of the conclusions available from conductance data.

solvent, acetonitrile, itself is 0.060 eV, the temperature range being 313–275°K. The latter value is not too different from the activation energy of the viscosity[9] of pure CH_3CN, viz., 0.069 eV. The activation energy of the conductivity of the complex solution is seen to be far below what would be expected from the effect of the viscosity on the mobility of the ions. It may be that higher temperatures favor recombination, so that the mean lifetime of the carriers drops with rising temperature, thus in part counteracting the rise in mobility to be expected from the viscosity change.

Alternatively, it is quite possible that the low activation energy might be due to an anomalous conductivity maximum such as has been observed[10] at certain temperatures in methylenediamine/aralkyl-Li complexes. The slope of the conductance vs. temperature curve in the vicinity of the maximum is much reduced, perhaps due to the thermal dissociation of molecular complexes; an entropy effect may also be involved.

One would be tempted to associate these very low activation energies with a Grotthus-chain-like charge transport via I_3^-, but the qualitatively similar transport mode of H_3O^+ in aqueous solutions involves[11] an activation energy of 0.09 eV at room temperatures, much in excess of that found in this case.

Another example[12] is the chlorpromazine (CPZ)/acetylcholine 1:2 complex dissolved in water. Both the donor (CPZ) and acceptor were used as the hydrochlorides; CPZ as free base is quite insoluble in water. The adduct was preprepared, dissolved in conductivity water at 10^{-2} M, and then diluted. The resulting equivalent conductivities Λ_{eq}, were found to be proportional to (concentration)$^{1/2}$ and extrapolate to a conductivity at infinite dilution Λ_∞ of 470 mho cm^2 eq^{-1} at 25°C. The Debye–Hückel–Onsager equation[13]

$$\Lambda_{eq} = \Lambda_\infty - (A + B\Lambda_\infty)C^{1/2} = \Lambda_\infty - \text{const } C^{1/2} \qquad (1)$$

is obeyed. The above Λ_∞ value is of the same order as that of known "strong" potential electrolytes (ionogens) such as HCl.[14] This behavior is in contrast to that exhibited in DMSO rather than water as the solvent. While the acetylcholine CPZ·HCl complex dissolved in water behaves like a classical "strong," or better "true," electrolyte,[12] the complex dissolved in DMSO shows the behavior typical of a "weak" "potential"[15] electrolyte: no linear extrapolation to yield a conductivity at infinite dilution is possible.

The CPZ free base/acetylcholine 1:1 pre-prepared complex, dissolved in DMSO, behaves like a typical liquid organic semiconductor[2]: its conductivity at 38°C rises linearly with concentration over a concentration range from 10^{-2} to 10^{-5} molar; at $10^{-2}\,M$ it is 3.33×10^{-4} (ohm cm)$^{-1}$ for the free base as the donor, and 6.98×10^{-4} (ohm cm)$^{-1}$ for CPZ·HCl as the donor.

Some CTC's exhibit quite unconventional behavior in the concentration dependence of their equivalent conductivity, if the latter involves solvent interactions; thus the CTC formed between DDT [1,1,1,-trichloro-2,2 bis (p-chlorophenyl) ethane] and iodine in an equimolar water–DMSO solvent appears[16] to obey an equation of the form

$$\Lambda_{eq} = \frac{\Lambda_\infty}{C} + \text{const.} \qquad (2)$$

rather than an Onsager equation; no extrapolation to infinite dilution is possible; see Fig. 1. The conductivity of the *Simonini* complex,[17] $AG^+(ArCO_2)I^-$, Ar = an aryl group, is also anomalous,[18] permitting no evaluation of Λ_∞.

Anomalous behavior, illustrated in Fig. 2, is also exhibited[19] by solutions of the 1:1 complex between anthracene and chloranil in methanol and in DMSO upon dilution with an inert solvent, CCl_4. Substantially straight lines, though of reverse slopes, result, i.e., the equivalent conductivity drops with increasing dilution. Deviations from the straight line are more pronounced in the high-permittivity solvent DMSO where the equivalent conductivity is less than expected. One might be tempted to ascribe this to a competition between the solvent, DMSO, and the diluent, CCl_4, in that the latter enters into a nondissociating side reaction. In fact, CCl_4 is capable of forming weak charge transfer complexes[20] with aromatic hydrocarbons, though the actual transfer of the electron in such weak charge transfer complexes has been questioned.[21] Since the diluent CCl_4 is inert, it appears that part of the conductivity is associated with formation of a complex involving methanol, which is known to be an electron donor.

Electrochemistry of Charge Transfer Complexes

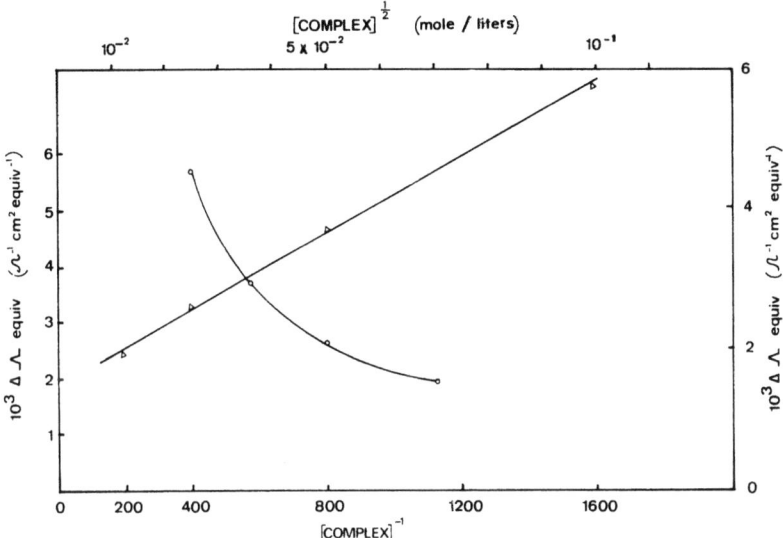

Figure 1. Dilution of the DDT/I_2 complex in equimolar DMSO/H_2O solvent, with the same solvent. (After A. Brau et al., Ref. 15.) The points marked ○ refer to the equivalent conductives scaled on the left; the corresponding abscissae are scaled on the top of the diagram. The points marked △ refer to the ordinates on the right of the diagram; the corresponding abscissae are scaled on the bottom. (After A. Brau et al., Ref. 15.)

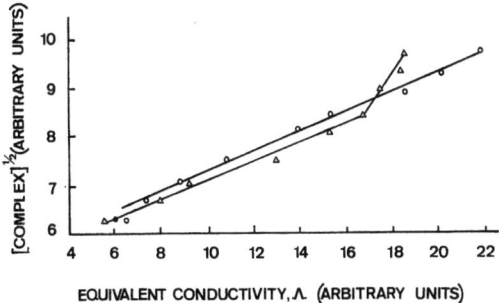

Figure 2. Dilution curve for the 1:1 anthracene/chloranil complex formed in methanol ○ and in dimethyl-sulfoxide △; the diluent in both cases is CCl_4. The solutions were aged for about a day before these readings were taken. The electrodes were gold. (After Gutmann and Keyzer, Ref. 19.)

The limiting value of Δ_{eq} for $C \to 0$, viz., Λ_∞, refers to a regime where the charge transport is limited only by the carrier mobilities and interionic interactions may be neglected; Eq. (2) would thus imply that in this case the converse applies, i.e., that at high concentrations the system tends toward an ionic, solvent-free solid in which Λ_{eq} is again limited by carrier mobilities only. The precision and range of the data at present available does not allow Eq. (2) to be distinguished from the form reported[22] to hold for some electrolytes:

$$\Delta \Lambda_{eq}/\Delta(C)^{1/2} = A + BC \qquad (3)$$

A and B being constants. It may well be that Eq. (2) is merely a short-range linear approximation to the more accurate form of Eq. (3).

The Onsager–Debye–Hückel equation employed here represents, of course, only a limiting case of the more accurate Fuoss–Kraus equation.[23] The latter, however, does not contain the variables in an explicit form and evaluation requires an iterative method. In the cases so far discussed, the precision of the data has not been deemed sufficient for consideration of these refinements. In a series of highly accurate measurements[24] on trimethyl-tin-iodide complexes with various donors and in different solvents, however, departures from the Fuoss–Kraus as well as from the limiting Onsager relations have been reported,[24] while others, e.g., the pyridine complex in DMSO, exhibit Onsager behavior, Λ_∞ for this system being about 40 $(\Omega\ cm)^{-1}\ eq^{-1}$.

Following Christov,[25] an ion transfer reaction can be written

$$AX^+ + B \to A + X^+B \qquad (4)$$

which can be represented by a potential energy surface in a many-dimensional manifold, taking into account all possible ion–ion and ion–solvent interactions. The problem may be greatly simplified by approximating it by a two-dimensional barrier; for details the literature should be consulted.[25,26] Matters are further complicated by the contributions arising from proton tunneling, especially at lower temperatures and in protonic solvents, such as water.

3. Charge Carriers

(i) Dissociation and Association

The conductivity of a CTC solution is due to the free ions produced by the dissociation of the complex; a mechanism similar to that proposed[27] for pure organic liquids. Dissociation in organic, complexing solvents may be followed by association,[28] leading to the formation of dimers and even higher-order entities. Multiple ions may be formed upon increasing dilution resulting in a relatively large change in the effective permittivity ε and causing a drop in conductivity. An increase in the value of the association constant with decreasing (ε) is well known[29] and quantitatively expressed by the Denison and Ramsay[30] or by the Fuoss and Kraus[23] relation; in addition, one would expect that the lowering of (ε) would tend to repress the dissociation of the complex.

Sometimes, conductivity titration yields a negative interaction: the conductivity as well as the capacitance exhibit a minimum at a certain stoichiometry.[31] Again, this must be associated with a double-layer reaction occurring within a certain active space fronting the electrode(s). At least in some cases, it appears to be due to the formation of a CTC which, however, fails to dissociate so that the carrier concentration drops. An example for this is the interaction[97] between heparin* and chlorpromazine (CPZ): the conductivity as well as the capacitance show a minimum at 83% heparin concentration. Since the heparin solution employed was $9.63 \times 10^{-3} N$ while the CPZ·HCl solution was $10^{-3} N$, this corresponds to a stoichiometry of about 2 heparin: 1 CPZ. Thus, each molecule of CPZ·HCl takes up two of the six negative charges associated with each tetrasaccharide unit of heparin so that three molecules of CPZ inactivate one tetrasaccharide unit. This matter is at present the subject of further clinical investigation.[215]

Another example[32] for such a negative interaction is that

---

* Heparin is a highly sulfated dextrorotatory mucopolysaccharide with a molecular weight, depending on source and methodology, of 6000–20,000. It is an important anticoagulant.

between CPZ·HCl and dilantin*; the latter behaves as a (weak) acceptor against CPZ which is known to be a strong donor. In this case, the stoichiometry is 1:1. The complex formed, apparently, fails to dissociate. Adrenalin (acid tartrate) and CPZ·HCl show a similar behavior; again a minimum is observed at a 1:1 stoichiometry. However, the same occurs with TMPD† instead of CPZ as partner; TMPD, however, is a powerful electron acceptor. Adrenalin thus appears to act amphoterically, as either donor or acceptor, depending on the partner. These effects are at present being further studied.

Janz and Danyluk proposed[33] the following empirical equation for the equivalent conductivity, of "systems in which the interactions give rise to solute–solvent species of molecular nature (cf. compounds) that ultimately contribute to the conductivity through additional ionic entities":

$$\Lambda = A + BM^{-1/2} \qquad (5)$$

here A and B are constants and M the molar concentration of the solute.

Breaks in a straight line representing Eq. (5) were thought likely to be associated with unilateral triple-ion formation. A Janz–Danyluk plot is shown in Fig. 3 for the anthracene–chloranil complex in methanol, following dilution with CCl_4. It is seen that Eq. (5) is indeed obeyed and that, as in Janz and Danyluk's original cases, a sharp break occurs. Cations in DMSO are thought to be solvated.[34] Thus, increasing dilution with CCl_4, assisted by the relatively large change in ε, could favor the formation of multiple ions with a resulting sharp drop in conductivity upon dilution.

The phenothiazine–iodine 1:2 complex dissolved in acetonitrile has been shown[8] (Section II.2) to be completely dissociated even at $3 \times 10^{-3} M$ concentration: the equivalent conductivity remains concentration independent. The conductance, thus, rises proportionally to concentration as in the case of a typical liquid semiconductor though the system is not an electronic, but an ionic conductor.

---

* Dilantin is one of the common names for diphenylhydantoin sodium, $C_{15}H_{11}N_2NaO_2$, mol. wt. 274. It is an important anticonvulsant and antiepileptic.
† TMPD stands for tetramethyl-p-phenylene diamine, also known as Wurster's blue.

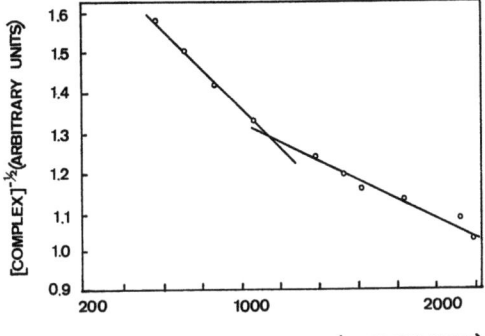

Figure 3. Janz–Danyluk plot [cf. Eq. (2.5)] for the anthracene/chloranil complex in methanol and diluted with CCl_4. Au electrodes. The data are the same as those used in Figure 2. (After Gutmann and Keyzer, Ref. 19.)

The proportionality holds[8] even if the solution is diluted not with CH_3CN but with CCl_4 (see Fig. 4), which is a solvent of very much lower permittivity; the slope of the curve is seen to change at about 3×10^{-4} M. This is considered to be due to the formation of larger ionic aggregrates caused by the drop in the permittivity of the solvent.[8]

The freshly prepared complex in solution is yellow, changing rapidly to green and eventually, after several days, to blue-violet. This aging effect may be speeded up by illumination and/or by addition of a few vol. % of water. The latter, in larger concentration, also causes about 60% further increase in the conductivity.

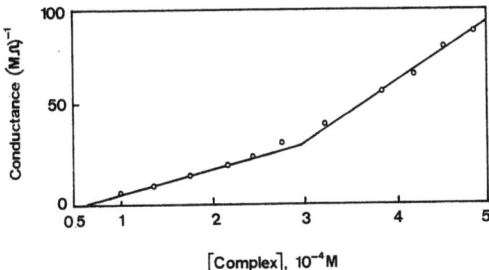

Figure 4. Dilution of a 5×10^{-4} M solution of the phenothiazine/I_2 complex in acetonitrile using again CCl_4 as the diluent. (After A. Brau et al., Ref. 8.)

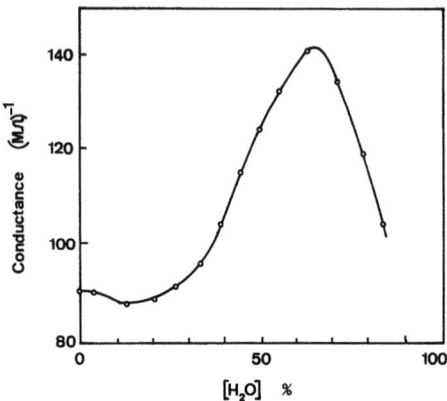

Figure 5. Dilution of the same complex, i.e., phenothiazine/I_2 formed in CH_3CN, with water. (After A. Brau et al., Ref. 8.)

This is illustrated in Fig. 5, which shows conductance values of the prepared complex at 25°C in CH_3CN as a function of added water.

The conductance remains substantially constant until about 20% water has been added; it then rises to a peak whence it decreases to approach the very low conductance of "pure" water. The acetonitrile used certainly contained traces of water, and thus further small additions are unlikely to produce any large effects; in fact it has been suggested[35] that "over the entire solvent range it is water molecules alone that are in the immediate neighborhood of the ions."

The pH at 25% water was 2.64 and rises approximately linearly to 3.4 at about 80% water, whence it starts to increase rapidly toward neutrality.

The pronounced rise in conductivity upon addition of about 36% water was considered[8] to be due to the further dissociation of the carriers caused by the rise in the permittivity of the medium.

An example for a complex solution that at room temperature is only partially dissociated is the aqueous solution of the chlorpromazine-acetylcholine 1:2 CTC,[12] discussed in Section II.2.

Assuming, as a first approximation, that the conductance ratio $\Lambda_{eq}/\Lambda_\infty$ can be equated to the coefficient of dissociation α, at

25°C its values result[12] as about 0.66 to 0.61 for the concentration range of 10^{-2} to 10^{-1} M.

The lifetime τ of the free ions can be calculated from the equation [cf. Eq. (7)]

$$\tau = \frac{\varepsilon}{4\pi\sigma} \tag{6}$$

Thus, the poorer conducting, only partially ionic complexes should be expected to yield more persistent ions. Free ions in organic solvents generally have rather long lifetimes, of the order of 10–100 msec.

Solvent interactions will be further discussed in Section II.4.

(ii) Majority Carriers and Mobilities

If the value of α is known, the total carrier concentration may be obtained. Taking the phenothiazine–iodine system in acetonitrile discussed in Section II.2, it is seen[8] that a concentration of the complex of 2.6×10^{-4} M upon complete dissociation yields a carrier concentration of 3.2×10^{17} carriers cm^{-3}, assuming equal contributions to the conduction current from both the 1:2 and 2:3 complexes, yielding a mean stoichiometry of 1.05:2.

We therefore shall simplify matters by assuming the presence only of the 1:2 complex. Thus, its concentration of 2.6×10^{-4} M results in 3.2×10^{17} carriers cm^{-3}.

From the conductivity equation

$$\sigma = e \sum_i n_i z_i \mu_i \tag{7}$$

where σ is the conductivity, $z_i e$ the charge carried by the ith carrier species of mobility μ_i and concentration n_i, e being the charge of the electron, the mean ionic mobility μ may be calculated as 6.2×10^{-4} cm^2 V^{-1} sec^{-1} at 20°C.

The transport numbers were determined[8] from electrolysis in a Hittorf-type cell as $t^- = 0.83$ for the anion, and $t^+ = 0.17$ for the cation. Thus, the individual ionic mobilities are $\mu^- = 1.0 \times 10^{-3}$ and $\mu^+ = 2.1 \times 10^{-4}$ cm^2 V^{-1} sec^{-1}. The mobility of the majority carrier, viz., the anion, is seen to be relatively high— approximately one-third that of the (hydrated) proton in aqueous solutions.[36]

The mobilities of most organic ions in solution are of the order of 10^{-4}–10^{-3} cm^2 V^{-1} sec^{-1}; for benzene values between 4.5 and 2.1×10^{-4} cm^2 V^{-1} sec^{-1} have been quoted.[37] Correcting for the rise in mobility to be expected from the smaller viscosity of acetonitrile compared to that of benzene—0.375 cp at 15° as against 0.705 for benzene—and using Walden's rule, the mobility of a hypothetical benzene ion in CH$_3$CN results as about 8.5 to 4.2×10^{-4} cm^2 V^{-1} sec^{-1} as against an experimental value for μ^+ of about 2.1×10^{-4} cm^2 V^{-1} sec^{-1}. Thus the mean cation radius is about 2 to 4 times larger than that of the hypothetical benzene ion.

The increased carrier size of the complex ion compared to that of a benzene ion is compatible with the carriers being solvated entities of the form $I_2 \cdot PH^+ \cdot I_2$, $I_2^- \cdot PH^+ \cdot I_2^-$ or $I_2 \cdot PH \cdot I_2^-$.[8]

The 2:3 complex has been suggested[38] as having a ground-state configuration of $I_2^- \cdot PH^+ \cdot I_2 \cdot PH^+ \cdot I_2^-$, which then perhaps dissociates thus:

$$I_2^- \cdot PH^+ \cdot I_2 \cdot PH^+ \cdot I_2^- \leftrightarrows \begin{cases} I_2^- \cdot PH^+ I_2^- + PH^+ \cdot I_2 \\ I_2 \cdot PH^+ + I_2^- \cdot PH^+ \cdot I_2 \end{cases} \quad (8a)$$
$$ 2I_3^- + 2PH^+ \quad (8b)$$

The 1:2 complex can be considered as a resonating structure with the negative charge being shared, viz., the aromatic intermediary, by the two iodine molecules, $I_2^- \cdot PH^+ \cdot I_2^-$. A collisional dissociation then should give rise to

$$2I_2 \cdot PH^+ \cdot I_2^- \leftrightarrows \begin{cases} I_2^- \cdot PH^+ \cdot I_2^- + I_2 \cdot PH^+ \cdot I_2 & (9a) \\ 2I_3^- + I_2 + PH^+ & (9b) \end{cases}$$

In a medium of higher permittivity it is expected that the processes (b) would be favored; however, in time more and more of the smaller entities would form if the permittivity of the medium is not too low. The above conjectures are supported by some spectroscopic evidence, since the I_3^- ion in CH$_3$CN is known[39] to give rise to an absorption line at 335 nm, evident in the spectra of all the PH/I$_2$ complexes. It becomes stronger with time and/or with addition of water.

For the CPZ:acetylcholine complex[12] already mentioned, assuming a valency $z = 1$, $\bar{\mu}$ results as 0.86×10^{-3} cm^2 V^{-1} sec^{-1}

at infinite dilution as the mean of four determinations on platinized and on bright Pt electrodes, the values ranging from 0.62 to 1.15×10^{-3} cm^2 V^{-1} sec^{-1}. The majority carrier mobility, $\mu' = 2\bar{\mu}$ in a single carrier model, then has a mean value of 1.7×10^{-3} cm^2 V^{-1} sec^{-1} at infinite dilution. Calculating μ' from the slope of the dilution curve[12] under the assumption that the coefficient of dissociation α remains constant, yields a value of 4.3×10^{-3} cm^2 V^{-1} sec^{-1}. Since α does alter with dilution, the agreement is not unsatisfactory. This relatively high mobility is comparable[40] with that of the (hydrated) proton, viz., 3.6×10^{-3}. The step that determines the rate of charge transfer in aqueous solutions of the CPZ·HCl acetylcholine Cl complexes thus is proton transfer, supporting the assumption of a univalent ($z = 1$) majority carrier made initially.

Since, for CPZ free base/acetylcholine-Cl in DMSO, a plot of conductivity versus concentration is linear[12] and very nearly passes through the origin, one may assume that the mobility as well as the coefficient of dissociation remain independent of concentration. On a one-carrier model, the (majority) carrier mobility calculated from the slope of the dilution curve results as $\mu' = 3.36 \times 10^{-4}$ cm^2 V^{-1} sec^{-1}, which is thus about one order of magnitude below the value reported above for the CPZ·HCl/acetylcholine in water system. It is a mobility of the order to be expected if the carrier is a relatively large molecular ion.[41]

The CPZ·HCl/acetylcholine complex in DMSO, and diluted with DMSO, yields a nonlinear relation between conductivity and concentration.[12] Therefore, α and/or μ depend on concentration. If one neglects this dependence and assumes $\alpha = 1$, a mobility value of 6.9×10^{-4} cm^2 V^{-1} sec^{-1} results, which is quite compatible with the other mobility values reported here.

Granting the validity of Eq. (2) for the DDT–iodine in equimolar DMSO/water solution already mentioned,[15] one may derive a mean mobility by the following reasoning: Λ_∞, obtained by extrapolation to infinite dilution, $C \to 0$ refers to a regime in which, in the absence of interionic interactions, the charge transfer is limited only by mobility so that

$$\Lambda_{\text{eq}} \underset{C \to 0}{=} \Lambda_\infty = F(\mu_+^\circ + \mu_-^\circ) \tag{10}$$

the superscript indicating that the μ values refer to infinite dilution. Equation (2) suggests extrapolation to the converse limit, viz., $C \to \infty$: this now assumes an infinite supply of carriers; the conductivity is again limited only by mobility so that

$$\Lambda_{eq} \underset{C \to \infty}{=} \text{const} = F(\mu_+ + \mu_-) \tag{11}$$

yielding for $\mu_+ + \mu_-$, at 25°C, in the 5×10^{-3} M solution the value of 8.6×10^{-6} $cm^2 V^{-1} sec^{-1}$. This corresponds to 2.9×10^{18} carriers cm^{-3}; the background conductivity of the solvent has been subtracted from the conductivity measured. If each complex molecule contributes one carrier, then for 5×10^{-3} M, 3×10^{18} carriers should be available, in excellent agreement with the value of 2.9×10^{18} cm^{-3} obtained above. The assumptions made thus are borne out and a one-carrier model is seen to be applicable to this system. It also appears that, again, the dissociation is virtually complete, i.e., $\alpha = 1$, varying but little with concentration. The low mobility value suggests that the carrier is a rather large, molecular ion.

The electrochemistry of several complexes between trimethyl tin iodine in aprotic donor solvents has been studied.[24] While these complexes are primarily coordination complexes rather than CTCs, it is of interest to note that the order of magnitude of the Λ_∞ values reported agrees with those reported above for charge transfer complexes: depending on the donor strength of the solvent, Λ_∞ at 25°C for several complexes varies from 23.8 to 82.3 $(ohm\ cm)^{-1}$. Charge transfer contributes only in a minor degree to the bonding of these complexes so that their properties fall outside the scope of this review.

4. Solvent Interactions

CTC solutions exhibit such interactions to a high degree: there is no really inert solvent. Every solvent is liable to act as either an electron donor or an acceptor and the most common of all, viz., water, because of its amphoteric nature, may act as either.[24] The solvent thus competes with the donor or the acceptor; a solvated ion may become an even more complicated entity. The anomalous behavior of the conduction properties of some CTS's is due to solvent interactions; it is quite likely, e.g., that in the case of the DDT/I_2 complex dissolved in equimolar DMSO/water, the complex initially forms solvent-shared ion pairs which then dissociate.

Solvation generally lowers the free enthalpy of complex formation as well as the entropy of the reaction. Especially in polymeric CTC's,[42] ion pairs and radical ions are solvated to a higher degree than the neutral complex. Being therefore energetically favored, this results in increased dissociation into solvated ion pairs and solvated radical ions.[42]

However, part of the solvent effect, notably anisotropies in the distribution of solvent molecules about one another, are due to[43] interactions which disturb the intrinsic polarizabilities of the molecules.

The condition that free ions can result from an electron transfer from a donor D to an acceptor A in a solvent can be written[44] as

$$|E_s + \beta E_c| > I - |E_A| \quad (12)$$

where E_s is the sum of the solvation energies of the individual ions and of any ion pairs, or higher aggregates, E_c the Coulombic energy to form an ion pair present in the solution at a concentration βM, M being the molarity of the complex, I is the ionization energy of the donor, and E_A the electron affinity of the acceptor. Since, in an organic CTC, I is likely to be low and E_A high, this condition is frequently fulfilled.

Thus, e.g., in acetonitrile, anions are quite stable in the absence of water and of oxidants, though the cations have lifetimes of only a few milliseconds until they react with the solvent.[45] However, competition by the solvent should also be allowed for,[46] i.e., the formation of a solvate complex following

$$A + nS \rightleftarrows AS_n^s \quad (13)$$

Assuming that the dissolved complex exists in solution exclusively in the form of such complexes,[46] the exchange of the, say, acceptor A between the competing molecules of the donor D and of the solvent S proceeds as

$$AS_n + D = AD + nS$$

$$K_e = \frac{[AD][S]^n}{[AS_n][D]} \quad (14)$$

The equilibrium $A + D \rightleftarrows AD$ is governed by a constant K_S which should vary linearly with $[S]^{-n}$ as long as the donor is present in excess, while K_e should remain independent of both $[D]$ and $[S]$. So far, these relations have been tested only for the benzene/I_2 in CCl_4 system where they appear to hold to about 5%.

The charge may be distributed over several molecules giving rise to the complicated multiple ions already encountered. Most aprotic solvents are quite good electron donors,[47] though, e.g., ethanol or acetone are relatively weak. In such cases, the solvent may behave amphoterically as does water.[24] The frequently used solvents DMSO and DMF are known to form CTCs with, e.g., chloranil in CCl_4[48]; the latter can act as an acceptor.[49]

The effect of anion solvation on complex formation is said[50] to follow the series

$$H_2O > ROH \gg DMSO \cong ES > CH_3CN$$
$$\cong TMS \cong NM \cong DMF > DMA$$

where ROH stands for methylalcohol, DMSO for dimethylsulfoxide, ES for ethylensulfite, TMS for tetramethylsulfone, NM for nitromethane, DMF for dimethylformamide, and DMA for dimethylacetamide.

The effect of 32 solvents on the acenaphthene-tetrachlorophthalic anhydride CTC has been studied[51]; from these effects, solvents may be divided into protic, dipolar aprotic, aromatic with or without functional groups, and n-donors. For a review on ionic solvation in nonaqueous solvents, see Ref. 52.

Yomosa defines[53] a critical energy of a complex as that value of an *isolated* complex molecule at which the ionic and nonbinding contributions to the ground state are equal. If now the potential energy of the complex molecule in the presence of a local field due to a material environment, in which the now no longer isolated molecule finds itself, exceeds this critical value, then the charge transfer band will show a blue shift. Its magnitude will rise with the value of the local field as well as with the permittivity of the medium.[51] Moreover, if thus the critical energy has been exceeded, the ground state of the complex will be predominantly ionic and its electronic structure will resemble that of the excited state of the complex in the absence of the local field.[54] No activation energy is needed to thus produce the ionic ground state, as has been observed[53] with several complexes. The blue shift predicted has been found,[55] e.g., for the TMPD (N,N,N',N'-tetramethyl-p-phenylenediamine)/chloranil complex upon changing the solvent from cyclohexane to acetonitrile, which has a much higher permittivity. Other blue shifts have been reported[8] for iodine complexes. Conversely, red shifts occur, e.g., for chloranil complexes with strong donors.[56] Again, it is surmised[57] that these shifts are due to contributions from ionized forms of the complex.

If a CTC is recrystallized from a solvent, then the nature of that solvent may considerably affect the properties of the (solid) CTC: thus, e.g., the 1,6-diaminopyrene/chloranil complex is dark blue when recrystallized from benzene.[58] These are two completely different chemical identities with different physical properties such as spectra and resistivity. This may be due to the formation of an inclusion compound[59] between the CTC and the solvent; the host lattice—here that of the (solid) CTC—has cavities which are occupied by the guest molecules, viz., those of the solvent. Such CTC inclusion compounds are known for several benzidine and TCNQ adducts.[60]

Polar solvent molecules will tend to become oriented[61] and thus will stabilize the CTC by means of a stereospecific orientation effect, giving rise to a local field such as has been discussed by Yomosa.[53]

5. Colloid and Surface Complexations

Complexes are known in which at least one component exists in colloidal form: thus, chlorpromazine as well as other phenothiazine derivatives are colloidal[62] and act as powerful electron donors[63]: melanin is colloidal as well as electron accepting[64] and many biologically active large molecules, such as acetylcholine,[12] are capable of forming CTC's.

Suspensions of polymers such as PVC may be coated with a donor or an acceptor and then irradiated.[65] A charge transfer reaction follows in which charge is transferred from or to the additive; the system then exhibits[65] a greatly enhanced conductivity. The surface CTC between polyethylene and SF_6 formed on the inner surfaces of the voids which are unavoidably present in the extruded polymeric insulation for high-voltage cables, improves[66] the dielectric strength of the cable by about 50%.

Some TCNQ complexes are colloidal and their critical micelle concentrations have been determined spectroscopically[67] as well as by surface tension measurements[67]; the latter method has been used to identify CTCs.[68] The complex appears to form above the critical micelle concentration between solubilized TCNQ and the surfactant[67] and has a palisade type of arrangement of molecules. However, very little work has been done on the colloidal properties of CTC's, though these must be of biological importance.

Organic donors, if exposed to an electron-accepting environment, may form surface charge transfer complexes[69]; the converse holds for acceptors. Examples are oxide surfaces, such as MgO exposed, e.g., to I_2.[70] Adsorbed Mg-phthalocyanine, without additional ligands, is reported[71] to have an electron affinity of 13–14 eV.

Since counterions interact with, and are attracted by the, say, hydrophilic heads of the colloidal particles, micellar assemblages of such particles should be catalytically active. This, in general, has been found[72] to be the case, and should be a rewarding field for the study of micellar CTCs as well as for electrocatalysis.

Surface CTCs appear to play a role in the observed[73] changes in the electrode potentials of merocyanine dyes in aqueous solutions of anionic, cationic, and nonionic surfactants; there are also changes in permittivity, as well as in the optical spectra.

Addition of surfactants stabilizes free radical ions in, e.g., anthraquinone:anthrahydroquinone complexes; lifetimes of several weeks have been reported.[213] The resulting (ternary?) adduct appears to have a micellar structure; if its concentration exceeds the critical micelle concentration, a spectral blue shift is observed.[213] The complex is said[214] to raise the solubilizing power of the surfactant. Addition of a dye yields a ternary adduct.

Chronoamperometry has been applied to the study of surface charge transfer complex formation at electrodes in which the reaction product is dissolved after formation, thus liberating active sites for further reaction.[76]

Many gases form surface CTCs with organic semiconductors,

thereby altering their surface properties to a great extent.[77] This effect can be employed to detect and identify ambient gases.

The surface complexes between donors and oxygen are of considerable importance in the design and operation of fuel cells. Substrate intermolecular CTCs are of importance in some enzymatic reactions,[78] e.g., in those of the substrate-porphyrin ionized complex, Subst:$^+$–Porph;$^-$ they also involve excited states. Other examples are[78] intermolecular complexes between cytochromes c and b.

The effect of the substrate carrying a catalyst on its activity has recently been reviewed.[78]

6. Stochastic Processes

The very nature of a charge transfer interaction is based upon a stochastic process, i.e., the probability that the energy levels of the two interacting particles remain the same for the time required for the electron exchange, say 10^{-15} sec. In solution, thermal fluctuations of the solvent dipoles—whether permanent or induced—change the interaction energy of the electron with the polarization field and thus the probability of the reaction.[79] In other words, the Franck–Condon principle, which requires the equalization of the energy levels of the donor and acceptor, permits the electron exchange only if fluctuations of the polarization field result in an instantaneous "activated complex" so as to fulfil this requirement.

The frequency of these fluctuations is in the optical or ultraviolet range, giving rise to the charge transfer spectra; in this section we shall be concerned with stochastic processes affecting ionic charge transport. These yield measurable noise voltages and currents which have been shown[80] to be related to the conductivity of the solution.

Consider an electrically neutral particle of mass m, free to move in one linear dimension in a liquid in which it is subject to Brownian motion. Let its macroscopic position as a function of time t be $x(t)$. The Brownian motion is governed by the Langevin stochastic differential equation:

$$m\frac{dv}{dt} + fv(t) = F(t) \qquad (15)$$

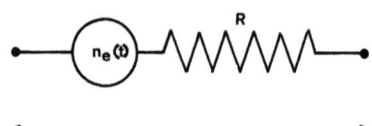

Figure 6. The equivalent circuit of a resistive circuit element, having a resistance R. The circuit element is seen to be equivalent to an ideal, noise-free resistance R in series with a white noise generator. (After D. Vasilescu et al., Ref. 80.)

Here, f is a friction coefficient and $F(t)$ a random function of time which includes the effect of all collisional processes between the particle and the solvent. The random variable $v(t)$ has a mean value $\langle v \rangle$ and an autocorrelation function

$$R_v(\tau) = \langle v(t+\tau)v(t) \rangle \tag{16}$$

The symbol $\langle \rangle$ denotes mean values.

The stochastic process considered here is stationary, i.e., $v(t+\tau)$ follows the same statistics as $v(t)$ for any τ. One may write

$$\langle v \rangle = 0$$
$$R_v(0) = \langle v^2 \rangle \tag{17}$$

In this notation,[81] a Fourier transform of $R_v(\tau)$ yields the power spectrum $S_v(\omega)$ of the velocity:

$$S_v(\omega) = \int_{-\infty}^{+\infty} R_v(\tau) e^{-j\omega\tau} \, d\tau \tag{18}$$

where $\omega = 2\pi\nu$, ν being the frequency of a noise component.

The autocorrelation function $R_v(\tau)$ is the inverse Fourier transform of the power spectrum

$$R_v(\tau) = (2\pi)^{-1} \int_{-\infty}^{+\infty} S_v(\omega) e^{j\omega\tau} \, d\omega \tag{19}$$

The behavior of the function $F(t)$ may then be closely approximated by comparing it with the white noise produced by a generator having an internal ideal and noise-free resistance R as in the equivalent circuit shown in Fig. 6. The spectral density (or power spectrum) of its thermal noise is constant, i.e., frequency independent, and only a function of temperature T:

$$S_{n_e}(\omega) = 2KTR \tag{20}$$

where k is Boltzmann's constant. This is known as the Nyquist–Johnson theorem.[82] The mean value $\langle e^2(t) \rangle$, where $e(t)$ is the instantaneous terminal noise voltage appearing across the resistance R, is given by

$$\langle e^2(t) \rangle = R_{n_e}(0) = (2\pi)^{-1} \int_{-\infty}^{+\infty} S_{n_e}(\omega)\, d\omega$$
$$= (2\pi)^{-1} \int_0^\infty 2S_{n_e}(\omega)\, d\omega \tag{21}$$

This last expression has a limiting value if the frequencies are not too high (the Planck condition $h\nu \ll kT$ requires $\nu \ll 10^{13}$ Hz). With this restriction, a good approximation of $\langle e^2(t) \rangle$ is given by

$$\langle e^2(t) \rangle = 4kTR\, \Delta\nu \tag{22}$$

where $\Delta\nu$ is the passband of the amplifier or other device used to measure $\langle e^2(t) \rangle$.

The Nyquist–Johnson theorem is usually quoted[83] in the form of Eq. (21). The Langevin equation (15) can also be written as

$$\frac{dv}{dt} + \beta v = n(t) \tag{23}$$

where $\beta = f/m$. It is seen that $F(t)/m$ has been replaced by the noise function $n(t) = n_e(t)$. This means that

$$\langle n_e(t) \rangle = 0$$
$$S_{n_e}(\omega) = \alpha = \text{const} \tag{24}$$

Equation (23) can be solved using the methods of the general theory of linear stochastic differential equations,[81] yielding

$$R_v(\tau) = \frac{kT}{m} \exp\left(-\tau \frac{f}{m}\right) \tag{25}$$

If the relaxation time $\tau \gg (f/m)^{-1}$, then the acceleration term dv/dt in Eq. (23) is negligible and the displacement of the particle is given by

$$x(t) = \frac{m}{f} \int_0^t n(t)\, dt \tag{26}$$

This is a case of a Wiener–Levy process[81] from which the mean square displacement of the particle position follows as

$$\langle x^2 \rangle = \frac{m^2}{f^2} \alpha t = 2Dt \tag{27}$$

since the diffusion coefficient D is given by

$$D = \frac{m^2}{2r^2} \alpha \tag{28}$$

One can now define a mechanical mobility u by

$$u = f^{-1} \tag{29}$$

u has the dimension time/mass: t/m. The relaxation time $mu \equiv \tau_0$ is of the order of 10^{-8} sec for a colloidal particle 1000 Å in diameter in water. For comparison, for Na$^+$ ions in water, τ_0 is of the order of 10^{-14} sec.

Large, highly solvated ionic entities such as those involved, e.g., in the CTCs formed from the phenothiazines, have a value of τ_0 in between these extremes, viz., of the order of 10^{-11} sec. The mechanical mobility should be measurable by ultrasonic techniques, i.e., by determining the electrochemical Debye voltages[83]: if an ultrasonic sound field is set up in a liquid which contains ions of different mass, then an electrical potential appears because of the heavier ions lagging behind the lighter ones; the effect is of the order of millivolts per unit velocity (1 cm sec^{-1}). No work appears to have been done in this field except that by Yeager's group with simple electrolytes.

Considering now a monovalent ion in Brownian motion, its electrical mobility μ_n is given by

$$\mu_n = eu \tag{30}$$

where e is the electronic charge. This "noise mobility" is the equivalent of the conventional mobility derived from the motion of an ion under the influence of an external electric field.

For a system of N identical monovalent ions, an equivalent noise conductivity σ_n may be defined by

$$\sigma_n = Ne\mu_n \tag{31}$$

assuming the ions to be acted on only by the same random forces of thermal origin, which also cause the noise phenomena; coulombic ion–ion interactions are supposed to be negligible. This means that the ions behave like a dilute Lifschitz–Landau[84] plasma, i.e., like an assembly of perfect gas molecules, which have a kinetic energy determined by thermal agitation only: their electrostatic energy of interaction, by comparison, is negligible.

The noise emission spectra obtained are linear and, within the accuracy of the measurements,[80] there is no evidence for any relaxation or resonance phenomena. There is no noise component varying with $1/f$: the electrolyte produces a white noise spectrum justifying the use of the Langevin equation (23). The equivalent noise resistance values agree, within the accuracy of the measurement,[80] with the resistance values obtained by means of conventional bridge methods; there is no excess noise arising from the electrodes. This is remarkable because in solid semiconductors the contact noise frequently exceeds the thermal noise. The absence of contact noise is probably due to the immobilization of the charges within the rigid part of the double layer because of the high electric fields in that region; the energy of the carrier due to the field is much larger than kT.

However, a $1/f$ noise component has been reported by other workers[85]; one would expect that such a noise component would appear whenever the number of carriers n is small as it is in some semiconductors.[86] Another source of a f^{-1} noise is the energy barrier at the contacts.[87] Empirically, it has been shown[88] that metals as well as semiconductors yield an f^{-1} noise obeying the relation

$$\left(\frac{\Delta R}{R}\right)^2 = \frac{a}{n}\frac{\Delta f}{f} \tag{32}$$

where R is the resistance, a an empirical constant having a value of about 0.002, and n the number—not the concentration—of carriers present in the sample. For ionic solutions, a has been found to be several thousand times larger, and it has been claimed that the f^{-1} noise is a bulk effect and thus not associated with surfaces.[88] It is clear, however, that this noise is not thermal in origin; it is most likely to be associated with the stochastic nature of the diffusion processes and thus is a "mobility noise."

In organic liquids, such as pure acetone, the noise current is said[89] to vary with $f^{-1.68}$ though the bulk of the noise power occurs in the very low frequency region. It would be most interesting to measure the noise spectra of CTCs in the solid state as well as in solution.

Thermal noise voltages $\bar{V}^2$ are also generated across the terminals of a capacitance C containing the system under test as a dielectric, as predicted by Einstein[90,91]:

$$\frac{C\bar{V}^2}{2} = \frac{kT}{2} \tag{33}$$

Given the Debye model of a dielectric, obeying a single relaxation time τ, so that

$$\varepsilon = \varepsilon_\infty + \frac{\varepsilon_0 - \varepsilon_\infty}{1 + j\omega\tau} \qquad \omega = 2\pi f \tag{34}$$

where ε is the effective premittivity, ε_∞ that for infinite frequency, and ε_0 that for zero frequency, an equivalent circuit as shown in Fig. 7 results yielding[92] a noise voltage given by

$$\langle V^2 \rangle = \frac{2kT}{\pi} \frac{a}{s} \frac{\varepsilon_0 - \varepsilon_\infty}{\varepsilon_\infty \varepsilon_0} \frac{\Lambda u}{1+u^2} \left(\frac{C_2}{C_1+C_2}\right)^{-2} \left[1 + a^2 \left(\frac{C_1 C_2}{C_1+C_2}\right)^e w^2\right] \tag{35}$$

where a is the distance between the plane parallel plates of the capacitor, each plate having an area s, and u is $(\varepsilon_\infty/\varepsilon_0)\omega\tau$.

Detailed studies of noise phenomena in CTCs should be of special value because they might yield a method for the direct measurement of the ionization–recombination equilibria involved in the complex formation. Noise arising from fluctuations in the ionization state of impurities in semiconductors has been observed,[93] as well as that connected with electrochemical reac-

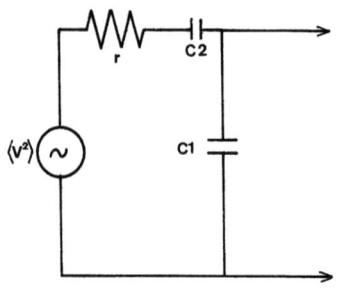

Figure 7. The equivalent circuit of a capacitive circuit element having a capacitance C_1, the dielectric of which obeys a single Debye-type relaxation time, τ; cf. Eq. (2.34). (After F. Micheron and L. Godefroy, Ref. 92.)

tions[94] such as H_2 evolution,[95] or those in primary electrochemical cells.[96] It is of interest to note that these fluctuations, while white (frequency independent), are considerably above purely thermal noise; this excess noise is claimed to be associated with the energy band structure of the noise generator.[97] The relaxation processes giving rise to such carrier generation and recombination are not Markovian in nature.[98]

CTCs might even offer a way to realize a system exhibiting an effective negative absolute temperature, as evidenced by the noise spectrum; CTC ions aligned with their mean magnetic moments opposite to an applied magnetic field by means of an applied electric field would fulfill this condition.[99]

III. ELECTRODICS

1. Complex Formation as an Electrode Reaction

The complex formation–dissociation equilibria may depend on electrocatalytic activation on an electrode surface, in other words would be affected by reaction at the electrode. Thus, conductivity titrations using platinized platinum electrodes are often less reproducible than those obtained on less catalytically active surfaces such as bright Pt; probably due to differences in adsorption and to the many different side reactions occurring at this highly catalytically active electrode.

Capacitance peaks have been observed in the course of several conductivity titrations of charge transfer complexes. They cannot possibly be caused by any reasonable increase in the bulk permittivity of the solutions which were only of the order of 10^{-3} M, and which is determined by that of the solvent. These changes thus must be due to processes occurring within an "active space"[100] or reaction zone associated with the double layer. During the titration, the electrochemical reaction proceeds within that reaction zone and concentrations therein may differ greatly from those in the bulk. Thus, e.g.,[101] acetylcholine adsorbs on the electrode to a lesser degree than the more surface-active CPZ (chlorpromazine) and fails to displace the CPZ from the double-layer region in proportion to its bulk concentration. During the titrations, the capacitances show considerable changes, peaking in both titration directions at the stoichiometry indicated by the conductance values, viz., at 1:2 CPZ·HCl: acetylcholine. The peak capacitance values were 12.8×10^{-4} μF when adding CPZ and 34.6×10^{-4} μF for the converse, i.e., when adding acetylcholine.[101]

In DMSO as the solvent, there is little evidence for such preferential adsorption and both branches of the titration curve coincide quite well.

CPZ·HCl forms a number of adducts if combined with serotonin in water; conductometrically, stoichiometries of 1:3, 1:2, 1:1, and 2:1 are reported.[101] Preferential adsorption of CPZ is even more evident than in the case of acetylcholine; serotonin is far less active[102] than acetylcholine. The interaction leading to complex formation is much weaker than that between CPZ and acetylcholine and may at least partially be due to the serotonin being kept away from the electrode surface by adsorbed CPZ.

The capacitance values remained[101] essentially constant at about 1.7×10^{-3} μF, after a slight initial rise using platinized platinum electrodes. The two branches of the titration curve fail to match. With bright platinum electrodes, however, a definite capacitance peak at 1 CPZ·HCl:3 serotonin is observed, the values rising from 5 to 10×10^{-3} μF at the peak.

The measured electron transfer rate is markedly affected by the material of which the supposedly inert electrode is made. This is due to the extent of orbital overlap between electroactive species and the electrode, adsorption, and other surface effects in respect to the transition state. The resulting differences in electron transfer rate for the same couple are evident from the data for the iron-oxalate complexes, where the rate constant has been found[103] to vary over three orders of magnitude, depending on the nature of the electrode.

The relation between specific adsorption and the donor-acceptor ligand interaction with the substrate has recently been reviewed.[103]

2. Electrocatalysis and Heterogeneous Catalysis

This discussion will be confined, for reasons of space, to charge transfer complexes other than coordination complexes, though these, at least as far as their catalytic activity is concerned, have been considered as CTCs because the electrons that are transfer-

red in catalytic reactions involve delocalized, mobile electrons of the complex.[104]
The subject has recently been reviewed.[105]

(i) *The Semiconductor Electrode*[105]

In general, electron transfer to or from an electrode requires that the energy level of the donor or the acceptor can attain a position within the valency or conduction band of the electrode. Neutral molecules as well as ions may be chemisorbed in the inner Helmholtz layer. The electrode potential E will then be determined by the dipole moment of the adsorbate, its charge, and its transfer rate. These should affect E in a linear fashion while capacitance effects should exhibit a parabolic relationship.[106] The charge transfer rate from electrochemisorption, under potentiostatic conditions, can be expressed by[106]

$$\left(\frac{\partial Q}{\partial \Gamma}\right)_E = -jzF \tag{36}$$

where F is the Faraday constant, Q the charge, Γ the surface charge excess, z the valency, and $j < 1$ stands for a factor describing the electrochemisorption process. Relatively large values of j, depending but little on the potential E, are an indication of adsorption of ions, while relatively small and highly potential-dependent values of j relate to the adsorption of neutral particles.

A neutral chemisorbed species increases the reaction rate of anions because of the predominance of electrostatic interactions, while cationic complexes show the opposite behavior.[107] Chemisorption of charged species onto an electrocatalytic surface thus provides a means of varying the effective field in the interfacial region without altering the electrode potential; this also holds, *mutatis mutandis*, for adsorbed free radicals, which indeed do enhance the electrocatalytic activity; cf. Section III.2(ii).

The main difference between a metallic electrode and a semiconducting one is the lower carrier density in the latter case, plus the fact that a semiconducting surface may be either p-type or n-type, i.e., that the majority carriers in the surface region may be either free electrons or free holes, i.e., electron vacancies.

The presence of an electrolyte causes the formation of an electric double layer in which an electric charge on the electrode is compensated by an equal, but opposite, charge on the solution side. In a metal, where there is a very high concentration of free electrons, there is virtually no space-charge region on the solid side. Application of an external potential causes a rearrangement of the electrons in a surface region extending at the most a few angstroms into the solid. In a semiconductor, the very much lower carrier density results in a space-charge region extending over considerable distances, of the order of 100 Å or so, into the solid; in relatively poorly conducting solids, the charge injected from the solution may be very much greater than that originally present. Depending on the majority carrier (electrons or holes), the space-charge region in the solid will be either an accumulation or an exhaustion region and thus, generally, rectification will occur. In other words, the energy bands in the solid will be bent upwards or downward, the depth of this perturbed region within the solid being considerable.

These considerations* are the result of the thermodynamic requirement that *at equilibrium*, the Fermi levels on both the electrode and the solution side must equalize.

Only those electrochemical reactions are possible in which either electrons are injected into the conduction band of the solid or holes into its valency band. The electrochemistry and physics of semiconducting electrodes have been extensively discussed by several authors,[108] and been recently reviewed[108]; let it thus suffice to give merely the most important results: The exchange current density, i_0 at either an n-type or a p-type electrode of a redox reaction subject to the Franck–Condon principle and involving solvated ions is given by[108]

$$i_0^n = Fv[\text{Red}]\exp\left\{-\frac{\lambda}{4kT}\left[\left(\zeta - \frac{\Delta\varepsilon_s - kT\ln[\text{Ox}]/[\text{Red}]}{\zeta\lambda}\right)^2 + \frac{4\Delta\varepsilon}{\lambda}\right]\right\}$$

$$i_o^p = Fv[\text{Ox}]\left[-\frac{\lambda}{4kT}\left(\zeta + \frac{\Delta\varepsilon + kT\ln[\text{Ox}]/[\text{Red}]}{\zeta\lambda}\right)^2\right] \tag{37}$$

---

* There is yet another difference between metal and semiconductor electrodes, which is associated with the width of the energy bands of the solid. In an ideal metal, the energy bands do not only overlap, but are also very wide, while in semiconductors, especially in poorly conducting ones, these may be narrow (compared to kT), CTCs, however, and particularly those having at least partially ionic ground states, which are the main concern of this review, have energy bands at least comparable to or wider than kT. The band width, being related to the carrier mobility, affects primarily the transmission coefficient of the equations governing the exchange current densities.

v stands for the velocity along the reaction coordinate, related to the transmission coefficient, the concentrations [Ox] and [Red] refer to those of the oxidised and reduced form at the equilibrium potential and in the outer Helmholtz plane, and ζ is the number of electrons involved in the charge transfer. λ refers to the energy required to reorient the solvation shell around the oxidized species so that at equilibrium it becomes the same as that formed around the reduced species, and $\Delta\varepsilon$ is the value of the energy gap in the surface, referred to the Fermi level, i.e., the distance between the bottom of the conduction band and the Fermi level for the n-type solid and that between the top of the valency band and the Fermi level for p-type material. Unless the semiconductor is intrinsic, $\Delta\varepsilon \simeq -\Delta\varepsilon_p$.

In the absence of surface states on the solid side—a rather questionable assumption—and assuming that nearly all the field due to the space charges is concentrated in the solid rather than in the electric double layer on the solution side, these equations may be simplified to yield

$$I_0^n = Fv[\text{Red}]\exp\left\{\frac{-[\zeta^2\lambda + \Delta\varepsilon_0 + e(E_0 - E_{\text{FB}})]^2}{4^2\lambda kT}\right\}$$

$$I_0^p = Fv[\text{Ox}]\exp\left\{\frac{-[\zeta^2\lambda + \Delta\varepsilon_0 - e(E_0 - E_{\text{FB}})]^2}{4^2\lambda kT}\right\}$$

(38)

where E_0 is the standard potential of the redox couple and E_{FB} the flat-band potential of the system, i.e., that potential at which there is no band bending across the solid-solution interface: in other words, it is the value of the Fermi level of the solid. The zero level of energy is at infinity. e stands for the electronic charge. $\Delta\varepsilon_0$ now refers to the bulk of the solid.

In contradistinction to metals, where $i_0 \propto ([\text{Ox}][\text{Red}])^{1/2}$, i_0 now is proportional to either [Red] or [Ox], and depends on the E_0 value of the redox couple. This relation has been shown to hold, e.g., for Ge electrodes, as well as for the wide gap, very narrow band, semiconductor anthracene. In very poorly conducting solids—such as anthracene—the carrier concentration is so low that one would expect the exchange current density to be limited by the current-carrying capability of the bulk solid. This, however, is not the case because the concentration of carriers injected from the solution completely swamps the very low concentration of carriers "intrinsically" present. It can be shown[108] that the hole and electron partial currents, in the case of anodic and cathodic reactions, respectively, cannot exceed the values of the exchange currents.

This limiting current, determined by electrochemical processes at the interface, must not be confused with a diffusion limited current; it is assumed that no diffusion limitations exist.

It is seen from Eq. (38) that for a strongly oxidizing couple, where

$$(E_0 - E_{FB}) \geq \frac{\Delta \varepsilon^p}{e} \gg 0 \qquad (39)$$

holds so that $i_0^p \gg i_0^n$, i_0 may attain values of the same order as observed with the same redox couple and a metal electrode. This has been confirmed experimentally.[108]

If the inequality (39) holds, then[108] the dependence of i_0 on the redox potential may be approximated by

$$\partial \log_{10} i_0 / \partial E_0 \simeq \frac{2.3 eF}{RT} = \frac{1}{120 mV} \qquad (40)$$

Any differences between the i_0 attained for a given redox couple and a metal, say Pt, electrode and a semiconducting electrode are due to lattice polarization effects and the preexponential factors involving, e.g., the transmission coefficient.[109] Moreover, in well-conducting semiconductors the assumption that the penetration depth of the space-charge region within the solid greatly exceeds the thickness of the double layer on the liquid side may not hold, giving rise to an additional overvoltage for practical currents; an additional voltage drop across the electric double layer on the solution side has to be supplied.

In the limit, the semiconductor surface may become degenerate due, say, to the conduction band being bent to such a degree that it contacts the Fermi level. There can then exist no potential drop across the solid side of the interface and the entire potential drop observed must appear across the double layer on the liquid side. The semiconductor electrode then will behave like a metal, and all metal electrode electrochemistry will be applicable. These considerations may even apply to metal electrodes which present a "demetalized" surface to the solution, due to the formation of a semiconducting film on the electrode surface,[110] as has been reported for some cases of oxygen reduction.[110]

(ii) Free Radical Reactions

Many of these reactions occur via excited states, often triplet states; they can be studied electrochemically by means of measuring the anodic photocurrent in oxidation reactions, and v.v.[111] In some cases of heterogeneous catalysis,[112] e.g., on ferroporphyrin complexes, the transition state involves a surface charge transfer complex exhibiting an extended and common energy level;[112] this requires the relaxation time for its formation to be shorter than the time constant τ of the catalyst–reagent interaction. This means $\tau \gg 10^{-12}$ sec, and thus longer than a molecular vibration period, which is of the order of 10^{-13} sec.

Another method for the detection of excited intermediates involved in electrocatalytic reactions consists in carrying out the reaction within the cavity of an ESR spectroscope; this method has been refined[113] by incorporating a coulometer placed within the cavity.

It is a truism to say that catalytic, and a fortiori electrocatalytic reactions occur via an excited state, because the transition state is necessarily a state of a potential energy higher than either the initial or the final one occupied by the system in the course of the reaction. However electrocatalytic reactions may also involve excitonic states of the electrode, such as have been reported,[114] e.g., for anthracene electrodes. Such reactions can be considered as reactions of a different kind[114] and may be written thus:

$$A_s^* + A \rightleftarrows A_s + \oplus + A^-$$
$$A_s + \oplus + A^- \rightleftarrows A_s + A \tag{41}$$

where the acceptor A, chemisorbed on the surface and being the locus of a localized excition, is denoted by A_s^*. The second reaction is a surface redox reaction resulting in the (dark) injection of positive holes. Due to the presence of traps in the surface layer, it is likely that the excition dissociates really in a layer somewhat below the surface, the carrier arriving at the interface by diffusion. The dissociation of the exciton yields a positive hole which is injected into the electrode and an electron which is available for an electrochemical reaction; the opposite sequence is also possible. The electrochemical reaction then becomes determined at least in part by the concentration and depth of traps in the semiconductor electrode surface and the rate of the above reaction sequence. Somewhat indirect support of this reasoning may be derived from the reported dependence of the electron transfer rate from free radicals to acceptors on the redox potential of the latter.[115]

In many catalytic reactions the main function of the catalyst can be considered to be to raise the concentration of free radicals in the interface, because less free energy is required to transform a free radical to a valency saturated ion. Ordinarily, the active centers acting as free radicals are free surface valencies, i.e., surface states of the solid.[116] Adsorbed free radicals, or localized regions of excitation, viz., excitons, may play the role of such active centres. It may well be that the high catalytic activity of Pt is associated at least partially with its strong paramagnetism. Thus, addition of a free radical should raise the catalytic activity: it has been shown that addition of the free radical of chlorpromazine to the acid electrolyte raises the potential of an oxygen electrode by about 50 mV.[117] An applied electric field affects the dissociation of excitons and thus the electrocatalytic reaction rate.[118]

Excitons may be readily introduced into a solid by irradiation; at least some of the resulting electron–hole pairs are likely to become localized at suitable centers and especially in the surface. The effect has been employed in photocatalytic reactions on semiconductor surfaces.[119] No work on CTCs in this field has been reported to the writers' knowledge, though such studies would be interesting and rewarding.

Charge transfer complexes are active in photopolymerization reactions, e.g., of styrene[120]; some CTCs homopolymerize by a free radical mechanism yielding the copolymers.[121]

(iii) Redox Reactions

A redox reaction can be defined as an electron exchange between two quantum states of equal energy, via an electron or a hole transfer. Redox catalysis thus involves reactions based on such reversible exchanges. The catalyst then reversibly and periodically changes from a reduced to an oxidized state.

Many organic semiconductors have been shown to be able to catalyze redox-type reactions.[122] This has been shown to be an intrinsic property of the "pure" solid, not associated[123] with impurities, though the presence of hetero atoms and of complexed, or chelated, metal atoms greatly affects the activity. While good electronic—be it p-type of n-type—conductivity appears to be a necessary condition for catalytic activity, it is certainly not a sufficient condition: it is the nature of the chemisorption bond between catalyst and reagent which also determines the activity. Since the nature of this bond depends so much on details of the surface state structure of the solid, these do affect the activity, as does the nature of the catalyst substrate, i.e., the carrier material which supports the catalytically active surface.[124]

This is likely to be due to a charge transfer interaction between catalyst and substrate, as has been proposed for metallic complexes[125] and, specifically, for the phthalocyanines.[126] Potentiodynamic current vs. voltage measurements in the absence of oxygen, though in an acid electrolyte, show a current maximum at a certain potential and the position of this peak, it has been suggested, can be correlated with catalytic activity in the electrocatalytic reduction of oxygen.[126] O_2 redox reactions will be further discussed later in this section.

Redox reactions of quinones such as benzoquinone, at semiconductor electrodes appear to be two-step processes[127] in which first an electron is transferred to the conduction band followed by a positive hole being transferred into the valency band of the solid. The electron transfer results in an excited state of the reactant,[127] followed by a further charge transfer from the valency band. Such reactions are likely to occur also with charge transfer complex electrodes.

It has been proposed that in redox reactions the change of the Fermi level across the interface is solely due to a change in the charge distribution.[128] Considering now a charge transfer dq associated with a change in the Fermi level dE_F, one can define a quantity C as

$$C \equiv dq/dE_F \qquad (42)$$

which characterizes a capability to donate or accept electric charge and which thus can be termed the redox capacity of the catalyst, or reagent. For a given reaction, that catalyst will give optimum performance when[128]

$$C_{\text{catalyst}} \gg C_{\text{reagent}} \tag{43}$$

Quite generally, for any redox reaction, one can say that the exchange current density is primarily governed by the height of the potential barrier and thus by the rate-determining step of the reaction path.[129] Any changes that leave that step unaltered will leave the barrier height unaltered. However, it should be pointed out that the statement[130] that there is no correlation between the Fermi level of the electrode and thus its work function, and heat of activation of a nonbonding reaction, holds only for materials in which the electron affinity is equal to the work function, as is the case, by definition, for metals. Nearly all organic materials do not exhibit such an equality.

Introduction of an organic intermediate into the surface—say a charge transfer complex—can yield significant increases in electrocatalytic activity though no drastic improvement can be expected; thus, e.g., adsorption of a free radical improves the performance of the aqueous electrode by about 20%.

In the electrocatalytic reduction of oxygen in an acidic electrolyte O_2^- is weakly adsorbed.[131] For electron transfer, the electron affinity A of O_2 must exceed that of the surface at the adsorption site. A dipole is then formed resulting in a weak bonding of the ion to the surface; the resulting entity resembles a weak, localized, charge transfer complex. There is experimental evidence[132] that such transient CTCs do actually exist in the case of oxygen adsorbed on a donor dye substrate; in effect an ion pair $D^+ \cdots O_2^-$ is formed.

Several coordination complexes and metal chelates have been tested[126] successfully for use as electrocatalysts in redox reactions, especially for the reduction of oxygen which is of utmost importance in fuel-cell technology. These complexes exhibit at least partial charge transfer character[133]; thus chloranil has been shown to form CTCs with several coordination complexes of 8-quinolinol and its Cu, Pd, and Ni chelates. Several other CTCs involving metal coordination complexes have been reported. These systems are interesting because of the enhanced catalytic activity shown by the central metal atom under such conditions, e.g., for the oxygen discharge.[133]

The dependence of electrocatalytic activity on the nature of the solvent from which the catalyst is precipitated again suggests that a charge transfer interaction may be involved: if, e.g., Co(II)-porphyrin catalysts are precipitated from pyrene rather than from dioxane, the current available at 600 mV potential is reported[126] to be more than doubled. Now, pyrene is likely to withdraw electrons from the porphyrin; the result would be a complex between the porphyrin and the pyridine solvent, though there is no published evidence for this effect.

Redox reactions involving methylene blue frequently involve the formation of a CTC; thus, e.g., the reduction rate of the dye by ascorbic acid was raised 50-fold by complexing dimethylalloxazine[134] or aromatic hydrocarbons.[135] These studies have important biological implications because of their applicability to some enzyme reactions.

Ferrocene complexes with, e.g., quinones or chloranil are also reported to enhance redox reactions.[136] The function of redox polymers is also likely to be associated with the formation of CTCs at least as intermediates.[137]

(iv) Heterogeneous Catalysis of Hydrogen Exchange Reactions

This is relatively the best explored field of CTC catalysis; especially the hydrogen–deuterium exchange recation catalyzed by charge transfer complexes has been extensively studied. The CTCs employed as catalysts were aromatic hydrocarbons complexed with Na,[138] and other alkali metals,[139] phthalocyanine-Li,[140] and Na–organic acceptor[141] complexes, anthracene-trinitrobenzene,[142] phthalocyanine-Na,[143] violanthrene-B-Cs,[144] and Ba–naphthacene[145] complexes.

At temperatures in excess of about 190°K *ortho/para* (*o/p*) hydrogen conversion takes place, on a suitable catalyst, parallel to the deuterium exchange.[146,147] The *o/p* hydrogen conversion is symmetry forbidden because the elementary step requires a change of spin; thus a material containing paramagnetic centers is capable of catalyzing it, though at low temperatures the H_2/D_2

exchange is very much slower.[148] The reaction thus appears to be a chemisorption one at higher temperatures, and mainly a physical one at low temperature.[148] It appears[139] that the *para*-to-*ortho* hydrogen conversion takes place through both these mechanisms while the H_2–D_2 exchange reaction occurs only by means of the first step.

The activation energy for the *o/p* hydrogen conversion was 4.4 kcal mole^{-1} with phthalonitrile–Na as the catalyst, 10 kcal mole^{-1} with violanthrene-B–Ca, and 0.6 kcal mole^{-1} with graphite–Cs.[149] This has been interpreted[149] as suggesting that the strongly activated hydrogen adsorption involves a charge transfer from the alkali metal to the acceptor component of the complex. There is some evidence, though not entirely convincing,[149] that the catalytic activity of CTCs in hydrogen exchanges depends strongly on the electron affinity of the acceptor; there is indication of an activity peak in the region of $A = 0$–0.6 eV. An excited state—perhaps a triplet state?—may be involved.[142] The decrease in the deuterium concentration, apparently a first-order reaction, for perylene–Cs as the catalyst had an activation energy of 8–9 kcal mole^{-1}.[144]

It is not clear what exactly is the role of the metal atom: no activity is shown[138] by the metal by itself, nor by the organic acceptor by itself, though the anthracene–Na complex behaves reversibly for the H_2–D_2 exchange reaction, with an activation energy of about 12 kcal mole^{-1} for the 1:2 adduct; the 1:1 required about 16 kcal mole^{-1}. There appears to be a connection, though, with the amount of, e.g., Na present in the complex, and its activity in the hydrogen exchange reaction.[146] The violanthrene complexes are said[147] to exhibit an activity dependence in that the apparent activation energy for the exchange reaction decreases with increasing atomic number of the donor atom: Na < K < Rb < Cs.

ESR and NMR techniques suggest that the locus of the exchange reaction is that having the lowest localization energy within the acceptor; azulene seems to be an exception.[138]

The tetracyanopyrene–Cs complex is highly active for the hydrogen exchange reaction though the tetranitropyrene–Cs adduct performs but poorly.[150] Depending mainly on the identity of the organic acceptor, CTCs may be classified[150] into two groups as far as their catalytic activity for hydrogen exchange is concerned.

Group "A," exemplified by tetracyanopyrene–Cs, maintains thermodynamic equilibrium throughout the reaction; complexes of group "B" such as tetranitropyrene–Cs, attain equilibrium only after several months. It is suggested[150] that H_2 is readily chemisorbed on the surface of a group "A" complex and equilibration involves an adsorption–desorption reaction step a in Eq. (44). In the case of group "B" complexes, however, there is very little H_2 chemisorbed and the reaction occurs slowly via step (b) in Eq. (44). The "B" group complexes are far tighter bonded than those of group "A"; in fact, the electrical conductivity of group "A" is about 40 times that of group "B," concurrent with a lowered activation energy for conduction from 0.94 to 0.7 eV. It is concluded[150] that the activity is due to an excess charge on the aromatic anions formed in the solid complex.

This is supported by some results[140] on the H_2–D_2 activity of metal phthalocyanine complexes, largely independent of the identity of the metal atom and said[140] to be mainly determined by the number of delocalized electrons in the porphyrin ligand. However, no, or only a very poor, catalytic activity has been found for aromatic–aromatic CTCs such as diaminodurene–TCNQ.[151]

How the electronic configuration of the CTC catalyst affects the catalytic activity is evident from Table 2, which refers to various phthalocyanines.[152]

Generally the relative activity for the hydrogen exchange reaction depends on the electronic properties of the CTC, especially on the reduction potential of its anion[153]: CTCs containing electron-donating anions with high reduction potential react readily with molecular hydrogen to cause its dissociative adsorption. On an aromatic hydrocarbon acceptor/Na complex surface,[154] an electron is transferred from the anion to H_2 followed by the formation of a monohydroanion complex[153]:

$$(A^{2-} - 2Na^+) + H_2 \rightleftarrows (A^-Na^+) + NaH + H \quad \text{(a)}$$
$$(A^-Na^+) + H \rightleftarrows (AH^-Na^+) \quad \text{(b)}$$
$$(A^-Na^+) + H_2 \rightleftarrows A + NaH + H \quad \text{(c)}$$
$$(A^-Na^+) + H \rightleftarrows (AH^-Na^+) \quad \text{(d)}$$

(44)

Table 2
The Electronic Configurations and Activities of Several Polynegative Ions of Charge Transfer Complexes of Phthalocyanines for the Hydrogen Exchange Reactions[a]

Complex	Electron configuration	D_2–HZ[*b] $(3+\log k)$	H_2–D_2^{**}[c] $(3+\log k)$
FePc	d^6	—	—
LiFePc	d^7	—	—
Li_2FePc	d^8	0.08	0.21
Li_3FePc	$d^8+\pi$	0.56	0.92
Li_4FePc	$d^8+\pi^2$	1.98	2.08
CoPc	d^7	—	—
LiCoPc	d^8	—	—
Li_2CoPc	$d^8+\pi$	0.12	0.24
Li_3CoPc	$d^8+\pi^2$	0.58	0.86
Li_4CoPc	$d^8+\pi^3$	1.46	1.82
Li_5CoPc	$d^8+\pi^4$	1.80	2.10
NiPc	d^8	—	—
LiNiPc	$d^8+\pi$	0.43	0.30
Li_2NiPc	$d^8+\pi^2$	1.28	1.26
Li_3NiPc	$d^8+\pi^3$	1.65	1.98
Li_4NiPc	$d^8+\pi^4$	2.12	2.46

[a] After J. Manassen, Ref. 105.
[b] $k(hr^{-1})$ at 60°, $P_{D_2} = 12.5$ cm Hg.
[c] $k(hr^{-1})$ at 60°, $P_{H_2} = 15$ cm Hg, $H_2:D_2 = 1:1$.

The relation between the catalytic activity of metal-phthalocyanines and their oxidation potentials is shown in Fig. 8. For a full discussion the reader is referred to the original papers.[105,139,152]

There does not appear to exist any simple correlation between fundamental physical properties such as concentration of carriers in the complex, the electron affinity or ionization potential of its components and its catalytic activity,[155] though a linear relation between at least initial activity and the energy of electron localization of the organic anion has been reported.[156]

(v) Other Catalytic Reactions on Charge Transfer Complexes

The catalytic activity of CTCs has also been demonstrated for a few other reactions: e.g., for the condensation reaction

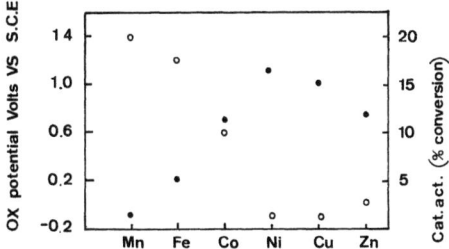

Figure 8. Oxidation potentials and catalytic activities of several metal-phthalocyanines as a function of the central metal atom. ○, Oxidation potential; ●, catalytic activity. (After J. Manassen, Ref. 105.)

between CO and H_2 (Fischer–Tropsch synthesis),[157] for the 1,2-butene isomerization,[158] for the Rosenmund–von Braun nitrile synthesis[159] and related reactions,[159] and polymerization reactions.[160] The complexes employed were alkali-metal–metal-phthalocyanine for Fischer–Tropsch syntheses,[157] violanthrene–iodine for the butene isomerization,[158] Cu(I)Cl–pyridine for the nitrile syntheses,[159] and maleic-anhydride–amide complexes for polymerizations.[160] The latter required additional energy in the form of illumination, but needed no initiator.[160]

Halogen exchange reactions were observed in the reactions between bromoferrocene and Cu(I)halide–pyridine complexes.[161] The mechanisms of these reactions appear to resemble those discussed for the hydrogen exchange reactions, Section III.2(iv).

3. Electrochemical Methods[162]

(i) Polarography of Donors and of Acceptors

The polarographic half-wave potential $E_{1/2}$ associated with the transfer of an electron from a metal electrode to a molecule in solution may be written

$$E_{1/2} = W - A_g - S \qquad (45)$$

where A_g is the (gaseous) electron affinity of the molecule, W the work function of the metal, and S the solvation energy, which can be estimated from

$$S \simeq z^2 e^2 \left(1 - \frac{1}{\varepsilon}\right)\frac{1}{2a} \qquad (46)$$

It is assumed that the ion being discharged can be approximated by a spherical cavity of radius a embedded in a continuum of permittivity ε; z is the valency and e the electronic charge. To a first approximation, a is the average radius obtained from the molecular volume. Alternatively, S may be deduced from a modified Born cycle using well-known electrochemical methods.[163] A_g then can be obtained. $E_{1/2}$ values for acceptors and donors thus yield the (gaseous) A_g and ionization energy I_g values for oxidation and reduction reactions, using

$$\begin{aligned} E_{1/2}^{ox} &= I_g - S^{ox} - c \\ E_{1/2}^{red} &= A_g - S^{red} - c \end{aligned} \qquad (47)$$

where c is a constant depending on the electrode. Then

$$E_{1/2}^{ox} - E_{1/2}^{red} = I_g - A_g - (S^{ox} - S^{red}) \qquad (48)$$

These equations have been used to estimate A_g for several acceptors[164] since I_g can be obtained, e.g., from charge transfer spectra or even calculated theoretically with reasonable accuracy.[165] Generally, donor and acceptor strengths rise as $E_{1/2}^{ox}$ diminishes and $E_{1/2}^{red}$ becomes less negative. The plots of $E_{1/2}^{ox}$ versus the peak of the charge transfer band are often linear if the same solvent is employed.[166]

While the converse of the method permits the evaluation of I_g if A_g is known, it is usually A_g that is required rather than I_g, which may be readily obtained, e.g., from the photoemission threshold.

Most aromatic acceptors may be reduced to their free radical ions, thus allowing potentiometric titrations[167]; these reductions, however, may readily and with greater convenience be carried out polarographically in aprotic, or nearly aprotic, solvents[168]:

$$\begin{aligned} A + e^- &\rightleftarrows A^- \\ D &\rightleftarrows D^+ + e^- \end{aligned} \qquad (49)$$

As long as the free radical ion is sufficiently stable not to affect the electrode potential, $E_{1/2}$ differs only very slightly from the reversible redox potential[169]; coulombic repulsion causes the reduction of a free radical anion, i.e., the addition of yet another electron, to require much more energy than the reduction of the neutral molecule, though exceptions, due to steric effects in nonplanar molecules, are known.[169]

The reversible redox potential E_0 is related to the frequency of the charge transfer band maximum, ν_{CT} by[169]

$$h\nu_{CT} = I_g - E_0 + (S - E_{ref}^0 - C) \qquad (50)$$

where C stands for the interaction energy of the charge transfer state containing coulombic and resonance contributions and E_{ref} to the potential of the reference electrode. S refers to the solvation energy of the free radical ion, assumed to remain constant. If this holds and C also remains constant, then the linear relationship illustrated in Fig. 9 should be expected.

The reversibility of the electrode processes can be estimated from ac polarography, which also shows up impedance changes due to tensammetric adsorption–desorption equilibria, if such are present. Many donors are strongly adsorbed at the electrode surface, e.g., chlorpromazine[152] or pyrene.[168] Differences in solvation can be studied by repeating the same polarogram in different solvents, though for many acceptors S appears to be remarkably constant.[168]

The redox potential in a donor solvent will be affected by the presence of anions. The effect becomes more prounced if the supporting electrolyte contains anions which have donor properties, allowing the anions to compete with the solvent molecules for coordination. In a protic solvent, hydrogen bonding affects the redox potential.[168]

(ii) Polarography and Voltametry of Charge Transfer Complexes

Consider a solution of uncomplexed acceptor A in the presence of a suitable, inert, supporting electrolyte.[168] Polarography then will yield a half-wave potential $E_{a/2}$ due to the reduction of A to A$^-$ at the dropping mercury cathode. If now donor D is added, in general several complexes of different stoichiometries form, having the composition D_jA, where j is an integer assum-

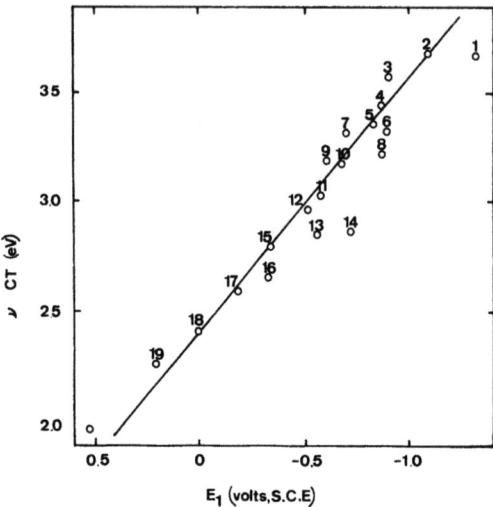

Figure 9. Dependence of the energy of the charge transfer band of charge transfer complexes of hexamethylbenzene as the donor, on the reduction potential of the acceptor. (After J. P. Williams, Ref. 177.) Key: 1, phthalic anhydride; 2, p-chloronitrobenzene; 3, m-dinitrobenzene; 4, 1,3-difluoro-2,4-dinitrobenzene; 5, 0-dinitrobenzene; 6, 1-fluoro-2,4-dinitrobenzene; 7, p-dinitrobenzene; 8, tetrachlorophthalic anhydride; 9, sym-trinitrobenzene; 10, 2,5-dimethyl-p-quinone; 11, methyl-p-quinone; 12, p-quinone; 13, pyromellitic dianhydride; 14, 1,2,4,5-tetracyanobenzene; 15, chloro-p-quinone; 16, dibromopyromellitic dianhydride; 17, 2,6-dichloro-p-quinone; 18, chloranil; 19, tetracyanoethylene; 20, 2,3-dichloro-5,6-dicyano-p-quinone.

ing values from unity to i; for a 1:1 complex $i = 1$. Each of the i stoichiometries is governed by an equilibrium constant K_{ij} corresponding to a complex D_jA. If now the activity of the donor remains constant throughout the solution because $[D] \gg [A]$, i.e., if there is a considerable donor excess, and introducing the Ilkovic equation, another half-wave potential, $Ec/2$ will be obtained, which will be shifted by $\Delta E_{1/2}$ relative to $E_{a/2}$. It can then be shown that[168]

$$\Delta E_{1/2} = \frac{RT}{F} \ln j_A \frac{I_c}{I_a} \sum_{j=0}^{j=i} \left(K_j[D]^j \bigg/ \sum_0^i j_{D_jA} \right) \quad (51)$$

j_A is the activity coefficient of the acceptor, I_c and I_a the polarographic diffusion currents of the complex and of the acceptor, respectively, and $[D]^j$ stands for the activity of the donor D in the jth complex. For a 1:1 complex, Eq. (51) reduces to

$$\Delta E_{1/2} = \frac{RT}{F} \ln j_A \frac{I_c}{I_a} \frac{K[D]}{j_{DA}} \qquad (52)$$

The values of the equilibrium constants K are similar to those obtained spectroscopically.[168] The solvent, as well as the supporting electrolyte, are assumed to be inert. It has also been assumed that no redox reactions occur of the form

$$D + A \rightleftarrows D^+ + A^- \qquad (a)$$
$$A^- + D \rightleftarrows D^- + A \qquad (b) \qquad (53)$$
$$D + e^- \rightleftarrows D^- \qquad (c)$$

Reactions like (a) are likely[168] with strong donors and acceptors, i.e., with strong complexes, and in a medium of relatively high permittivity. Otherwise, the resulting shifts in the half-wave potential are small relative to $\Delta E_{1/2}$ in Eq. (51).

If the dissociation of the complex is slow, then[168] $E_{1/2}$ shifts in a direction opposite to that derived from the free energy of complexation [Eqs. (51) and (52)], and a 1-mV shift corresponds to a drop in I_c by about 4%. The same reasoning holds for added acceptor to an excess of donor.

The relation between the oxidation potentials of several CTCs and their equilibrium constants is illustrated in Fig. 10.

The half-wave potentials, e.g., of phenothiazines, shift if oxygen is not expelled from the solution; this has been interpreted as evidence for the formation of a CTC.[170]

An inverse relation between the diffusion current and the molecular weight of the acceptor has been reported.[171]

Polarography has been applied to the determination of trace amounts of iodine by means of complexation with suitable donors and for following the process by means of ac polarography.[172] Several complexes such as pyrene:TCNQ have thus been studied[173]; for further discussion, the literature[174] should be consulted.

The technique of cyclic voltametry is also applicable to the study of charge transfer complexes; some preliminary studies

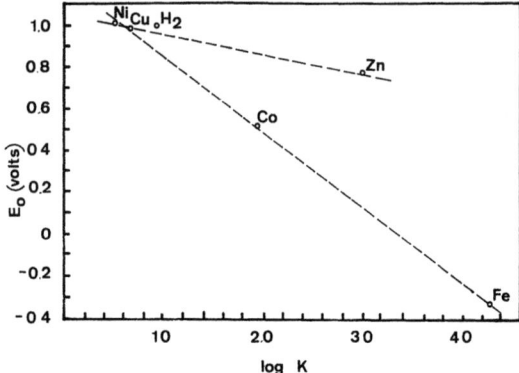

Figure 10. Relation between the oxidation potentials and the logarithm of the equilibrium constants of charge transfer complex formation for several metal-tetraphenylporphyrins. (After J. P. Williams, Ref. 177.)

have confirmed[31] this: new peaks are observed, e.g., in the well-known CPZ:I_2 interaction which are absent in the components. Likewise, the CPZ: dilantin interaction [see p. 370, Section II.3(i)] shows a new peak; these shift as a function of scan rate.

(iii) Other Electrochemical Methods

The pyridine:iodine complex has been studied[175] in a concentration cell Pt|[Pyridine-2I_2]$_{c_1}$ + CH_3CN | [Pyridine-2I_2]$_{c_2}$ + CH_3CN | Pt and by potentiometry; it appears that the 1:2 complex tends to decompose upon dilution with acetonitrile.

The phenothiazine:iodine complex has been subjected to prolonged electrolysis in a two-compartment cell with a cellulose separator and gold electrodes 6 cm apart.[8] 280 V applied resulted in the anolyte becoming brown (iodine) while the catholyte became pinkish-violet. The color changes are reversible upon repeated polarity reversal. While the value of the maximum current in each electrolysis experiment tends to drop approximately exponentially upon repetition, using the same solution, the charge transported stays very nearly constant. Upon cessation of the electrolysis, the anolyte and catholyte slowly recombine by diffusion until the original complex solution is reconstituted. It thus appears that the complex formation/dissociation equilibria are reversible. The discoloration of the anolyte indicates that the anions are iodine ions, I_x^-, molecular iodine being plated out at the anode. The polarization voltage measured between the electrodes, after electrolysis, was about 1.5 V: it decays slowly with time due to recombination of ions through the separator.[8] There always remains a small, constant current showing no signs of diminution with time. This current may well be electronic, such as has been observed, e.g., in the pyridine/iodine system, in contrasdistinction to the time-dependent ionic component.[169]

The formation of a CTC can also be followed by a potentiometric titration[31]; the electrode potential of, say, a Pt electrode measured against a reference electrode exhibits a maximum, or minimum, at the stoichiometry of complexation.

Potentiometry has also been employed to study what appears to be the formation of surface CTCs on membranes.[176]

IV. TERNARY AND PROTON TRANSFER COMPLEXES

1. Ternary Complexes

Such adducts arise in either solid or liquid solutions. In the solid state, they often involve guest molecules in a host matrix, where the host forms a charge transfer complex. The resulting electronic properties are quite different from those exhibited by the solvent-free CTC:[178] the resistivity typically drops from, say, 10^9 to 10^3–10^5 ohm cm, the activation energy of the conductivity drops from, say, 0.3 eV to 0.1–0.2 eV, and free radicals become evident from ESR as well as from spectroscopic studies. The new ternary complex thus formed has the structure D:A:S:S:D:A or, perhaps more likely, D:S:A:S:D, where S stands for a solvent molecule.

Thus, while the benzidine:TCNQ complex is[179] diamagnetic, the ternary B:TCNQ:S [S = e.g., dichloromethane (CH_2Cl_2) or chloroform ($CHCl_3$)] is strongly ESR active and the charge transfer band is shifted to longer wavelengths with a new shoulder appearing in the spectrum. The spectral changes depend on the relative donor/acceptor properties of the solvent.[180] The solvent dipoles try to align themselves with the host dipole moment; the ground state of the complex is less polar than the excited state.[180] The benzidine and the TCNQ molecules are alternately stacked on each other, forming molecular columns with intervening channels accommodating solvent molecules.[181] Energy and charge transport in such "doped" CTCs occurs, at room temperatures, predominantly by percolation, i.e., by variable range (as distinct from nearest-neighbor) hopping of excitons,[182] often triplet excitons. The mobility is thermally activated.[183] Excimers may be formed as, e.g., in pyrene in a naphthalene host, as evident from fluorescent spectra.[184] The ternary complex then is D–D*–A. Exciton trapping is important in these processes; the size of the trapping region depends critically on lattice distortion.[184] The energy transfer distance is about 30 Å for pyrene in naphthalene, 23 Å for anthracene in naphthalene, and only 12 Å for tetracene in naphthalene. Larger values, up to 130 Å, have been suggested.[184]

A particularly interesting class of what effectively are ternary complexes are electrolytically formed intercalation complexes such as the deep blue graphitic C_{24}^+ $(HSO_4)^- 2H_2SO_4$.[185] These lamellar complexes are very good catalysts for the formation of, e.g., formates and acetates. They are highly ESR active indicating a high concentration of free radicals and delocalized electrons. Upon contact with an acid, i.e., with an electron acceptor, considerable electron localization takes place. The graphitic compound as such has, apparently, a great many electron donating centers. Bromine,[186] $SbCl_5$,[186] chromic anhydride,[187] Al_2Cl_6[187] as well as other strong acids[185] such as $HClO_4$ or H_3PO_4 may also be intercalated. These materials should have interesting conduction properties. Graphite also forms intercalation ternary adducts stabilized by metal halides. Some of these are superconducting at very low temperatures,[188] though TaS_2 intercalation compounds exhibit relatively high transition temperatures.[189] The ternary adduct $TaS_2(pyridine)_{1/2}$ is claimed to exhibit two-dimensional superconductivity.[190]

Metal ion bridged CTCs too have been reported,[214] e.g., for Cu^{2+} adducts such as Cu–bipyridil $(ATP)^{2-}$. This has a plane parallel, lamellar structure with the metal forming a bridge between the two aromatic ring systems. Such charge transfer interactions are of considerable importance in biological systems[191];

thus metal ion bridged ternary purine–indole adducts,[192] as well as some nucleic-acid–protein interactions via an ionic bridge, have been reported.[193] The stability of the complex may be determined by potentiometric titration of the ternary and of the binary adducts, i.e., with and without, e.g., tryptophane; this refers to the system (metal)–(adenosine triphosphate)–(tryptophane). The most likely structure[192] of the adduct is shown in Fig. 11.

Iodine bridges may link the ether oxygens in different dioxane molecules forming crystalline iodine–dioxane complexes.[75]

Ternary CTCs produced by alkali metal doping of an anthracene:organic acceptor adduct—such as dibenzanthracene, tetracene, or pentacene—have been studied.[194] Considerable electron delocalization is reported,[194] amounting to 6–10 anthracene per dibenzanthracene molecule and 4×10^4 anthracenes per pentacene molecule in the K-anthracene–pentacene complex.

Ternary complexes may also arise as solvent inclusion compounds: the host lattice has cavities which are occupied by guest (solvent) molecules. Intermolecular forces of the order of 0.2–0.5 eV bind guest to host.[178] Thus, e.g., the p-tricyanovinyl-N,N-dimethylaniline auto-complex is deep blue. In chloroform, it forms a ternary complex which is deep red.[195]

Many solvents enter into ternary complexes; thus much of the work on complexes in acetonitrile as solvent is suspect because this solvent has been shown[196] to form CTCs with

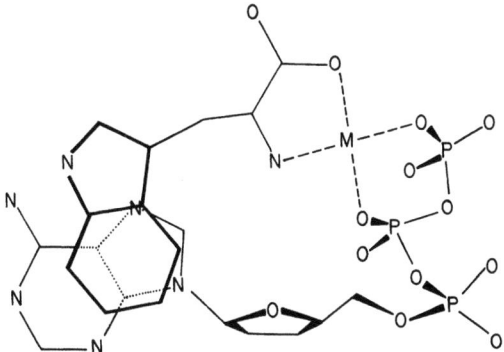

Figure 11. Proposed structure of the metal–adenosine-triphosphate–tryptophane adduct. (After H. Siegel and C. F. Naumann, Ref. 192.)

halogens which are frequently employed as acceptors. Many reported[197] irregularities in conductivity titrations using CH_3CN as solvent are probably due to the formation of ternary adducts. The red-colored serotonine–picrate monohydrate,[198] which forms alternating stacks of continuous columns much like TCNQ salts, is probably another case in point.

Of particular interest are the recently reported[199] inclusion complexes of mononuclear aromatic hosts with α-cyclodextrin: these form weakly coupled hydrophobic cavities capable of binding hydrophobic residues of compatible size. This small, water-soluble, "active site" catalyzes specific reactions with kinetics similar to those of enzymes. It thus acts as a model compound for specific enzymatic reactions. The electron donor–acceptor interaction, though, is reported[199] to be minor.

The entropy contributions to the formation, and stability, of ternary complexes must be quite complicated and do not appear to have been studied to any extent. CTC formation, generally, involves a desolvation and thus an increase of entropy because of the decrease of order due to the (at least partial) destruction of the solvation shell. In many cases, complexation in solution occurs because it involves a drop in the free energy of the system, though the enthalpy change may be endothermic. Ternary systems would thus be favored from the point of view of entropy production: there is a decrease of order because the bonds active in solvation are now engaged in interactions with the occluded solvent molecules. Ternary complexes are reported[213] to play a role in the formation of coal.

2. Proton Complexes

CTC formation may be followed, or be accompanied, by protonation[200]: thus, the orange-colored α-naphthylamine-picric acid 2:1 complex is really a ternary adduct consisting of picrate ions, protonated amine, and molecular α-naphthylamine. Several other complexes of that type have been reported[200]: e.g., picric acid and benzidine, or adducts between donors such as phenothiazine and the strong acceptor tetranitrobiphenyl 4,4'-diol (TNB). Other protonated complexes have been reported[201] with the strong acceptor $SbCl_5$ and solvents such as 1,2-dichloroethane. Adducts of the form of $(DH)^+SbCl_6^-$ have also been suggested.[202] At least part of the stability of some ternary, columnar, solvent inclusion complexes is due to a proton transition.[181]

It is, of course, difficult to decide whether the proton transfer discussed is not merely a case of conventional hydrogen bonding; however, there is considerable evidence for the simultaneous transfer of an electron and a proton giving rise to a new and different type of molecular adduct. As one extreme, one might consider the 1:1 complex between diphenylcyclopropenone and substituted acetic acids such as phenylacetic acid: the complexation appears[203] to involve a hydrogen bond between the oxygen of the polar carbonyl group of the propenone and the acidic hydrogen of the acetic acid. The infrared spectrum exhibits a red shift.[203]

At the other extreme, there are a number of complexes in which the CTC formation is due to the simultaneous transfer of electronic charge and a proton: the already mentioned orange-colored α-naphthylamine–picric acid 2:1 complex[204] or the orange/red 2:1 (or higher) complexes of tetranitrobiphenyl-4,4'-diol with aromatic monoamines[205] are examples of such adducts.

Thus, one part of a diamine molecule acts as an electron donor while another part acts as proton acceptor. Conversely, in 1:2 picric acid complexes, e.g., one picric acid functions as the proton donor and the other as the electron acceptor.[205]

The 1:1 and 1:2 complexes between oxalic acid and α-amino acids also involve a proton transfer from the carboxyl group of the oxalic acid to the carboxyl ion of the amino acid; there is a similar interaction in the 1:1 complex between malonic acid and glycine.[206] These adducts may well involve a great deal of proton delocalization resulting in the formation of proton energy bands rather than energy levels. Proton tunneling is a well-established fact and even halogen ions have been shown[207] to be able to tunnel between two suitable energy levels.

Ions are thermally excited from localized ionic states to free-ion-like states capable of migration with a velocity v determined by their energy E:

$$E = mv^2/2$$

Interaction with the rest of the solid then produces a finite lifetime τ corresponding to excitation of the free-ion-like state. This mechanism has been proposed[208] for MAg_4I_5, where M stands for NH_4 or an alkali metal, or for $Ag_4HgSe_2I_2$ adducts. Formation of protonic complexes is more likely in nonaqueous media, because the proton affinity of water, about 7.6 eV, is so very high; that of CH_4 is only 5.5 eV though ethers are reported[208] to have a proton affinity of 8.9 eV.

The now commercially available "proton sponges",[209] viz. 1,8-*bis*(dimethylamino)naphthalene compounds,[210] should be capable of forming new and interesting proton complexes. Formation of a protonic complex between aliphatic amines and phenols is reported[211] to result in a considerable increase in viscosity.

Ternary, protonic surface complexes arise[212] from the doping of a poly(*N*-vinyl carbazole): nitro aromatic (e.g., *o*-dinitrobenzene) CTC with strong organic acids such as trichloroacetic acid. The resulting films are highly photoconductive and exhibit a photoinduced ESR signal which might be used as a photoelectric memory.

V. CHARGE TRANSFER COMPLEXES AS ELECTROCHEMICAL ENERGY STORAGE DEVICES

An electrochemically active reactant may be stored in the form of a charge transfer complex.[214] Since the adduct is generally rather weakly bonded, the system may assume a state of lowered potential energy by the release of, say, the acceptor component to act in an electrochemical reaction. Thus, e.g., iodine may be stored as the iodine–phenothiazine complex, facing a magnesium anode. The complex acts as cathode. The cell is completed by an inert, say carbon or platinum, counter electrode. The device becomes practicable because complexation reduces the very high resistivity of elemental iodine by many orders of magnitude so that the cell shows a reasonably low internal resistance. Its open circuit voltage is about 1.5 to 2.5 V.[214] As current is drawn, more iodine is released from the solid complex. The resulting device may be used as a primary battery of long shelf life, lasting up to several years in low-power, intermittent operation such as in cardiac pacemakers.

ACKNOWLEDGEMENTS

We wish to thank Professor J. O'M Bockris of Texas A & M University for his patience, and our many friends and associates in the University of Nice, France, for their helpful comments. One of us (J.-P. F.) is indebted to the School of Chemistry, Macquarie University, for their hospitality and for a Visiting Fellowship.

REFERENCES

[1] J. O'M. Bockris and A. K. N. Reddy, *Modern Electrochemistry*, Plenum Press, New York, 1970, p. 2.
[2] F. Gutmann and L. E. Lyons, *Organic Semiconductors*, Wiley, New York, 1967; *Conduction in Low Mobility Materials*, Proceedings of the Second International Conference, Eilat, Israel, 1972; H. Fritzsche, *Ann. Rev. Mater. Sci.* **2** (1972 697; D. Adler, *Amorphous Semiconductors*, CRC Press Inc., West Palm Beach, Florida, 1971.
[3] V. Gutmann, *Chimia* **23** (1969) 285; V. Gutmann and A. Scherhaufer, *Monatsh. Chem.* **99** (1968) 335; V. Gutmann and U. Mayer, *ibid.* **98** (1967) 294; V. Gutmann, *Fortschr. Chem. Forsch.* **27** (1972) 59.
[4] M. J. Mantione, *Theor. Chim. Acta* **11** (1968) 119; J. P. Suchet, *Electrical Conductivity in Solid Molecules*, Pergamon Press, Oxford, 1975.
[5] V. Gutmann and U. Mayer, *Rev. Chim. Miner.* **8** (1971) 447.
[6] Y. Matsunaga, *Helv. Phys. Acta* **36** (1963) 800; F. Gutmann and H. Keyzer, *J. Chem. Phys.* **46** (1967) 1969; R. Foster and C. A. Fyfe, *Biochim. Biophys. Acta* **112** (1966) 460.
[7] B. Kratochvil and H. L. Yeager, *Fortschr. Chem. Forsch.* **27** (1972) 1.
[8] A. Brau et al., *Electrochim. Acta* **17** (1972) 1803.
[9] *Handbook of Physics and Chemistry*, 51st ed., Chemical Rubber Co. Cleveland, Ohio, 1970–71, p. F-37.
[10] N. Mataga, *Bussei Kenkyu* **18** (1973) A-37.
[11] See Ref. 1, p. 473.
[12] F. Gutmann et al., *Adv. Biochem. Psychopharmacol.* **9** (1974) 15.
[13] See Ref. 1, p. 434.
[14] See Ref. 1, p. 179.
[15] See Ref. 1, p. 472.
[16] A. Brau, J.-P. Farges, and F. Gutmann, unpublished results.
[17] See H. O. House, *Modern Synthetic Reactions*, 2nd ed. W. A. Benjamin Inc. Menlo Park, California, 1972, p. 438.

[18] E. J. Ariens and A. M. Simonis, *Afinidad* **33** (1976) 329; B. I. Lirova, *Khim. Vysokomol. Soedin Neftekim.* **1973** 106.
[19] F. Gutmann and H. Keyzer, *Electrochim. Acta* **11** (1966) 555.
[20] R. Anderson and J. M. Prausnitz, *J. Chem. Phys.* **39** (1963) 1225.
[21] H. O. Hooper, *J. Chem. Phys.* **41** (1964) 599.
[22] A. Brau, J.-P. Farges, and F. Gutmann, unpublished results.
[23] R. M. Fuoss and C. A. Kraus, *J. Am. Chem. Soc.* **55** (1963) 2387; H. S. Harned and B. B. Owen, *Physical Chemistry of Electrolyte Solutions*, Reinhold, New York, 1958.
[24] V. Gutmann and U. Mayer, *Monatsh. Chem.* **100** (1969) 2048; F. Frank, Ed., *Water*, Plenum Press, New York, 1973; T. Kagija, *Bull. Chem. Soc. Jpn.* **41** (1968) 767.
[25] S. G. Christov, 22nd meeting ISE Dubrovnik 1971; *Croatica Chim. Acta* **44** (1972) 67.
[26] S. G. Christov and Z. L. Georgiev, *J. Phys. Chem.* **75** (1971) 1748; S. G. Christov and S. Ikonopisov, *J. Electrochem. Soc.* **116** (1969) 56.
[27] H. Bassler and N. Riehl, *Z. Naturforsch.* **19A** (1964) 1070; **20A** (1965) 85, 394, 401, 587.
[28] M. Z. Swarc, *Z. Elektrochem.* **67** (1963) 763.
[29] F. Accascina, in *Electrolytes*, Ed. by B. Pesce, Pergamon Press, Oxford, 1962, p. 293.
[30] J. T. Denison and J. B. Ramsey, *J. Am. Chem. Soc.* **77** (1955) 2615.
[31] G. Eckert and F. Gutmann, *J. Electroanal. Interfacial Electrochem.* **62** (1975) 267.
[32] G. Eckert and F. Gutmann, *J. Biol. Phys.* **6** (1978).
[33] G. J. Janz and S. S. Danyluk, in *Electrolytes*, Ed. by B. Pesce, Pergamon Press, Oxford, 1962, p. 280.
[34] H. N. Parton, *Proceedings of the first Australian Conference on Electrochemistry*, Ed. by J. A. Friend and F. Gutmann, Pergamon Press, Oxford, 1964, p. 464.
[35] J. E. Prue, in *Chemical Physics of Ionic Solutions*, Ed. by B. E. Conway and R. G. Barradas, Wiley, New York, 1966, p. 163.
[36] See Ref. 1, p. 473.
[37] K. Tsuji et al., *J. Chem. Phys.* **46** (1967) 2808; **45** (1966) 2894; L. I. Boguslavskii and A. V. Vannikov, *Organic Semiconductors and Biopolymers*, Plenum Press, New York, 1970; F. Gutmann and L. E. Lyons, Ref. 2.
[38] F. Gutmann and H. Keyzer, *J. Chem. Phys.* **46** (1967) 1969.
[39] A. I. Popov and R. F. Swensen, *J. Am. Chem. Soc.* **77** (1955) 3724.
[40] See Ref. 1, p. 472.
[41] G. B. Sergeev et al., *Russ. J. Chem.* **47** (1973) 396; K. De Groot et al., *J. Chem. Phys.* **47** (1967) 3084.
[42] D. Braun and H. J. Sterzel, *Ber. Bunsenges. Phys. Chem.* **76** (1972) 551.
[43] A. D. Buckingham et al., *Trans. Faraday Soc.* **67** (1971) 577; L. Jansen and P. Mazur, *Physica* **21** (1955) 193, 208.
[44] G. Briegleb, *Elektronen-Donator-Akzeptor Komplexe*, Springer, Berlin, 1961, p. 182; D. Braun and H. J. Sterzel, *Ber. Bunsenges. Phys. Chem.* **76** (1972) 551.
[45] B. Case et al., *J. Electroanal. Chem.* **10** (1965) 360.
[46] G. B. Sergeev et al., *Russ. J. Chem.* **47** (1973) 396.
[47] M. A. Slifkin, *Charge Transfer Interactions of Biomolecules*, Academic Press, London, 1971, p. 47.

[48] M. A. Slifkin, *Spectrochim. Acta* **25A** (1969) 1037; R. Kh. Ibragimova *et al.*, *Izv. Akad. Nauk Uzb. SSR Ser. Fiz. Mat. Nauk* **17** (1973) 55.
[49] O. B. Nagy, J. B. Nagy, and A. Bruylants, *J. Chem. Soc. Perkin Trans. II* **1972** 968; V. Gutmann and U. Mayer, *Monatsh. Chem.* **99** (1968) 1383; M. A. Slifkin, *Charge Transfer Interactions in Biomolecules*, Academic Press, London, cf. also *Non-Aqueous Solutions*, Ed. by V. Gutmann, Pergamon Press, Oxford, 1976; cf. also S. Ahrland, *Solvation and Complex Formation in Protic and Aprotic Solvents*, Ed. by J. J. Lagowski, Academic Press, New York, 1978.
[50] U. Mayer and V. Gutmann, *Monatsh. Chem.* **101** (1970) 912.
[51] O. Nagy *et al.*, *J. Chem. Soc. Perkin Trans.* **1972** 968.
[52] J. I. Padova, *Modern Aspects of Electrochemistry*, No. 7, Ed. by J. O'M. Bockris and B. E. Conway, Plenum Press, New York, 1972, p. 1.
[53] S. Yomosa, *Progr. Theoret. Phys. Suppl.* **40** (1967) 249.
[54] R. S. Mulliken, *J. Am. Chem. Soc.* **74** (1952) 811.
[55] B. G. Anex and E. B. Hill, *J. Am. Chem. Soc.* **88** (1966) 3648.
[56] A. Brau *et al.*, *Electrochim. Acta* **17** (1972) 1803.
[57] See Ref. 47, p. 20.
[58] S. Koizumi and Y. Matsunaga, *Bull. Chem. Soc. Jpn.* **45** (1972) 423.
[59] L. Mandelcorn, *Chem. Rev.* **59** (1959) 827; J. H. v. d. Waals and J. C. Platteeum, *Adv. Chem. Phys.* **2** (1959) 1.
[60] M. Ohmasa *et al.*, *Bull. Chem. Soc. Jpn.* **44** (1971) 391.
[61] J. I. Aihara *et al.*, *Bull. Chem. Soc. Jpn.* **42** (1969) 1824; A. K. Covington and T. Dickinson, *Physical Chemistry of Organic Solvent Systems*, Plenum Press, New York, 1973.
[62] W. Scholtan, *Kolloid Z.* **142** (1955) 84; M. E. Kitler and P. Lamy, *Pharm. Acta Helv.* **46** (1971) 1483; P. M. Seeman and H. S. Bialy, *Biochem. Pharmacol.* **12** (1963) 1181.
[63] See Ref. 3, 6, 7, and 8.
[64] J. Forrest *et al.*, *Rev. Aggressol. (Paris)* **7** (1966) 147; A. H. Ademan and C. M. Verber, *J. Chem. Phys.* **39** (1963) 931; R. L. Ward, *J. Chem. Phys.* **39** (1963) 852.
[65] H. Veda and D. Yashiro, *Bull. Chem. Soc. Jpn.* **44** (1971) 391.
[66] F. Gutmann, unpublished results.
[67] K. Deguchi and K. Meguro, *J. Colloid Interface Sci.* **38** (1972) 596.
[68] I. Blei, *Arch. Biochem. Biophys.* **109** (1965) 321.
[69] R. Memming and F. Möllers, *Faraday Soc. Symp. No.* 4 (1970) 1140; R. Memming and G. Kürsten, *Ber. Bunsenges. Phys. Chem.* **75** (1971) 1140; S. A. Alkaitis, G. Beck, and M. Grätzel, *J. Amer. Chem. Soc.* **97** (1975) 5723; M. Chen, M. Grätzel, and J. K. Thomas, *J. Amer. Chem. Soc.* **97** (1975) 2052; A. J. Frank *et al.*, *Ber. Bunsenges. Phys. Chem.* **80** (1976) 547; M. Grätzel, *Ber. Bunsenges. Phys. Chem.* **79** (1975) 475.
[70] J. F. J. Kibblewhite and A. J. Tench, *J. Chem. Soc. Faraday Soc. Trans. I* **70** (1974) 72.
[71] J. Robillard, private communication 1976; see also L. I. Ahmed, *J. Phys. Chem. Solids* **29** (1968) 1653.
[72] C. A. Bunton, *Progr. Solid State Chem.* **8** (1973) 167.
[73] F. Gutmann, unpublished results.
[74] S. Oka, *Proc. Phys. Math. Soc. Jpn. III* **15** (1933) 247, 413; P. Debye, *J. Chem. Phys.* **1** (1933) 13.
[75] F. Tokiwa and K. Tsuji, *Bull. Chem. Soc. Jpn.* **46** (1973) 2684; H. Lange, *Kolloid Z.—Z. Polymer* **243** (1971) 101; M. L. Fishman and F. R. Eirich, *J.*

Phys. Chem. **75** (1971) 3135; K. Kano and T. Matsuo, Bull. Chem. Soc. Jpn. **47** (1974) 2836.
[76] A.-M. Baticle et al., C. R. Acad. Sci. **273** (1971) 1589.
[77] Th. G. J. van Oirschot et al., J. Electroanal. Chem. **37** (1972) 373; M. J. Sparnaay, Surf. Sci. **13** (1969) 99; H. Inokuchi et al., J. Catalysis **8** (1967) 383; F. Gutmann and L. E. Lyons, Ref. 2.
[78] J. P. Williams, 5th Keilin Memorial Lecture, Biochem. Soc. Trans. **1** (1973) 1; G. M. Schwab, Fortschr. Chem. Forsch. **25** (1972) 105; M. E. Peover and J. D. Davies, Trans. Faraday Soc. **60** (1964) 476.
[79] R. R. Dogonadze, Ber. Bunsenges. Physik. Chem. **75** (1971) 628.
[80] D. Vasilescu et al., Electrochim. Acta **19** (1974) 181; V. A. Tyagai, Electrochim. Acta **18** (1973) 229; H. Kranck et al., Electrochim. Acta **23** (1978) 891.
[81] A. Papoulis, Probability, Random Variables and Stochastic Processes, McGraw Hill, New York, 1965; R. Fürth and A. J. Allnutt, Physica **32** (1966) 869.
[82] J. B. Johnson, Phys. Rev. **32** (1928) 97; H. Nyquist, Phys. Rev. **32** (1928) 110.
[83] F. H. Lange, Correlation Techniques, Iliffe Books, London, 1967, p. 167.
[84] L. Landau and E. Lifschitz, Physique Statistique, Mir Publ. Co., Moscow, 1967, p. 349.
[85] F. N. Hooge, Phys. Lett. **33A** (1970) 169; N. M. Hosseini and B. K. Jones, Phys. Status Solidi **A40** (1977) k-185; G. P. Vasileu et al., Pisma Zh. Tekh. Fiz. **2** (1976) 604; K. Klason and J. Kubat, J. Appl. Phys. **47** (1976) 1970; G. Blanc et al., Electrochim. Acta **20** (1975) 687; J. Electroanal. Chem. Interfacial Electrochem. **75** (1977) 97.
[86] E. Groschwitz, Phys. Bl. **11** (1955) 121; A. M. H. Hoppenbrouwers and F. N. Hooge, Philips Res. Rep. **25** (1970) 69; G. Ya. Kolbasov and V. A. Tyagai, Fiz. Tekh. Poluprok. **6** (1972) 946.
[87] R. E. Burgess, Fluctuation Phenomena in Solids, Academic Press, New York, 1965; N. M. Hosseini and B. K. Jones, Phys. Status Solidi **A40** (1977) K-185.
[88] F. N. Hooge, Phys. Lett. **29A** (1969) 139; **33A** (1970) 169; J. H. Lorteije and A. M. H. Hoppenbrouwers, Philips Res. Rep. **26** (1971) 29; M. Weissmann and G. Feher, J. Chem. Phys. **63** (1975) 586; G. Blanc et al., Electrochim. Acta **20** (1975) 687; H. Kranck et al., Electrochim. Acta, **23** (1978) 891; K. Klason and J. Kubat, J. Appl. Phys. **47** (1976) 1970.
[89] W. F. Pickard, Nature (London) **201** (1964) 283.
[90] A. Einstein, Ann. Phys. (Leipzig) **IV-22** (1907) 569; cf. also: B. Breyer and F. Gutmann, J. Proc. Roy. Soc. NSW **83** (1950) 66; L. Davis, J. Appl. Phys. **35** (1964) 2004.
[91] B. K. P. Scaipe, Progr. Dielectrics **5** (1963) 143.
[92] F. Micheron and L. Godefroy, Rev Phys. Appl. **7** (1972) 279.
[93] L. J. Giacoletto, Proc. IEEE **49** (1961) 921; see also Ref. 85.
[94] F. Cardon, Physica **57** (1972) 390; see also Refs. 80 and 85.
[95] G. C. Barker, J. Electroanal. Chem. Interfacial Electrochem. **39** (1972) 484.
[96] K. J. Euler, Naturwissenschaften **58** (1971) 621.
[97] V. A. Shenderovskii, Ukr. Fiz. Zh. **16** (1971) 1907.
[98] V. L. Bonch-Bruevich, Sov. Phys.-Solid State **7** (1966) 1728.
[99] V. Vysin and V. Janku, Phys. Lett. **9** (1964) 19.
[100] B. Breyer and H. H. Bauer, AC Polarography and Tensammetry, Interscience, New York, 1963; B. Breyer and F. Gutmann, Aust. J. Sci. **8** (1946) 163; B. Breyer and S. Hacobian, Aust. J. Chem. **7** (1954) 225; D. Smith, Crit. Rev. Anal. Chem. **2** (1971) 247.

[101] F. Gutmann et al., Adv. Biochem. Psychopharmacol. **9** (1974) 15.
[102] J. Kurk, Zesz. Nauk Uniw. Jagiellon. Phys. Chem. **16** (1971) 101.
[103] J. E. B. Randles and K. W. Somerton, Trans. Faraday Soc. **48** (1952) 937.
[104] A. Goudot, Wave Mech. Molec. Biol. **1966** 28.
[105] J. Manassen, Fortschr. Chem. Forsch. **25** (1972) 1; W. Hanke, Z. Chem. **9** (1969) 1; S. Toshima, Progr. Surf. Membrane Sci. **4** (1971) 250; M. Spiro, in Essays in Chemistry, Vol. 5, Ed. by J. N. Bradley, R. D. Gillard, and R. F. Hudson, Academic Press, New York, 1973.
[106] K. J. Vetter and J. W. Schultze, Ber. Bunsenges. Phys. Chem. **76** (1972) 920, 927, 1134.
[107] R. F. Lane and R. T. Hubbard, J. Phys. Chem. **77** (1973) 1411.
[108] W. Mehl and J. M. Hale, in Advances in Electrochemistry, Vol. 6, Ed. by P. Delahay, 1967, p. 399; W. Mehl and F. Lohmann, Electrochim. Acta **13** (1968) 1459; V. A. Myamlin and Y. V. Pleskov, Electrochemistry of Semiconductors, Plenum Press, New York, 1973; J. O'M. Bockris et al., Nature London Phys. Sci. **240** (1972) 143.
[109] W. Mehl in Reactions of Molecules at Electrodes, Ed. by N. S. Hush, Interscience, New York, 1971 p. 305.
[110] A. K. Vijh, J. Electrochem. Soc. **119** (1972) 1498.
[111] R. Memming and G. Kürsten, Ber. Bunsenges. Phys. Chem. **76** (1972) 4.
[112] I. B. Beruske and S. S. Budnikov, Fourth International Congress on Catalysis, Moscow 1968, paper No. 3 published as revised 1970, Ed. by Ya. T. Eidus, Nauka Publishing House, Moscow, 1970; A. Goudot, Ref. 104.
[113] J. K. Dohrmann and F. Gaulluser, Ber. Bunsenges. Phys. Chem. **75** (1971) 430.
[114] L. I. Boguslavsky and B. T. Lozhkin, Surf. Sci. **38** (1973) 413; W. Mehl and F. Lohmann, Ref. 108.
[115] P. S. Rao and E. Hayon, Nature (London) **243** (1973) 344.
[116] F. Gutmann and L. E. Lyons, Ref. 2. R. Memming and G. Kürsten, Ber. Bunsenges. Phys. Chem. **75** (1971) 1140.
[117] F. Gutmann, unpublished.
[118] P. Durand and R. Founie, Conference on Dielectric Materials; Measurements and Applications, 1970, Publ. No. 67, IEE, London, 1970.
[119] Th. Wolkenstein, Adv. Catal. **24** (1973) 157; T. Freund et al., Fourth International Congress on Catalysis, Moscow 1968, Ed. by Ya. T. Eidus, published as revised, Nauka Publishing House, Moscow, 1970, p. 94; W. Mehl. Ref. 109.
[120] T. Kodaira et al., Polymer J. **4** (1973) 1.
[121] S. T. Bashkatova et al., Vysokomol. Soedin **A14** (1972) 2640.
[122] S. Toshima, Progr. Surface Membrane Sci. **4** (1971) 231; J. Manassen, Ref. 105; F. Gutmann and L. E. Lyons, Ref. 2.
[123] S. Z. Roginsky and M. M. Sakharov, J. Phys. Chem. USSR **42** (1968) 1331.
[124] G. M. Schwab, Fortschr. Chem. Forsch. **25** (1972) 105.
[125] R. B. Hall and F. Williams, J. Chem. Phys. **58** (1973) 1036; E. H. Cordes and R. B. Dunlap, Accounts Chem. Res. **1969**, 2329; S. I. Ahmad and S. Friberg, J. Am. Chem. Soc. **94** (1972) 5196; E. J. Fendler and J. H. Fendler, Adv. Phys. Org. Chem. **8** (1970) 271; O. Inacher, Chem. Phys. Lett. **27** (1974) 317.
[126] H. Alt et al., Electrocatalysis and Fuel Cells, Ed. by G. Sandstede, University Press, Washington D.C., 1972, p. 113; W. Beyer and F. von Sturm, Agnew. Chem. **84** (1972) 154; A. Andro et al., C.R. Acad. Sci. **272C** (1971) 366; H. Tanaka et al., Denki Kagaku **39** (1971) 596; H. Jahnke and M. Schönborn, Ber. Bunsenges. Phys. Chem. **74** (1970) 944.

[127] R. Memming and F. Möllers, *Ber. Bunsenges. Phys. Chem.* **76** (1972) 609.
[128] I. B. Bersuker and S. S. Budnikov, Ref. 112.
[129] J. O'M. Bockris and S. Srinivasan, *Fuel Cells: Their Electrochemistry*, McGraw-Hill, New York, 1969, p. 1146.
[130] See Ref. 129, p. 351.
[131] Th. Wolkenstein, *Fourth International Congress on Catalysis*, Moscow 1968, published as revised, Ed. by Ya. T. Eidus, Nauka Publishing House, Moscow, 1970, paper No. 2; Th. Wolkenstein, *Elektronen Theorie der Katalyse an Halbleitern*, VEB Deutscher Verlag der Wissenschaften, Berlin, 1964; *The Electronic Theory of Catalysis*, Pergamon, Oxford, 1963.
[132] Y. Usui, *Bull. Chem. Soc. Jpn.* **42** (1969) 1231.
[133] Y. Iida, *Bull. Chem. Soc. Jpn.* **44** (1971) 2564; S. Koizumi and Y. Iida, *Bull. Chem. Soc. Jpn.* **43** (1970) 1436; Y. Usui, Ref. 132; K. Gollwick and G. O. Schenck, *Pure Appl. Chem.* **9** (1964) 507.
[134] Y. Iwasawa et al., *Bull. Chem. Soc. Jpn.* **43** (1970) 720, 2656.
[135] N. Wakayama and H. Inokuchi, *J. Catal.* **11** (1968) 143.
[136] Y. Omote et al., *Bull. Chem. Soc. Jpn* **44** (1971) 3463; J. C. Coan et al., *J. Org. Chem.* **29** (1964) 975.
[137] D. A. Seanor, "Electrical Properties of Polymers," Chap. 17 in *Polymer Science*, Ed. by A. D. Jenkins, North Holland Publ. Co., Amsterdam, 1972; I. M. Barkalov et al., *Dokl. Akad. Nauk SSSR* **169** (1966) 1111; F. Beck, *Ber. Bunsenges. Phys. Chem.* **71** (1973) 353.
[138] S. Tanaka et al., *Bull. Chem. Soc. Jpn.* **41** (1968) 1278.
[139] M. Tsuda, *Bull. Chem. Soc. Jpn.* **43** (1970) 3415; M. Ichikawa et al., *Bull. Chem. Soc. Jpn.* **43** (1970) 3672; D. D. Eley and H. Inokuchi, *Z. Elektrochem.* **63** (1959) 29; D. D. Eley et al., *Disc. Faraday Soc.* **28** (1959) 54.
[140] M. Ichikawa et al., *Bull. Chem. Soc. Jpn.* **41** (1968) 1739.
[141] M. Ichikawa et al., *Bull. Chem. Soc. Jpn.* **40** (1967) 1294.
[142] H. Inokuchi et al., *J. Catal.* **8** (1967) 91.
[143] M. Ichikawa et al., *J. Catalysis* **6** (1966) 336.
[144] H. Inokuchi et al., *J. Catal.* **8** (1967) 288.
[145] N. Wakayama and H. Inokuchi, *J. Catal.* **15** (1969) 417.
[146] N. Wakayama and H. Inokuchi, *J. Catal.* **12** (1968) 15.
[147] Y. Mori et al., *J. Catal.* **14** (1969) 1.
[148] T. Kondow et al., *J. Chem. Phys.* **43** (1965) 3766.
[149] H. Inokuchi et al., *J. Chem. Phys.* **46** (1967) 837.
[150] N. Wakayama and H. Inokuchi, *J. Catal.* **11** (1968) 143.
[151] M. Tsuda and H. Inokuchi, *Bull. Chem. Soc. Jpn.* **43** (1970) 3410.
[152] J. Manassen, *Fortschr. Chem. Forsch.* **25** (1972) 2; F. Gutmann et al., *Adv. Biochem. Psychopharmacol.* **9** (1974) 15.
[153] M. Ichikawa et al., *Bull. Chem. Soc. Jpn.* **43** (1970) 3672.
[154] S. Bank et al., *J. Am. Chem. Soc.* **91** (1969) 5407.
[155] H. Inokuchi, *Faraday Soc. Disc.* **51** (1971) 183.
[156] M. Ichikawa and K. Tamaru, *Z. Phys. Chem.* **84** (1973) 217.
[157] K. Tamaru et al., Japan Pat. 72 08 284, March 9, 1972.
[158] M. Tsuda et al., *J. Catal.* **11** (1968) 81.
[159] M. Sato et al., *Bull. Chem. Soc. Jpn.* **43** (1970) 2972.
[160] N. G. Gaylord, *Nuova Chim.* **49** (1973) 81; G. P. Belov et al., *Nuova Chim.* **48** (1972) 73; E. Tsuchida et al., *Nippon Kagaku Kaishi* **1972**, 2416; R. F. Tarvin, University of Iowa Dissertation 1972, Diss. Abst. Int. (B1972) **33** 3030.

[161] M. Sato et al., *Bull. Chem. Soc. Jpn.* **42** (1969) 1976.
[162] *Topics in Organic Polarography*, Ed. by P. Zuman, Plenum Press, New York, 1973; T. Kambara, *Modern Aspects of Polarography*, Plenum Press, New York, 1973; A. N. Frumkin and A. B. Ershler, *Progress in Electrochemistry of Organic Compounds*, Plenum Press, New York, 1973.
[163] See Ref. 1, Vol. 1.
[164] Sh. Shata, *J. Bioenerg.* **1** (1970) 325; F. Gutmann and L. E. Lyons, Ref. 2.
[165] F. Gutmann and L. E. Lyons, Ref. 2, Chap. VI.
[166] M. A. Slifkin, *Charge Transfer Interactions of Biomolecules*, Academic Press, London, 1971, p. 186.
[167] G. J. Hoijtink et al., *Rec. Trav. Chim. Pays Bas* **75** (1956) 487.
[168] B. Kastening, *Progr. Polarogr.* **3** (1973) 195; M. E. Peover, *Trans. Faraday Soc.* **60** (1964) 417, 479; Yu. A. Karbainov et al., Nov. Polyarogr. Tezisy Dokl. Vses Soveshla Polyargogr. 6th 1975, Ed. by Ya. Stradyn, p. 49.
[169] N. E. Wisdom and E. O. Forster, *J. Polymer Sci.* **C-17** (1967) 125.
[170] Ref. 166, p. 207.
[171] H. Bartelt and H. Skilandat, *J. Electroanal. Chem. Interfacial Electrochem.* **24** (1970) 207.
[172] L. V. Mirovich, Ostsillogr. Peremenotok Polyarogr. **1971**, 126.
[173] K. G. Boto and F. G. Thomas, *Aust. J. Chem.* **26** (1973) 1669; V. F. Toropova et al., *Zh. Obsch. Khim.* **43** (1973) 711; M. Barigandi et al., *Bull. Soc. Chim. Belges* **79** (1970) 625; A. Delsaut, Memoire de License en Sci. Centre Univ. Mons, Belguim, 1969.
[174] J. Wolf et al., *Melliand Textil Ber. Int. Ed.* **54** (1973) 61; N. Nagy et al., *J. Chem. Soc. Perkins Trans. 2* **1972**, 2048; E. J. Rudd and B. E. Conway, *Trans. Faraday Soc.* **67** (1971) 440; E. G. Chikayzova, Usp. Perspekt. Razv. Polyarogr. Metoda **1972**, 164; R. N. Adams, *Acc. Chem. Res.* **2** (1969) 175; G. Dayhurst and P. J. Elving, *Talanta* **16** (1969) 885.
[175] A. Ya. Gorenbein and A. K. Trofimchuk, *Zh. Obshch. Khim.* **37** (1967) 1422.
[176] M. E. Starzak, *J. Biol. Phys.* **2** (1974) 57; D. V. Lamsweerde-Gallez and A. Meessen, *J. Biol. Phys.* **2** (1974) 75; R. A. Llenado, *Anal. Chem.* **47** (1975) 2243; cf. also P. Groll and F. Grass, *Electrochim. Acta* **16** (1971) 31.
[177] J. P. Williams, *Diss. Abstr. Int.* **36** (8) (B-1976) 3950.
[178] H. Bretschneider, *Wiss. Z. Tech. Hochschule Karl Marxstadt* **17**(2) (1975) 281; F. Cramer, *Einschlussverbindungen*, Springer, Heidelberg, 1974.
[179] M. Ohmasa et al., *Bull. Chem. Soc. Jpn.* **44** (1971) 391.
[180] L. Fredin and B. Nelander, *J. Molec. Struct.* **16** (1973) 205; B. Nelander, *Theoret. Chim. Acta* **25** (1972) 382.
[181] M. Mandelkorn, *Chem. Rev.* **59** (1959) 827; M. Ohmasa et al., *Bull. Chem. Soc. Jpn.* **44** (1971) 391, 395; K. Deguchi and K. Meguro, *J. Colloid Interface Sci.* **38** (1972) 596; J. B. Torrance and B. D. Silverman, *Phys. Rev. B* **15** (1977) 788.
[182] R. C. Powell, *J. Chem. Phys.* **58** (1973) 920.
[183] See, e.g., H. Scher and M. Lax, *Phys. Rev. B* **7** (1973) 4491; G. G. Roberts and J. I. Polanco, *Phys. Status Solidi A* **1** (1970) 409; A. J. Grant and E. E. Davis, *Solid State Commun.* **15** (1974) 563; cf. also F. Gutmann et al., *Nature (London)* **221** (1969) 1237; A. M. Hermann, *Electrical Properties of Polymers*, Ed. by K. C. Frisch, Technomic, Westport, Conn., 1972, p. 103.
[184] R. C. Powell, *J. Chem. Phys.* **58** (1973) 920; L. E. Lyons, *Search* **7** (1976) 339.
[185] Jean Bertin et al., *J. Am. Chem. Soc.* **96** (1974) 8113.

[186] Jean Bertin et al., *Tetrahedron Lett.* **1974**, 763.
[187] J. M. Lalancette et al., *Can. J. Chem.* **52** (1974) 589; *Can. J. Chem.* **50** (1972) 3058.
[188] M. E. Volpin and Yu. N. Novikov, *Topics in Non-Benzenoid Aromatic Chemistry*, Vol. 1, Ed. by T. Nozoe et al., Wiley, New York, 1973, p. 269.
[189] F. R. Gamble et al., *Science* **168** (1970) 568; **174** (1971) 493; M. M. Labes, *Nature (London) Phys. Sci.* **246** (1973) 122.
[190] A. H. Thompson, *Solid State Commun.* **13** (1973) 1911.
[191] E. C. Johnson et al., *Can. J. Chem.* **56** (1978) 1381; P. R. Mitchell and H. Sigel, *J. Am. Chem. Soc.* **100** (1978) 1564; C. F. Naumann and H. Sigel, *FEBS Lett.* **47** (1974) 122; P. Chaudhuri and H. Sigel, *J. Am. Chem. Soc.* **97** (1975) 3209; H. Sigel et al., *Europ. J. Biochem.* **41** (1974) 209; G. Cilento, *Quart. Rev. Biophys.* **6** (1973) 488; Y. Fukuda, P. R. Mitchell, and H. Sigel, *Helv. Chim. Acta* **61** (1978) 638.
[192] H. Siegel and C. F. Naumann, *J. Am. Chem. Soc.* **98** (1976) 737.
[193] A. V. Heuvelen, *J. Biol. Phys.* **1** (1973) 215.
[194] H. Moehwald, *Chem. Phys. Lett.* **26** (1974) 509.
[195] Y. Matsunaga et al., *Bull. Chem. Soc. Jpn.* **47** (1974) 2926.
[196] H. Negita et al., *Bull. Chem. Soc. Jpn.* **46** (1973) 2662.
[197] F. Gutmann, *J. Sci. Indust. Res.* **26** (1967) 19.
[198] U. Thewalt and C. E. Brugg, *Acta Crystallogr.* **B-28** (1972) 82.
[199] F. Cramer, *Einschluss-Verbindungen* Springer, Heidelberg, 1974; J. P. Behr and J. M. Lehn, *J. Am. Chem. Soc.* **98** (1976) 1743.
[200] Y. Matsunaga and G. Saito, *Bull. Chem. Soc. Jpn.* **45** (1972) 963; **47** (1974) 2873; Y. Matsunaga and R. Osawa, *Bull Chem. Soc. Jpn.* **47** (1974) 1589; G. Saito and Y. Matsunaga, *Bull Chem. Soc. Jpn.* **47** (1974) 1020; **46** (1973) 714; Y. Matsunaga et al., *Bull Chem. Soc. Jpn.* **48** (1975) 37; J. M. Dumai et al., *J. Chim. Phys. Phys.-Chim. Biol.* **72** (1975) 1185; H. Ratajcak et al., *Chem. Phys.* **17** (1976) 197; J. M. Dumai et al., *J. Chim. Phys. Phys.-Chim. Biol.* **72** (1975) 1185; H. Ratajcak et al., *Chem. Phys.* **17** (1976) 197; G. Saito and Y. Matsunaga, *Bull. Chem. Soc. Jpn.* **46** (1973) 1609; A. Kofler, *Z. Elecktrochem.* **50** (1944) 200.
[201] P. Stilbs and G. Olofsson, *Acta Chem. Scand.* **A28**, (1974) 647; G. Olofsson and I. Olofsson, *Tetrahedron* **29** (1973) 1711.
[202] Q. Appleton et al., *Tetrahedron* **27** (1971) 5921.
[203] I. Agranat and S. Cohen, *Bull. Chem. Soc. Jpn.* **47** (1974) 723; cf. also N. Inoue and Y. Matsunaga, *Bull. Chem. Soc. Jpn.* **45** (1972) 3478; G. N. Felix and P. L. Huyskens, *J. Phys. Chem.* **79** (1975) 2316.
[204] G. Saito and Y. Matsunaga, *Bull. Chem. Soc. Jpn.* **46** (1973) 1609.
[205] G. Saito and Y. Matsunaga, *Bull. Chem. Soc. Jpn.* **46** (1973) 714, 1609.
[206] J. Nishijo, *Bull. Chem. Soc. Jpn.* **47** (1974) 1539.
[207] P. W. Anderson et al., *Phil. Mag.* **25** (1972) 1; J. Tauc, *Phys. Today* **Oct. 1976**, 27; S. G. Christov, *Contemp. Phys.* **13** (1972) 199; *Phys. Status Solidi* **7** (1971) 371; *Croatica Chim. Acta* **44** (1972) 67; R. R. Dogonadze and A. M. Kuznetsov, *J. Res. Inst. Catalysis Hokkaido Univ.* **22** (1974) 93; G. Gusman and R. Deltour, *Solid State Commun.* **15** (1974) 1597.
[208] M. J. Rice and W. L. Roth, *J. Solid State Chem.* **4** (1972) 294; see also L. J. Gagliardi, *J. Chem. Phys.* **58** (1973) 2193.
[209] Trade Name registered to Aldrich Chem. Corp. Milwaukee, Wisconsin 53233.
[210] R. W. Bowman et al., *Chem. Commun.* **1968**, 723.

[211] G. N. Felixand P. L. Huyskens, *J. Phys. Chem.* **79** (1975) 2316.
[212] K. Morimoto and Y. Murakami, *Appl. Opt. Supl. No. 3* (*Electrophotogr.*) **1969**, 50; G. Pfister, M. Abkowitz, and D. J. Williams, *J. Chem. Phys.* **57** (1972) 2979.
[213] W. Zeichmann and M. Wakill, *Erdöl, Kohle, Erdgas, Petrochem.* **29** (1976) 297.
[214] F. Gutmann, A. M. Hermann, and A. Rembaum, *J. Electrochem. Soc.* **114** (1967) 323; *J. Electrochem. Soc.* **115** (1968) 359; M. Pampallona *et al.*, *J. Appl. Electrochem.* **6** (1976) 269; H. Sigel *et al.*, *Eur. J. Biochem.* **41** (1974) 209.
[215] G. Eckert, *Proceedings of the Australia–United States Symposium on Bio-Electrochemistry, Los Angeles, 1979*, Ed. by H. Keyzer and F. Gutmann, Plenum Press, New York, 1979.

Index

Acetonitrile, capacity behavior of, 181
Activation energies and conductance, 63
   tabulated, 64
Adsorbed films, lateral interactions in, 225
Adsorption
   and free energies
      for organic substances at metal-aqueous-solution interfaces, 120
   of transfer, 229
   on mineral solids, 208
   on minerals
      chemical state of, 233
      current research, 242
Adsorption equations for minerals, 222
Adsorption isotherms for surfactants on minerals, 221
Adsorption kinetics for minerals, 240
Adsorption potential shifts, 100
   of alcohols, 105
   at mercury and bismuth electrodes, 105
   and solvent orientation, 102
Adsorption studies and characterization of electrode surfaces, 330
Adsorption of surfactants on minerals, 209
Air-solvent interface, surface potentials at, 186
Air-water interface, surface potential at, 105

Alcohol adsorption and capacity, 114
Aliphatic alcohols, capacity behavior of, 179
Aluminum and x-ray photoelectron spectroscopy (XPS), 335
Anisotropy, of water dielectric properties, 113
Apatite, isoelectric point, leaching effect on, 219
Apparatus
   for precise conductance measurement, 7
   for purification of organic solvents, 11
Association constants, for ion-pair formation, 20, 46
Au: *see* Gold
Auger electron spectroscopy
   and corrosion, 331
   and passivity, 331
Auger electrons, x-ray excited, 297
Auger line shape, 265
Auger spectra
   and background subtraction, 269
   chemical effects, 263
   chemical shifts, 263
   deconvolution, 269
   electron beam effects, 274
   examples, 256
   quantitative analysis, 258
Auger spectroscopy, 251
   details, 253
   instrumentation, 255

Band density of states, 295
Batteries, high-energy, 1
Binding energies of 1s electron in
 nitrogen compounds, 340
Bizmuth
 adsorption potential shifts at
  electrode, 105
 surface potential of water at
  interface, 134
Bjerrum treatment, in conductance, 17
Boyd and Zwanzig effect, 26

Cadmium, surface potential of water
 at interface, 134
Calcite, point of zero charge, 212
Capacity
 and adsorption of alcohols, 114
 double-layer, for mercury, 85
 experimental data and solvent
  orientation, 133
 at gallium electrode, 86
 hump, 149
 inner layer
  for mercury, 86, 132
  for metals, 123
  minimum, 135
 and solvent adsorption, 85
 and solvent orientation, 173
Capacity behavior
 of acetonitrile, 181
 of aliphatic alcohols, 179
 of dimethylformamide, 180
 of dimethylsulfoxide, 182
 of ethylene carbonate, 184
 of formamide, 175
 of formic acid, 184
 of methanol, 178
 of sulfolane, 183
Catalysis for hydrogen exchange, 397
Cell, for permittivity measurement,
 13
Cell constant, 8
 temperature dependence of, 9
Cells, for conductance, 6
Charge carriers
 and charge transfer complexes, 369
 mobility of, 373

Charge for zero dipole orientation, 151
Charge generation
 by dissociation on minerals, 210
 in surfactant adsorption, 209
Charge-potential curves
 general, 92
 for water orientation, 146
Charge transfer complexation and
 conductivity, 364
Charge transfer complexes
 and charge carriers, 369
 dissociation of, 374
 for electrochemical energy storage,
  412
 in electrochemistry, 361
 polarographic study of, 403
 solvent interactions, 376
Chemical effects and photoelectron
 spectroscopy, 287
Chemical forces, in adsorption on
 minerals, 233
Chemical shift and photoelectron
 spectroscopy, 291
Chemical shifts, in Auger spectra, 263
Chemical state plots, 290
Chemisorption
 and photoelectron spectroscopy, 293
 on minerals, 233
Chlorine adsorption and ruthenium
 dioxide. 329
Chlorine electron binding energies, 339
Clusters, of water at interfaces, 142
Colloid and surface complexes, 379
Complex formation, as electrode
 reaction, 387
Complexation
 charge transfer and conductivity, 364
 and donicity number, 363
 stochastic processes in, 381
Complexes
 colloid and surface, 379
 ternary and proton transfer, 407
Concentrated electrolyte solutions, 52
 conductance of, 2
 experimental data on, 55
Conductance
 activation energies for, 63
 advanced setup for, 7

Index

Conductance (cont.)
  cell constant in measurements of, 8
  cells, 6
  and Debye-Hückel theory, 24
  and dielectric relaxation, 26
  of electrolytes in organic solvents, 55
  experimental aspects, 4
  general monographs on, 71
  of individual ions, 25
  ion pairing in, 17
  ion size parameters for, 45
  limiting, 25
  maximum, 55
  measurements, 6
  organic solvents for, 10
  requirements for accuracy, 2
  results of data analysis, 30
  statistical-mechanical treatment, 3
  survey of publications, 38
  temperature dependence of, 6
Conductance data
  analysis of, 27
  empirical equation fits for, 32
  for tetraalkylammonium salts, 34, 36
Conductance equations, 2
  for dilute solutions, 14
  empirical, 4
  for high concentrations, 3, 4
  of various forms, 16
Conductivity and charge transfer
    complexation, 364
Constants and conversion factors, 1, 2
Contact angle and mineral surfaces, 239
Conversion factors and constants, 1, 2
Corrosion and photoelectron
    spectroscopy, 331
Critical micelle concentration, 231

Debye-Huckel theory, 24
Density, measurement by dilatometry, 14
Density of organic solvents, 12
Density of states, in valence band, 295
Destructive techniques, in
    photoelectron spectroscopy, 305

Dielectric constant
  and hydrophilicity, 112
  of inner-layer, 136
  of water molecules, 140
Dielectric properties
  and interfaces, 122
  of water at interfaces, 113
Dielectric relaxation, in
    conductance, 26
Dilatometer, for solvent density, 14
Dimethylformamide, capacity
    behavior of, 180
Dipole moment, of water at
    interfaces, 138
Dipole orientation
  at electrodes, temperature effects, 155
  at metals other than Hg, 161
  model prediction, 162
Dissociation, in organic solvents, 47
Dissociation processes, for minerals, 210
Donicity, 362
Donicity numbers in complexation, 363
Double layer
  dielectric aspects, 122
  of minerals, Stern model, 218
  models for, 87
  and solvent adsorption, 81
  water structure in, 126

Electrocatalysis, in relation
    to heterogeneous catalysis, 388
Electrocatalysis
  photoelectron surface analysis in, 312
  on ruthenium dioxide electrodes, 325
Electrochemical energy storage,
    and charge transfer complexes, 412
Electrochemistry of charge transfer
    complexes, 361
Electrode reactions and complexes, 387
Electrode surfaces, characterization
    for adsorption studies, 330

Index

Electrodes
  Auger spectroscopy at, 251
  general surface analysis techniques for, 346
  photoelectron spectroscopy at, 251
  solvent adsorption at, 81
Electrolyte solutions, concentrated, 52
Electron beam effects, in Auger spectra, 274
Electron binding energies, for chlorine, 339
Electron spectroscopy and surface analysis, 252
Electronic configuration of phthalocyanine complexes, 400
Electrostatic adsorption on minerals, 219
Elemental analysis, by photoelectron spectroscopy, 284
Energies of adsorption, of organic substances at electrodes, 120
Energy storage, electrochemical, 412
Entropy, surface excess, at interfaces, 164
Escape depth and photoelectron spectroscopy, 300
Excess entropy, theoretical calculations of, 165
Experimental data on capacity, 133

Field effects, on water orientation, 145
Flotation
  and ion adsorption, 225
  and mineral surfaces, 207
  with oleic acid, 237
Formamide, capacity behavior of, 176
Free energy of transfer, of CH_2 groups from aqueous solution, 229
Free-radical processes and electrochemistry, 393
Fuoss–Onsager equations, 16

Gallium, capacity, 86
Gold
  hydrophobicity of, 118

Gold (cont.)
  photoelectron spectroscopy of, 318

Hg: see Mercury
High-energy batteries, 1
Hump, in capacity curves, 149
Hydrodynamics, in ion conductance, 25
Hydrogen exchange
  catalysis for, 397
  and phthalocyanine complexes, 400
Hydrophilicity
  and dielectric constant, 112
  of gold, 118
  of metals, 109
  and organic adsorption, 121
  scale of, 110, 117

Indium, chemical state plots from photoelectron spectroscopy, 290
Inner layer
  dielectric constant of, 136
  surface excess entropy of, 164
  water population in, 141
Inner-layer capacity
  for mercury, 132
  minimum, 135
  and potential of zero charge, 123
  and solvent orientation, 173
Interfaces
  dipole moment of water at, 138
  models for water structure at, 127
  water clusters at, 142
  water structure at, 126
Interfacial entropy, calculation of, 165
Interfacial tension and mineral surfaces, 239
Ion adsorption and flotation, 224
Ion association energy, noncoulombic contributions, 49
Ion-pair formation
  consecutive steps in, 23
  thermodynamic data for, 50
  thermodynamics of, 18

Index

Ion pairs
 association constants for, 20
 in conductance, 17
 contributions to energy of, 20
 degrees of freedom for, 22
Ion size parameters, in
  conductance, 45
Ionic conductance, 25
 models for, 25
Ionic radii
 and conductance, 45
 and specific conductance, 59
Interaction parameters, for metal
  surfaces, 115
Iron surfaces, passivation, 334
Isoelectric point
 for leached apatite, 219
 for various minerals, 215, 218
 leaching effect, 219
Isologous sections, 6
Isotherms, adsorption, for minerals, 221

Justice treatment, in
  conductance, 18

Kinetics of adsorption at minerals, 240

Lateral interaction, in adsorbed
  films, 225
Leaching effect, and isoelectric
  point, for minerals, 219
Limiting conductance, 25
 and ion size, 45
 temperature dependence of, 42
Limiting orientation of water
  molecules, 141

Maximum specific conductance, 68
Mercury
 adsorption potential shifts at, 105
 capacity of electrode, 85
 charge for zero water
  orientation at, 151

Mercury (cont.)
 features of capacity curve, 85
 inner-layer capacity for, 86, 132
 interface, basic information
  from, 190
 surface potential of water for,
  100, 134
 temperature dependence of dipole
  orientation for, 155
 temperature dependence of dipole
  orientation, tabulated, 159
 water-orientation at pzc, 88
Metal surfaces, interaction parameters
  of, 115
Metals
 adsorption of organic substances
  at interfaces, 120
 hydrophilicity of, 109
 water surface potentials for, 95
Methanol, capacity behavior of, 178
Micelle formation, 231
Mineral solids, adsorption on, 207
Mineral surfaces
 charge generation in adsorption on,
  209
 and contact angles, 239
 and flotation, 207
 and interfacial tension, 239
Minerals
 adsorption equations for, 222
 adsorption kinetics for, 240
 adsorption of oleates on, 235
 charge generation by dissociation,
  210
 chemical adsorption effects, 233
 current research on adsorption on,
  242
 dissociation processes of, 210
 electrostatic adsorption on, 219
 isoelectric points of, 215
 isoelectric points for, tabulated,
  218
 lateral interaction in films, 225
 potentials of zero charge for, 214,
  218
 solubility constants for, 214
 surface potentials for, 217

Mixed solvents, 51
  viscosity of, 70
Mobilities of charge carriers, 373
Models
  cluster type, for water, 142
  for double-layer, 87
  for solvent orientation, 168
Moving boundary method, 41

Nitrogen electron binding energies, 340
Nonaqueous electrolyte solutions,
  previous literature, 1
Nonaqueous solvents
  models for orientation at
    electrodes, 175
  orientation at electrodes, 174
  surface potential at air solvent
    interface, 186
  surface potentials for, 187
  surface potentials and potentials of
    zero charge, 187

Oleates, adsorption on minerals, 235
Oleic acid and flotation efficiency, 237
Onsager equation, 16
Organic adsorption, and
  hydrophilicity, 121
Organic solvent mixtures, 66
Organic solvents
  apparatus for purification of, 11
  for conductance, 10
  data on, 12
  dissociation of salts in, 47
  mixed, 51
  pure electrolytes in, 55
  purification of, 10
  survey of publications on
    conductance, 38
Organic substances, energies of
  adsorption at metals, 120
Orientation
  field effects on, 145
  of nonaqueous solvents at electrodes,
    174
Oxide layers on noble metals,
  photoelectron spectroscopy of,
  313

Pair distribution function, 21
Palladium, photoelectron
  spectroscopy of, 321
Passivation, of iron surfaces, 334
Passive films on aluminum,
  photoelectron spectroscopy of,
  342
Passivity and photoelectron
  spectroscopy, 331
Permittivity
  measurement of, 13
  of organic solvents, 12
Photoelectron spectra, satellites
  in, 279
Photoelectron spectroscopy, 251
  and band density of states, 295
  and chemical effects, 287
  and chemical shift, 291
  and chemisorption, 293
  elemental analysis by, 284
  of gold, 318
  instrumentation for, 283
  nondestructive techniques, 304
  of palladium, 321
  of passive films on aluminum, 342
  of platinum, 313
  of ruthenium dioxide electrodes, 325
  x-ray, 277
Physical constants
  conversion factors, 2
  of organic solvents, 12
Platinum, photoelectron
  spectroscopy of, 313
Platinum surface oxide, x-ray
  photoelectron spectra of, 313
Point of zero charge
  of calcite, 212
  and solvent adsorption, 83, 90; see
    also Potentials of zero charge
  and surface potentials, 187
  for various minerals, 218
Polarography
  and charge transfer complexes, 403
  of donors and acceptors, 401
Potential drop and solvent adsorption,
  81
Potentials of zero charge
  factors in, 82

Index

Potentials of zero charge (cont.)
  and inner-layer capacity, 123
  and organic adsorption, 120
  and solvent orientation, 83
  and surface potentials, 188
  and water orientation, 90
  and work functions, 99
Potentials of zero charge, for mercury, water orientation effects, 88
Proton complexes, 410
Pt: see Platinum
Publications, survey of, for conductance, 38
Pyrazine, as probe for water orientation, 103

Radii, ionic, and conductance, 45, 59
Redox reactions and semiconductor electrodes, 394
Redox systems, 390
RuO_2, 325, 329; see also Ruthenium dioxide
Ruthenium dioxide and chlorine adsorption, 329
Ruthenium dioxide electrodes and photoelectron spectroscopy, 325

Screening potential, 228
Semiconductor electrodes
  and complexes, 389
  and redox reactions, 394
Solvation of charge transfer complexes, 376
Solvent adsorption
  at electrodes, 81
  models for, 83
  previous reviews on, 81
Solvent dipole and potential of zero charge, 83
Solvent dipole potential, 81
Solvent orientation
  and adsorption potential shifts, 102
  and double-layer models, 90
  and inner-layer capacity, 173
  historical survey on electrodes, 84
  for various metals, 95

sp metals, standard potentials of, 129
Specific conductance
  and ionic radii, 59
  maximum, 68
  temperature dependence of, 63
Spectra, Auger, examples, 256
Spectroscopy
  Auger, 251
  details of, 253
  photoelectron, 251
Spectrum, photoelectron, 277
Standard potentials of sp metals, 129
Stern model, of double layer, 218
Stochastic processes in complexation, 381
Structure in photoelectron spectroscopy, 3
Structural models, for water at interfaces, 127
Surface analysis
  by Auger and x-ray photoelectron spectroscopy, 299
  in electrocatalysis, by photoelectron spectroscopy, 312
  by electron spectroscopy, 252
  techniques for electrodes, general, 346
Surface dipole potential, estimation of, 88
Surface excess entropy, inner-layer, 164
Surface potential
  and capacity, temperature dependence of, 173
  for nonaqueous solvents, 187
  and water adsorption, 108
  of water,
    at free surface, 105
    at mercury, 100
    tabulated, 134
Surface potentials
  for minerals, 217
  of water, for various metals, 107
  tabulated, 98, 108
  on various models, 148
Surface roughness, in photoelectron spectroscopy, 309

Surfactants
  adsorption isotherms for minerals, 221
  and free energies of transfer in adsorption, 229
  and mineral surfaces, 209

Temperature dependence
  of cell constant (conductance), 9
  of conductance, 6
  of inner-layer capacity, related to surface potential, 173
  of organic solvent properties, 12
  of specific conductance, 63
Temperature effect on Walden product, 62
Temperature effects, on dipole orientation, 155
Ternary and proton transfer complexes, 407
Tetraalkylammonium salts
  conductance data for, 34
  ion-pair formation, 50
Thermodynamic data, for ion-pair formation, 50
Thermodynamics of ion pairs, 20
Thiourea, probe for water adsorption 125
Transference numbers, 41
Transport numbers, 41

Viscosity
  of mixed solvents, 70
  of organic solvents, 12

Walden product, 62
Walden's rule, 26
Water
  dipole moment at interfaces, 138
  interfacial structure of, 126
  ordering in, 125
  surface potential of, 134

Water adsorption
  and work function, 118
  pyrazine probe for, 103
  thiourea probe for, 125
Water–air interface, surface potential at, 105
Water clusters, at interfaces, 142
Water dipole orientation and potential of zero charge, 90
Water dipole potential, for various metals, 95
Water molecules, dielectric constant of, 140
Water orientation
  field effects on, 145
  limiting, 141
  and pyrazine probe, 103
  and thiourea probe, 125
Water surface potentials, for various metals, 98
  tabulated, 108
Work function
  and potential of zero charge, 99
  and water adsorption, 118

X-ray-excited Auger electrons, 297
X-ray-induced damage, in photoelectron spectroscopy, 298
X-ray photoelectron spectra, of platinum oxide, 316
X-ray photoelectron spectroscopy, 277
  of aluminum surfaces, 337
  and corrosion, 331
  and passivity, 331
Xenon adsorption, and hydrophilicity, 116

Zero dipole orientation
  charge for, for Hg, 151
  charges for, tabulated, 160
Zeta potentials for apatite, 215
Zwanzig and Boyd effect, 26

MIX
Papier aus verantwortungsvollen Quellen
Paper from responsible sources
FSC® C105338

If you have any concerns about our products,
you can contact us on
ProductSafety@springernature.com

In case Publisher is established outside the EU,
the EU authorized representative is:
**Springer Nature Customer Service Center GmbH
Europaplatz 3, 69115 Heidelberg, Germany**

Printed by Libri Plureos GmbH
in Hamburg, Germany